THE IMPERIAL LABORATORY

EXPERIMENTAL PHYSIOLOGY AND CLINICAL MEDICINE IN POST-CRIMEAN RUSSIA

THE WELLCOME SERIES IN THE HISTORY OF MEDICINE

Forthcoming Title:

The Stepchildren of Science
Heather Wolffram

Please send all queries regarding the series to Michael Laycock,
The Wellcome Trust Centre for the History of Medicine at UCL,
183 Euston Road, London NW1 2BE, UK.

THE IMPERIAL LABORATORY

EXPERIMENTAL PHYSIOLOGY AND CLINICAL MEDICINE IN POST-CRIMEAN RUSSIA

Galina Kichigina

Amsterdam – New York, NY 2009

First published in 2009
by Editions Rodopi B.V., Amsterdam – New York, NY 2009.

Design and Typesetting by Michael Laycock,
The Wellcome Trust Centre for the History of Medicine at UCL.
Printed and bound in The Netherlands by Editions Rodopi B.V.,
Amsterdam – New York, NY 2009.

Index by Henry Mitchell.

British Library Cataloguing in Publication Data
A catalogue record for this book is available from the British Library

ISBN 978-90-420-2658-2
E-Book ISBN 978-90-420-2659-9

'The Imperial Laboratory:
Experimental Physiology and Clinical Medicine in
Post-Crimean Russia' –
Amsterdam – New York, NY:
Rodopi. – ill.
(Clio Medica 87 / ISSN 0045-7183;
The Wellcome Series in the History of Medicine)

Front cover:
Left: portrait of S.P. Botkin, courtesy of the Wellcome Library, London.
Right: portrait of I.M. Sechenov, courtesy of the Tretyakov Gallery, Moscow.
Background: map of Russia, courtesy of the Wellcome Library, London.

Printed in The Netherlands

All titles in the Clio Medica series (from 1999 onwards) are available to download from the IngentaConnect website: http://www.ingentaconnect.co.uk

Contents

List of Images

List of Tables

Acknowledgements

First off, I would like to express my deepest thanks to Professor Pauline M.H. Mazumdar at the University of Toronto's Institute for the History and Philosophy of Science and Technology for her generosity, enthusiastic support and encouragement on my historical path. Pauline read the entire manuscript, and her immense knowledge as a historian of medicine and expert advice as a physician–scientist made this book better. I am also grateful to the institute's scholars, Susan G. Solomon and Trevor H. Levere, for valuable suggestions on my PhD thesis in the history of nineteenth-century Russian and German medical science.

I am extremely thankful to the Russian historians of medicine, who were helpful in diverse ways during my research on this book, and most especially to Vladimir Olegovich Samoilov, Director of the State Centre for Pulmonology, St Petersburg, Nora Andronikovna Grigoryan, Doctor of Medical Sciences, Moscow, the Institute for the History of Science of the Russian Academy of Sciences, and to Iurii Petrovich Golikov, Head of the Museum of the Institute for Experimental Medicine of the Russian Academy of Medical Sciences, St Petersburg.

I heartily thank Professor Ingrid Kästner for hospitality and shared scholarship, and the staff of the Karl-Sudhoff-Institut für Geschichte der Medizin und der Naturwissenschaften at Leipzig University, where I have happily worked every summer for many years on this and other projects, and where I participated in the programme on Russian–German relations in medicine and science in the eighteenth and nineteenth centuries.

Since I worked in the archives and libraries of the most significant nineteenth-century European and Russian laboratory centres, I owe a debt to many archivists and librarians for the access to the historical laboratory material and intellectual culture, apparatus, textbooks, picture collections. In particular, I am thankful to the staff of the museums and libraries of the St Petersburg Military Medical Academy, the Moscow Medical Academy, the Vienna Josephinum, Berlin Humboldt University's Physiological Institute, Heidelberg University's Organic Chemistry Department, the Institute for the History of Science and Library at the Deutsches Museum in Munich, the Bibliothèque Interuniversitaire de Médecine in Paris, and Strasbourg University's Physiological Institute. I am also indebted to the staff of the

State Archive for War History in Moscow for the locating of and access to valuable sources.

I would like to express my deep thanks to Mike Laycock, publishing administrator at The Wellcome Trust Centre for the History of Medicine at the UCL, for his patience, superb editorial assistance and workmanship with the illustrations for the book. I am grateful to Cathy Doggrell and Benedict Flynn who also helped in editing the text. I am also thankful to my readers at the *Clio Medica* series for copious suggestions.

I extend my thanks to the cardiologist–scientists at the Toronto Hospital for Sick Children Reseach Institute and Heart Centre – where I pursued an experimental study on cardiac arrhythmias towards a degree in cardiovascular sciences – for a fresh insight into the past and present of medical research.

I offer my warmest thanks to Professor Valentina Fedorovna Novodranova at Moscow Medical Academy, supervisor of my first PhD thesis, for sustaining friendship and care.

To my family, my husband Gerasim and my son Karen, I owe love, nurturing, and support in all my research venues.

This book is dedicated
to the loving memory of my mother,
Valentina Roshchina.

Introduction

This book is about the rise of the laboratory in nineteenth-century Russian medicine. It discusses how one of the major innovations of nineteenth-century medicine, the teaching–research laboratory, was introduced to the St Petersburg Medico-Surgical Academy in the early 1860s and within two decades spread out to increasingly dominate in both medical research and practice. Why and how the laboratory became so important in medicine has received much attention from historians of science. A multifaceted picture of intellectual, political, institutional, scientific, and educational components of nineteenth-century European laboratory medicine has suggested that the transition to laboratory medicine was a revolution comparable in importance to the transition to hospital medicine, which preceded it.[1] The laboratory continues to be the major site for research and training in medicine worldwide, and in this sense its nineteenth-century counterpart exemplifies the modern laboratory. Medical scientists nowadays tend to think of advances in medicine as making their way from laboratory bench to bed-side. The physiological and clinical laboratories of the 1850s and 1860s were shaping their research programmes in remarkably similar ways, giving priority not only to extending knowledge but also to the relevance of this knowledge to the clinic. Although the advancement of science and scientific medicine had been widely recognised as the symbol of material and cultural progress, some of the great nineteenth-century thinkers, such as Fyodor M. Dostoyevsky, were more concerned with the dangers of medical science rather than its benefits.[2] This dilemma of nineteenth-century science is still an open question today.

Experimental sciences began to supersede an empirical tradition in medicine by the 1840s and within the next decade experimental physiology, medical chemistry, cellular pathology, and medical histology had evolved as clearly defined laboratory disciplines. Among the laboratory disciplines, physiology quickly assumed a dominant position in nineteenth-century medical research and teaching, largely due to its spectacular discoveries and their clinical implications. The interplay between research programs, medical practice and training, institutional forms, professional careers, and state or social interests has received critical attention in the important scholarship on nineteenth-century experimental physiology.[3] At the same time, substantial and well-integrated studies of institutional structures with some serious accounts of Russian laboratory scientists are very rare. Daniel Todes's work on Pavlov and his laboratory at the St Petersburg Imperial Institute of Experimental Medicine is an exception.[4]

Nineteenth-century Russian laboratory medicine is a challenging and rewarding topic for the historian of medicine. The early laboratory leaders at the St Petersburg Medico-Surgical Academy – Ivan M. Sechenov, Sergei P. Botkin, Elie de Cyon, Nikolai N. Zinin, and Alexander P. Borodin – are among most interesting and noteworthy figures representing both the laboratory disciplines they built and a diverse socio–scientific network of nineteenth-century Russian medical science. Popular teachers and capable scientists, they exerted powerful influence on the development of physiology and medicine in Russia. All of them were closely connected with the network of important European medical research institutions and its leaders, such as Justus Liebig, Emil du Bois-Reymond, Carl Ludwig, Rudolf Virchow, Claude Bernard. These Russian scientists are key figures for a study of the dissemination of scientific ideas, methods and instruments, and the ways innovative laboratory practices and culture were introduced and further developed in a new setting in Russia. Their perceptions, experiences, and participation in the ongoing research during their studies abroad and at home may add some important insights into science-in-the-making both at the leading European research institutions and at the newly established laboratories in Russia. The Russian scientific community of the 1860s appears to have been more sensitive to foreign influences and foreign models than has hitherto been appreciated.[5] It is one of the purposes of this book to discover why this was and how it happened.

The scientific and laboratory revolution arrived in Russia in the late 1850s. It had a profound impact on Russian culture and society, and made a strong impression on Russian literature. With the realisation that science was not only an accumulation of facts but also a way of looking at the world, scientific ideas and methods spread into areas of thought where they had been hitherto absent such as philosophy, political and social theory, and history. Science gave a new and powerful legitimisation to philosophical materialism and thus encouraged determinist and positivist ideas. Natural sciences, physiology in particular, captured the allegiance of young radicals, for whom the laws of validation, prediction, and calculation seemed to be transferable to the solution of social and political problems. New possibilities for adapting, developing, and applying advanced knowledge in all spheres of human activities, and growing awareness within Russian society that improvements in social life could be achieved through the development of scientific knowledge were responsible for the enormous prestige of science and science education. Sofia Kovalevskaia (1850–91), the future brilliant mathematician, wrote about that time:

> We were so exalted by all these new ideas, so convinced that the present state of society could not last long, that the glorious time of liberty and general knowledge was quite near, quite certain.[6]

Among Russia's institutions of higher learning, the Medico-Surgical Academy at St Petersburg, which came under the War Ministry, occupied a privileged position as a specialised school for training both military and civilian doctors. Its institutional history allows studies of important structural as well as functional changes underlying the organisation of teaching and research as well as the diffusion of knowledge. In Russia, governmental support for the development of sciences within the medical educational system acquired a pivotal role from the 1860s onwards. The same was true in Europe – in Saxony, the Minister of Education, Johann Paul Baron von Falkenstein, was responsible for substantial financial support in the construction of the famous physiological institute for Carl Ludwig in 1864 as well as some other research institutes at Leipzig University. In Baden, a liberal ministry, appointed by Grand Duke Frederick I, spent a substantial sum of money on laboratories for physiology, botany, and zoology, as well as on expanding the existing structures for chemistry and physics at the University of Heidelberg. Friedrich Althoff, head of the Higher Education Section of the Prussian Culture Ministry in the early 1880s, was responsible for the unprecedented growth of the new research institutes, laboratories, and clinics in Prussian universities.[7] In Russia, in the late 1850s, soon after the Crimean War, substantial governmental support was rendered not to the universities, but to military medical education. Through the 1860s, the liberal Minister of War, Count Dmitrii Alekseevich Miliutin (1816–1912) continued to provide crucial support for the development of the natural sciences at the Medico-Surgical Academy. Substantial funding allowed the construction of a number of new facilities, the Natural History and Sciences Institute, and the Anatomo-Physiological and Botanical Institutes, all with their own well-equipped laboratories, as well as new hospitals with teaching clinics. After the educational reform of 1864, the Government began slowly to support laboratory sciences in the universities, but the scale of expansion and construction of university research and laboratory facilities was consistently smaller than in the military educational institutions. The Ministry of Education was always financially constrained and dependent on contradictory changes in governmental policy towards universities. The War Ministry, in contrast, could count on consistent support and sufficient funding.

Whether the military medical authorities were convinced of the far-reaching benefits of the laboratory to medical or surgical practice is a question hard to verify. What is clearly evident from the official correspondence of the War Ministry is the attraction of the reformist administration of the Medico-Surgical Academy to the examples of the former enemies in the Crimea, France and Britain, and neutral Germany and Austria, and their continuous support for science and education. The reformist administration of the Medico-Surgical Academy presented to the high officials of the War Ministry an image of the laboratory as the major vehicle of modernisation in teaching and a core element in the development of scientific medicine which, by its very definition, implied modern advancement. Introduction of the laboratory denoted an important shift towards practicality of knowledge and opened up possibilities for the application of this knowledge in the hospital and on the battlefield. Finally, the shocking statistics of the heavy losses from diseases in the Russian army during the Crimean War gave a strong impetus to the urgent decision of the government to reform military medicine and education.

The 'national calamity' in the Crimea showed not only severe deficiencies in medical administration and service but, more importantly, it showed the basic deficiencies in Russia's power itself: an army system based on serfdom, the backwardness of her economy, and of her social structure. The emancipation of the serfs in 1861, followed by the advancement of reforms in most of the institutions of society, including educational, governmental, juridical, and military systems, ultimately moved Russia ahead towards modern economy and society.[8] Enlightened high officials and educational reformers interpreted the Crimean defeat as the unavoidable result of Russia's backwardness in science and technology. Inadequate teaching in natural sciences, total absence of any practical component in science training, and disproportionate emphasis on classical education were the underlying causes.[9] To gain from European scientific, technical, and educational advancement, the government began to permit a new freedom of access to studies abroad and mitigated its censorship, which brought a new flood of foreign literature into the country. The educational reforms liberalised academic life and made higher education more accessible for different social estates. These measures enormously facilitated growth of Russian science, medicine, and education.

The first generation of Russian medical scientists, like Sechenov, Botkin, Cyon, who studied abroad in the late 1850 and early 1860s, devoted themselves to the introduction of some ideas of German laboratory culture and practices to Russian medicine. These scientists are convenient vehicles for the study of important changes in the disciplines of experimental physiology and clinical medicine. Interdisciplinary practices of experimental physiology and physiologically oriented clinical medicine required a new generation of researchers that appeared on the scene by the mid-nineteenth century. These were young people who had a thorough background in physics, chemistry, and mathematics. The German physiologist–physicists Emil du Bois-Reymond, Ernst Brücke, Hermann Helmholtz, and Carl Ludwig, adopted a style of exact science and its instrumentation and methods into physiological theory and practice. From the dawn of their careers, they strove to transform physiology into an applied physics. The very nature of the problems they studied was related to the physics and chemistry of the living organism and required them to bring together traditions of research from these sciences. In Russia, this new generation was represented by Sechenov, whose engineering and science background made him well fitted to handle physiological problems as natural science problems distant from the clinic. Sechenov's researches embraced all the changes that were shaping the physiology of his day, the precision measurements, the graphical inscription, and the quantitative definition of bodily functions. His interest in the design and improvement of apparatus enabled him to cross the boundaries of several different fields of research, including electrophysiology, blood gases, and theory of solutions, and create a cross-disciplinary discourse that shared concepts, methods, and instruments. At the same time, Sechenov's attempt to translate a reductionist approach to studies of mental functions in his *Reflexes of the Brain* represented a broader transformation of physiology as a unified science of mind and body that used one single method, experimentation. This conception of physiology gave it a philosophical status and made it a key element in the ideology of materialism with its threat to the established cultural and spiritual values, norms and life-styles of nineteenth-century Russian society.

For the clinician Sergei P. Botkin, physiology along with pathology and therapeutics constituted integrative parts of scientific medicine. Although Botkin insisted that medicine is more science than art in defining the cause of illness, he admitted that when it comes to the cure, medicine, at least in part, is an art based on intuition, clinical skills, and bedside experiences. For him, however, the basis of scientific medicine was the clinical laboratory, where systematised and well-planned research and training in microscopic and chemical techniques and animal experimentation was directly linked to the clinic. Clinical application of the new diagnostic tools – microscope, thermometer, and ophthalmoscope – rendered practical medicine quantitative and transformed it into an exact science. But the gap between diagnostics and therapeutics, which remained mostly empirical, was not so easy to overcome. Botkin, however, consistently constructed his investigative strategies as experimental studies of physiological and pharmacological action of therapeutic means. Like Ludwig Traube, his close colleague and friend in Berlin, Botkin was a powerful proponent of the integration of the laboratory into the clinic: his major innovation was a specialised clinical laboratory that became increasingly important in training both a medical researcher and a scientifically educated physician. Botkin's clinical laboratory was one of the first in Europe.

As medicine increasingly relied on theoretical disciplines such as pathological anatomy, physiology, physics, and chemistry, their role in medical education began to rise in prominence. At the Medico-Surgical Academy, a simultaneous development of related disciplines – physiology, chemistry, and physics – and their possible fruitful interactions, was promoted by Nikolai Zinin and Alexander Borodin, the future composer and Zinin's close associate. As early as the 1840s, Zinin, by then Russia's leading organic chemist, introduced the Liebigian system of training chemists to Kazan University. In the late 1850s, as an influential member of the reformist administration at the Medico-Surgical Academy, Zinin strongly advocated practical classes in chemistry as an integrative and important part of the pre-clinical curriculum. He believed that laboratory training in special methods, such as elementary organic analysis, gas analysis, the technique of titrating solutions, and of reactions in sealed tubes under high pressure was important not only for chemists but for future medical researchers and doctors as well. The creation of the Medico-Surgical Academy's Natural History and Sciences Institute, equipped with laboratory facilities, enabled Zinin and Borodin to institutionalise practical classes in chemistry and physics – an important step in the evolution of medical curriculum. Zinin's consistent support of other laboratory disciplines, including experimental physiology and cellular pathology, was pivotal for the establishment of laboratories for Sechenov and Botkin.

The transition from a small practical laboratory to an institute with a spacious laboratory equipped for research in all areas of physiology at the Medico-Surgical Academy took place within a decade.[10] In the early 1870s, Elie de Cyon, one of the most successful students and collaborators of both Bernard and Ludwig, succeeded Sechenov at the Medico-Surgical Academy's newly opened Physiological Institute. Cyon's tenure denoted not only institutional and structural changes that included growing demand for funding and space due increasing number of investigators and undergraduates, but also important changes associated with the increasing sub-specialisation of physiology according to the methods applied, and a simultaneous shift from the earlier reductionist physico–chemical approach to more integrated biological approach. A dedicated vivisectionist and skillful in quantitative methods and devices, Cyon uniquely combined the two leading approaches in his research and teaching. His *Atlas zur Methodik der physiologischen Experimente und Vivisectionen* is one of the best illustrations to the material culture of nineteenth-century physiology and its analytical tools. The picture that emerges from the *Atlas* is not only a glimpse of how a standard physiological laboratory in Russia or Germany was equipped, it also presents another side of the physiological laboratory and its practices, the 'costly ghastly kitchen', as Latour calls it, versus the image of its fine and sophisticated tools that facilitated and controlled the experiment. Both Cyon and Sechenov continuously pointed to the importance of scientific instruments and techniques in medical teaching and research, and their appeals aimed at promoting the laboratory may be interpreted as practical as well as rhetorical and aesthetic accomplishment.[11]

Among the issues discussed above, this book has a biographical intent. It focuses on key events in the lives of Sechenov, Botkin, Cyon, Zinin, and Borodin relevant to the rise of the laboratory in Russian medicine. An appreciation of the participation of these men in the collective ideas and practices of the nineteenth-century laboratory, as well as in the intellectual, cultural, and social discourses of the time, may provide some important insights into what it meant to be a physiologist, a clinical scientist, or a chemist in nineteenth-century Russia. These figures also provide us with an entry into most of the significant developments in nineteenth-century Russian science and medicine. The research programmes pursued by Sechenov, Botkin, or Cyon were closely related to research programmes in Leipzig, Berlin, or Paris. The institutional forms in St Petersburg were fairly similar to those of their foreign counterparts. But the Russians' choice of research topics, their way of tackling them, as well as the broader social and cultural implications of, and responses to, scientific activities are unique in many ways to the Russian scene in which they worked.

Notes

1 A. Cunningham and P. Williams, 'Introduction', in A. Cunningham and P. Williams (eds), *Laboratory Revolution in Medicine* (Cambridge: Cambridge University Press, 1992), 1–13: 4.

2 F.M. Dostoevskii, *Notes from Underground* (London: Everyman, 1994), 127.

3 W. Coleman and F.L. Holmes, 'Introduction', in W. Coleman and F.L. Holmes (eds), *The Investigative Enterprise: Experimental Physiology in Nineteenth-Century Medicine* (Berkeley: University of California Press, 1988), 1–14: 4–5.

4 D. Todes, *Pavlov's Physiology Factory: Experiment, Interpretation, Laboratory Expertise* (Baltimore: Johns Hopkins University Press, 2002).

5 There is still no book-length study on the German impact on Russian medicine with the exception of S.A. Chesnokova, 'Osnovnye napravleniia i tendentsii razvitiia fiziologii v Rossii i Germanii v XIX veke i russko-nemetskie nauchnye sviazi', Doct. dissertation (Moscow: Institut normal'noi

fiziologii im. P.A. Anokhina Akademii Meditsinskikh Nauk SSSR, 1979). Expressions of European influence on American physiology and medicine have been explored, particularly the French influence on American medicine, see J.H. Warner, *Against the Spirit of System: The French Impulse in Nineteenth-Century American Medicine* (Princeton: Princeton University Press, 1998); on the German impact on American physiology, see R.G. Frank, 'American Physiologists in German Laboratories, 1865–1914', in G. Geison (ed.), *Physiology in the American Context, 1850–1940* (Baltimore: American Physiological Society, 1987), 11–46.

6 A. Leffler and S. Kowalewski, *Biography and Autobiography,* L. von Cossel (trans.), (New York: Macmillan, 1895), 11. On Kovalevskaia's biography, see A. Koblitz, *A Convergence of Lives. Sofia Kovalevskaia: Scientist, Writer, Revolutionary* (Basel: Birkhäuser, 1983). On Kovalevskaia's studies with Carl Weierstrass and their correspondence, see 'Master and Pupil', in E.T. Bell, *Men of Mathematics* (New York: Simon & Schuster, 1937), 423–32.

7 On the role of the governmental support at Leipzig University, see T. Lenoir, 'Science for the Clinic: Science Policy and the Formation of Carl Ludwig's Institute in Leipzig', in Coleman and Holmes, *The Investigative Enterprise, op. cit.* (note 3), 139–78: 139–45; at the University of Heidelberg, see A. Tuchman, *Science, Medicine, and the State in Germany: The Case of Baden, 1815–1871* (New York: Oxford University Press, 1993), 138–67; at Prussian universities, R. Kremer, 'Building Institutes for Physiology in Prussia, 1836–1846: Contexts, Interests and Rhetoric', in Cunningham and Williams, *op. cit.* (note 1), 72–109: 101–4; see also C.E. McClelland, *State, Society, and University in Germany, 1700–1914* (Cambridge: Cambridge University Press, 1980), 280–1.

8 On the Great Reforms, see A. Polunov, *Russia in the Nineteenth Century: Autocracy, Reform and Social Change, 1814–1914* (Armonk: M.E. Sharpe, 2005), 110–25. On Russia in the Crimean War and her foreign policy after the war, see W. Baumgart, *The Crimean War, 1853–1856* (London: Arnold, 1999), 213–28.

9 N.I. Pirogov, 'Voprosy zhizni', in N.I. Pirogov, *Izbrannye pedagogicheskie sochineniia* (Moscow: Akademiia Nauk SSSR, 1953), 56. On educational reforms of the 1860s, see A. Vucinich, *Science in Russian Culture* (Stanford, CA: Stanford University Press, Vol. I, 1963, Vol. II, 1970), Vol. II, 3–14.

10 Historians of nineteenth-century German physiology and chemistry have pointed to the importance of the small private-research university laboratories such as the laboratories of Johann Purkynje at Breslau, Johannes Müller, Gustav Magnus, and Heinrich Rose at Berlin, in the transition to the first large-scale university-supported institute laboratories that began to appear in the 1870s, see R.S. Turner, 'Justus Liebig versus Prussian Chemistry: Reflections on Early Institute Building in Germany,' *Historical*

Studies in the Physical Sciences, xxiii (1982), 129–62; W. Coleman, 'Prussian Pedagogy: Purkynje at Breslau, 1823–1839', Coleman and Holmes, *op. cit.* (note 3), 16–28; A. Rocke, *The Quiet Revolution: Hermann Kolbe and the Science of Organic Chemistry* (Berkeley: University of California Press, 1993), 28.

11 N. Jardine, 'The Laboratory Revolution in Medicine as Rhetorical and Aesthetic Accomplishment', in Cunningham and Williams, *op. cit.* (note 1), 307–9.

PART I

The German Laboratory and 'Scientific Medicine' in the Late 1850s and Early 1860s: A Russian View

1

The Old–New Tradition

> With the new reign and with the conclusion of peace, all obstacles at once disappeared. The doors were opened wide, and all Russia rushed abroad. It was as if an entirely new world opened, full of charm and poetry, presenting realization of all my ideals. Wonders of nature and art, the educated mode of life of the countries that had left us behind on the way of enlightenment, science, and freedom, people and things – I craved to see all these by myself: I wanted fresh new impressions presenting human life in its height.[1]

For the 1860s generation of young Russian scientists, the period of reform that followed Russia's defeat in the Crimean War was crucial to the unfolding of their scientific and academic careers. Abolition of serfdom, attraction of foreign capital to build a railway network that would develop industry and make the army mobile, the reforms of governmental and social structures – these were tremendous changes that marked a turning point in Russian history. For Russia in particular, the immediate aftermath of the Crimean War signified a great move to modernity, directly associated with industrialisation and developments in technology and science both pure and applied. The poor state of Russian education and science which reached its crisis point under Nicholas I (1796–1855), changed dramatically after the humiliation of the Crimean War, which provided an effective stimulus for reform.

There are a number of justifications for emphasising the magnitude of the changes that stimulated the development of natural sciences in Russia during the second half of the nineteenth century. The essential elements of these changes included liberalisation of academic and cultural life, which gave way to the moral sentiment within the intellectual community that long-awaited improvements in Russian social life could be achieved through developments in science and its practical application to all spheres of human activities. Intellectuals of the 1860s believed in the need to pursue scientific knowledge as part of Russia's openness to the influence of Western European thought. Such changes created favourable conditions for the introduction of modern scientific and medical ideas and laboratory practices, and the young

Image 1.1

St Petersburg c.*1850.*
Courtesy of the Siemens Forum and Archive, Munich.

Russian scientists of the 1860s came to be the vehicles of these advances to the Russian milieu.

These developments were rooted in the tradition of Russians' travelling to Europe to study, and the historical concerns of governmental policy in building its own Russian system of university education. The tradition of the 'travels for knowledge' reflects, in a curious way, both the attempts of the Russian rulers to maintain connections with enlightened Europe, and their old fears that the influx of Western liberal and humanistic ideas would undermine Russia's autocratic order. Introduced by Peter the Great (1672–1725), known for his consistent policy of the Westernisation of Russia, the tradition of getting advanced education abroad, as well as inviting foreign scientists to the newly founded Academy of Sciences became prominent during the reign of Catherine II (1729–96). Catherine II's early humanism and allegiance to the French philosophers of the Enlightenment intensified Western influence at all levels, notably in Russia's academic science. Great European scholars such as mathematician Leonard Euler (1707–83), naturalist and encyclopaedist Peter Simon Pallas (1741–1811), and embryologist Caspar Friedrich Wolff (1734–94) were invited to St Petersburg by the personal request of Catherine II. The eighteenth-century St Petersburg Academy of Sciences did not represent 'Russian science', it represented the high status of science in Russia.[2]

During the reign of Catherine II, Russia became the largest and one of the most powerful countries in Europe. However, as far as education was concerned, she remained backward. Founded in 1755, Moscow University was the only university in the country, while secondary education was inadequately provided and badly planned. Consequently, the expanded state machinery, the army and navy, trade and industry, were all in sore need of qualified personnel. Catherine II formed a committee to work out some sort of educational reform, but fearing the growth of revolutionary ideas in the West, she drew back from implementing reform in education. Moscow University and its two 'gymnasiums' could not satisfy the country's growing demand for professional manpower, and the government had to send advanced students to study law and natural sciences in Leipzig, at the time one of the largest and most prestigious universities in the German states. For medical training, students were sent to Edinburgh and Leyden, centres famous for their clinical teaching.[3]

During the short reign of Tsar Paul I (1754–1801), policy was directed towards obliterating all traces of the governance of his mother, Catherine II, which in his opinion had 'impaired the foundations of absolutist power'. In 1798, following the French Revolution and fearing the spread of its 'poisonous' ideas, Paul I banned foreign literature, including scientific literature, and even music from entering the country. He also refused funds to scholarly institutions. The St Petersburg Academy of Sciences, the pride of Catherine II's heart, was in a desperate state. 'In order to protect the immature minds of the Russian youth from unrestrained and corrupted reasoning', Paul I forbade any studies at foreign universities.[4] Instead, he initiated the foundation of the St Petersburg Medico-Surgical Academy, following the example of the Prussian Pépinière and the Austrian Josephinum, specialised military–medical institutions, which had emerged from the demands of mass armies during the French Revolutionary Wars.[5] In 1799, Paul I gave permission to the Baltic nobility, who traditionally sent their sons to German universities, to set up their own 'Protestant Dorpat University', the predecessor of which, the Akademia Gustaviana had been founded by the Swedish king Gustavus Adolphus (1594–1632) in 1632.[6]

By the time Alexander I (1777–1825) ascended the throne in 1801, education had become a forgotten land. Among the immediate educational measures were the abolition of his father's edict banning studies in foreign universities and generous support for the Academy of Sciences.[7] The young tsar continued Catherine II's policies, which had strengthened Russian science, and in particular had stressed the development of mathematical and natural sciences. The Academy of Sciences was to represent the Enlightenment tradition of publishing scholarly works for both scientific and educational purposes. At the same time, the foremost concern of Alexander I and his 'young friends', French-educated liberal aristocrats, was the foundation of a new education system. The major reason for the decision was the drastic alteration of the political world wrought by Napoleon. Russia needed to add cultural prestige to her growing military influence in Europe, and she needed a national, centralised university system to train officials, doctors, teachers, and engineers. The French influenced the shape of Russian higher education policy in the 1800s. Russia would adopt the centralising, highly hierarchical principles associated with the Napoleonic system of 'université impériale'.[8]

In 1802, following the course of moderate liberal reforms, Alexander I decreed a re-organisation of the governing body of state machinery into eight ministries. The first such ministry to function was a new Ministry of Education, which acquired, for the first time, a political significance as well as a cultural one. The Russian Empire was divided into six educational districts, each with a university, governing the work of the schools within the district. Three new imperial universities were founded – in Kazan in 1804, in Kharkov in 1805, and in St Petersburg in 1819 – as district centres of education. The two old ones, in Vilno and in Dorpat, were re-opened as Imperial Universities in 1802. The Central Board of Educational Establishments attached to the Ministry of Education was placed in charge of the Academy of Sciences, universities and other educational institutions. All of these establishments' charters conferred upon them a regular, uniform system of organisation and gave them the power to confer degrees, and prescribed that, irrespective of the social class of the recipient, those degrees should carry with them a right to enter the state service.[9] The state, however, held a dominant role. Universities had powerful curators appointed by the tsar and the government, usually from the military élite and aristocracy, and were dependent on the state budget for most of their capital and operating expenses.

Powerful supporters and authors of the educational reform, such as Prince Adam Czartoryski (1770–1861), took a strong personal interest in university matters. In the re-organisation of the re-opened universities, Vilno and Dorpat in particular, the planners of the reform were strongly and consciously influenced by the model of Göttingen University. Göttingen was a Protestant university, largely state funded and state controlled, modernising in outlook and curriculum, and designed to attract upper-class students.[10] Czartoryski became the curator of Vilno University, founded in 1578, in the Grand Duchy of Lithuania, which had fallen under Russian rule in 1793. Under Czartoryski, Vilno University began to flourish and notably attracted members of the Polish aristocracy. To implement reorganisation of the Vilno medical faculty, Czartoryski invited Johann Peter Frank (1745–1821), known for his reformist activities at the medical faculties of Göttingen in 1786, and Vienna in 1798. Before long, Frank was transferred to St Petersburg, and his son Joseph, graduate of both Göttingen and Padua continued the reformation of the medical faculty at Vilno.[11]

In 1802, on his way to Prussia, Alexander I visited Dorpat University. Greeting his august visitor in French, the Professor of Physics, Georg Friedrich Parrot (1767–1852) skilfully incorporated into his speech the ideas of the Enlightenment. Parrot managed to pinpoint the ambitions of the young tsar, to be 'a promoter of education, faithful to the ideals of enlightened monarchism and the lofty humane aims of science'. Presented in such an appealing manner, Parrot's speech immediately 'gained benevolence and favour of the tsar'. Parrot became Rector of Dorpat University, and later, Academician of the St Petersburg Academy of Sciences. His close relations with Alexander I proved decisive for the most generous funding of Dorpat University and a special attitude to its freedom and independence.[12]

The Dorpat University acquired an important and unique place in the newly established Russian university system. Its special status as the university for Baltic Germans and geographic closeness to Europe made it the most advanced among Russian universities in the first half of the nineteenth century, where German influences on education and science were especially noticeable. German universities had tended to train more specialists than the states needed, and many graduates – doctors, pastors, lawyers and teachers – left to work in the Baltic provinces and Russia. In the early nineteenth century, eighty-two percent of the teaching staff at Dorpat University were German, and many of them were highly qualified for their work. Professor of Chemistry and Pharmacy, Alexander Nicolas von Scherer (1771–1824), a notable scholar and editor of the *Allgemeines Zeitschrift der Chemie* contributed much to chemistry teaching and research at Dorpat.[13] The lavishly funded, excellent Dorpat University Observatory provided training for the future great astronomer Friedrich Georg Wilhelm von Struve (1793–1864), who later became its director. Struve's observatory and instrument collection was the best in Europe. The acquisition in 1824 of a twenty-four-centimetre refractor by Joseph Fraunhofer (1787–1826), the largest such instrument in the world, made possible the most astonishing astronomical study of the century, Struve's *Mensurae micrometricae*, which provided information on the positions, distances, and brightness of over three thousand double and multiple stars.[14]

The three newly founded Imperial Universities, in Kazan, Kharkov, and St Petersburg, bore some of the traits of Göttingen University.[15] Reformers at least attempted to introduce into the new universities the kind of freedom of teaching and research associated with the reformed German Protestant universities. Indeed, many of the new teaching staff brought in to inject new academic life in Kazan and Kharkov were German universities' alumni. In the 1810s, Kazan University, the most Eastern and remote university in the Empire, acquired the services of a group of distinguished German professors and, according to Friedrich Engel, its Faculty of Physical and Mathematical Sciences had more accomplished scholars than most German universities. Johan Martin Bartels, the teacher of Carl Friedrich Gauss (1777–1855), taught pure mathematics, Casper F. Renner taught applied mathematics, Felix C. Brunner taught physics, and Joseph J. Litton taught astronomy. All three taught the future creator of non-Euclidian geometry, Nikolai Lobachevskii (1792–1856), the 'Copernicus of geometry'. The German scholars brought to Kazan University up-to-date knowledge in exact sciences, modern educational ideas, and a great deal of philosophical erudition, Kant's ideas in particular. In great measure they were responsible for the conversion of Kazan University from an isolated intellectual oasis at the threshold of Siberia into a cultural centre alert to modern scientific developments.[16]

After Napoleon's defeat, the general political reaction in Europe manifested itself in a conscious effort to bolster the authority of monarchy and the church. In Russia, Alexander I drastically changed his educational policy after 1815. The newly established Ministry of Religious Affairs and National Education, which succeeded the Ministry of Education, brought to the fore the idea that the basis of all true enlightenment was 'a salutary concord between faith, knowledge, and authority'. The new direction was chiefly implemented by the Ministry's official Mikhail Magnitskii, who led the campaign to suppress academic freedom at the educational institutions. Kazan University, which Magnitskii considered a 'deplorable den of atheism and immorality', suffered the most. The effect on the Universities was ruinous: they lost many professors, mainly foreign, either through dismissal or departure. The universities as centres of tutorial free thought were 'amended' to be guided by Magnitskii's 1820 'Instruction', aimed at eradicating the modern scientific approach to teaching the philosophical, historical, and political sciences as well as mathematical and natural sciences. The 'Instruction' also imposed strict regulations on students personal behaviour. The students themselves were to be 'formed into a sort of quasi-monastical body.'[17] The suppression of the universities was the most glaring example of the type of control over intellectual life for which the Magnitskii era is notorious.

The move by Alexander I towards stricter control over domestic universities was provoked by the liberal agitation and radical activism of the student fraternities, Burschenschaften, in German universities. The Burschenschaft movement was a spontaneous accompaniment to the general reforms being carried out in universities and was the most influential of the organisations calling for constitution, national unity, and freedom. Neo-humanist ideas about the liberty of learning and general education reform, and the radical nationalism of the Burschenschaft, came to be seen as a threat to the policies of Prince Klemence von Metternich-Winneburg (1773–1853), the all-powerful Minister of Foreign Affairs of the Habsburg Empire and one of the founders of the Holy Alliance. The Wartburg Student Festival of 1817 and the assassination of the writer and informant to the Russian government, August von Kotzebue, by a student in 1819, frightened the German governments. They finally agreed to Metternich's demands for suppression measures and control of universities, adopted at a meeting of the German states in Carlsbad in 1819.[18]

Alexander I, another Holy Alliance originator, was forced to follow the decisions of the Carlsbad Conference in order to avoid undermining the existing order at home. After Carlsbad, access to international study for students was limited by the Russian government. In the 1820s, Russians were forbidden to attend universities in Jena, Giessen, Heidelberg, and Würzburg, where student unrest and dangerous liberalism were especially pronounced. However, in favour of Russia's intellectual and cultural connections with Europe, Alexander I was not equally suspicious of all German universities. He considered that the governments of Prussia and Hanover were on guard against pernicious and harmful influences on youth, so Berlin and Göttingen universities were not ranked as dangerous.[19]

To sum up, the Russian educational system in the first quarter of the nineteenth century underwent a series of unparalleled changes. Despite governmental inconsistencies and paradoxes in Alexander I's educational policy, Russia built an established network of Imperial Universities, and an adequate system of gymnasium education. Universities' student numbers increased to 51 at St Petersburg, 820 at Moscow, 332 at Kharkov, 118 at Kazan, 365 at Dorpat, and 900 students at Vilno University. Whereas this strengthening of the universities was quite slow, the professional schools developed into well-established training centres and scientific workshops. In addition to the handful of specialised schools, such as the Medico-Surgical Academy and the Mining Institute, founded in the late eighteenth century, there were seven more established between 1806 and 1824. The Main Pedagogical Institute became the major institution for training teachers for all school levels, for all parts of the country. The two other prestigious schools, the Communication Institute, similar to the Paris École des Ponts et Chaussées, and the Engineering School, placed special emphasis on an education in mathematical and natural sciences.[20] As a result, Alexander I's reign is remembered not only for historically significant events such as the campaigns of 1812–15 against Napoleon, but also for its contributions to the creation of new conditions that would profoundly affect Russian science and education.

The sudden death of Alexander I, in December 1825, was followed by the famous imbroglio over the Imperial succession. A group of the nobility belonging to the élite guard corps used the opportunity to stage an abortive military revolt against Russia's two pivotal institutions, autocracy and serfdom. The participants of the Decembrist uprising had acquired their conceptions of civil rights from reading the classical Western European political and social theorists. The Decembrists were convinced that Russia's progress was dependent on freedom of thought, which could not exist without constitutional government.[21] The new tsar, Nicholas I (1796–1855), with an intensified suspicion of the West, consolidated all means possible to suppress the forces opposing the power and social standing of the autocracy, and imposed a policy of political oppression and official nationalism. In foreign policy, Nicholas I maintained the role of 'gendarme of Europe'.

Censorship regulations in 1826 and 1828 strictly controlled all published literature, including scientific publications, and higher learning institutions fell under the restraint of official power. The 1830 Revolution in France interrupted Franco–Russian relations, which prevented Russian students from attending French universities. The Polish revolt of 1830–1 resulted the in restriction of any studies abroad. Seeking to bar the way to revolutionary ideas from Europe, Nicholas I resorted to drastic means in the re-organisation of the education in the western districts of the Russian Empire. Vilno University was abolished in 1832 and its medical faculty was converted into a specialised Academy of Medicine and Surgery, similar to the St Petersburg Medico-Surgical Academy. For the Polish aristocracy, Nicholas I founded in 1834 the University of St Vladimir in Kiev, intended to symbolise 'the close relation of intellectual strength to military power.'[22]

By contrast, Dorpat University, although affected by the government's nationalist policy, continued its existence under conditions of moderate reaction. Its former Rector, Georg Parrot, by then Academician of the St Petersburg Academy of Sciences, and Count Karl von Lieven, the influential Trustee of the St Petersburg Educational District, consistently promoted significance and supported independence of Dorpat University. Beginning with Magnitskii, the government policy of replacing foreign professors with Russians had seriously depleted the ranks of university instructors. The

Professors' Institute was founded in 1828 at Dorpat University with the purpose of training Russian academic personnel. Parrot, the founder of the Institute, suggested that outstanding students from Moscow, Kazan, and Kharkov Universities should continue their studies first in Dorpat and after graduation, specialise abroad. This scheme appeared quite efficient and, the Institute during the short period of its existence, prepared some twenty professors, six of them in medicine. The future distinguished surgeon Nikolai Ivanovich Pirogov (1810–81), one of the Institute's graduates, left revealing memoirs of his time studying at Dorpat under his mentor, Johann Christian Moier (1786–1858). A student of the great Pavia anatomist Antonio Scarpa (1747–1832), an able pianist and friend of the composer Ludwig van Beethoven (1770–1827), Moier was an eminent representative of Dorpat medical faculty.[23] Limited though it was, the Institute's programme of training was great help in mitigating the major difficulty in Russian education system, a lack of national academic cadres. Equally important was the role of Dorpat University as a centre for learning and scientific exchange between Russia and Europe. While Russia was secluded from any Western influence, Dorpat University remained the only 'window' open to the developments of European science for Russian scientists and students.

From 1833, under Count S.S. Uvarov (1786–1855), the educational policy at gymnasiums and universities shifted from humanistic and scientific to didactic formalism with a particular emphasis on mathematics and classics. Uvarov's doctrine as Minister of National Education was memorably simple: 'Orthodoxy, autocracy, and nationality'. His oppressive legislation culminated in the elimination of university autonomy in 1835. Teaching at the universities was placed under tight control, and a core of 'national' subjects, such as ancient and contemporary Russian history, Russian law, church Slavonic and Russian languages, and orthodox church history and theology began to acquire a prominent place in the curriculum. The new University Regulations made the university fully subordinate to the Trustee, the appointed official, usually of military rank.

Highly educated and gifted with respect to the applied sciences, Uvarov had been President of the St Petersburg Academy of Sciences since 1818. As a young man in diplomatic service, he had become personally acquainted with Goethe and Alexander von Humboldt, who, in 1829, was invited to undertake a systematic survey of the geography of Central Asia, the first study of Russia's physical environment based on modern scientific methods. To maintain the prestige of the Academy, Uvarov acquired the services of a group of renowned scientists from Germany and Dorpat: the embryologist Karl Ernst von Baer (1792–1876), the physicist Heinrich F. Lenz (1804–65), the analytical chemist Hermann Hess (1802–50), and the astronomer Struve. Uvarov placed the resources of the Academy's observatory, which was opened in 1839 in Pulkovo near St Petersburg and contained the most complete and splendid collection of instruments in Europe, at the disposal of Struve. The collection included a fifteen-inch refractor by Merz & Mahler of Munich (Fraunhofer's successors), which exceeded in size and quality the famous Dorpat telescope, which, for a long time, had no rival in the world. As before, the Academy of Sciences more than compensated for the meagre development of scientific research at the universities.[24]

Although an ardent proponent of official nationalism, Uvarov, however, rescinded the ban on foreign studies immediately after assuming his ministerial post in 1833. He realised that the Dorpat Professors' Institute alone could not meet the demands of even the slowly growing national university system, especially in natural sciences. The shortage of instructors and lack of adequate libraries, research and teaching facilities made the European universities the only places for Russian students to get postgraduate training in natural sciences and medicine. Besides, German universities traditionally remained places for education of the Russian nobility and of those preparing for governmental services in such fields as jurisprudence and law. In the 1830s, Russian graduates were allowed to study in Berlin, Göttingen, and Heidelberg. According to Pirogov, in 1833:

> [A]ll the medical students were supposed to go to Berlin, all natural scientists to Vienna, and all others (jurists, philologists, and historians) also to Berlin. For some reason, no one was permitted to go to France and England.[25]

Russians students abroad were under vigilant surveillance of the Russian officials. A.A. Voskresenskii (1809–80), the future distinguished chemistry teacher, did not get permission from Count de Ribopierre, the Russian ambassador in Prussia, to leave Berlin for Giessen to study with Justus Liebig. Darmstadt principality still had a suspicious reputation amongst Russian high officials. Liebig had to write a letter to the embassy that Voskresenskii would be engaged only in the study of chemistry.[26] Pirogov recalled an episode when Nicholas I during his visit to Prussia in 1833 ordered all Russians studying in Berlin to the embassy. At the meeting, the Emperor reprimanded those students wearing moustaches, as it was a privilege of the military, and commanded that these be shaved off immediately.[27]

The revolution of 1848 in Western Europe led to the most severe restrictions within Russian universities. By Nicholas I's decree, the enrolment in any university was restricted to three hundred students, only medical faculties did not fall under that rule. Access to the universities was limited to the nobility, and the total university enrolment was considerably reduced: from 4,016 students in 1848, to 3,018 in 1850. Western literature on science and philosophy again was banned from entering the country. All studies abroad were prohibited.[28] B.N. Chicherin, the jurist and historian, and member of the 'Moscow circle of Westerners' recalled:

> Since 1848 the Russian government had made all possible difficulties for those daring to cross the holy borders of the fatherland... A traveller abroad was considered a man who had partaken of the fruits of enlightenment.[29]

In the early 1850s, however, the educational policy began to change, influenced by the example of the German government, which consistently supported science education in gymnasiums and universities. These policies were aimed at strengthening scientific education, which had been seriously undermined by the classical curriculum. The teaching of natural sciences, however, did not make any gains due to a lack of modern teaching facilities and laboratories, adequate science libraries and qualified instructors.[30] Nevertheless, the turn in policy was conducive for the development of natural sciences in the following decade.

After the death of Nicholas I in 1855 and Russia's defeat in the Crimean War in 1856, the situation in the country changed drastically. The repercussions of the Crimean War on the policy of the European powers and on their mutual relations were also most profound. The 'Concert of Europe,' which had been formed after the Napoleonic Wars in order to subdue revolutionary movements, had to a large extent broken down. Russia entered the new era, the period of 'great reforms' of the 1860s. As mentioned earlier, these reforms powerfully affected most institutions, including the entire educational system. Among the first measures of the government of Alexander II (1818–81) was the annulment of all restrictions for studies abroad. The Ministry of Foreign Affairs facilitated the procedure for getting travel documents and reduced the high prices for issuing foreign passports.[31] This sparked an unprecedented move to European centres of learning by Russian graduates, such as Sechenov, Botkin, and Borodin. Their experiences as young men reveal the overall enthusiasm for advances in experimental sciences that were being made in the late 1850s in Germany. Their aspiration to bring back the scientific ideas and methods, and their belief that science would help to overcome Russia's backwardness was representative of their age and generation.

By looking at the developments in experimental physiology, 'scientific' medicine, and chemistry through the eyes of Sechenov, Botkin, and Borodin, the transformations in research practice and pedagogy become clear from the perspective of those in the midst of events both in Germany and in Russia. By extention, much can be learnt from these perceptions and images about the rise of the German laboratory, its and societal implications.[32] The focus on the cross-national transfer of ideas, methods, apparatus, and people from laboratory to laboratory opens a window on a critical stage in the development of the German laboratory, that became a model widely emulated, particularly in Russia.

Notes

1. B.N. Chicherin, *Vospominaniia: puteshestvie za granitsu* (Moscow: Sever, 1935), 21. Boris N. Chicherin (1828–1904), Professor of Law at Moscow University and a prolific writer on the history of law and political science, was one of the most notable representatives of the liberal trend in Russian philosophic, juridical, and historiographical school of thought. His *Vospominaniia* offers a comprehensive overview of political, cultural, and academic life in mid-nineteenth century Britain, Italy, France, and Germany. An aristocrat and admirer of the constitutional monarchy system, Chicherin visited sittings of the British Parliament, and communicated with some prominent historians and jurists in Berlin, Heidelberg, Munich, and Vienna. Chicherin was a close associate of the Grand Duchess Elena Pavlovna (1806–73), known for her salon, where all distinguished figures of the St Petersburg literary, musical, scientific, and liberal political circles were welcome. Notably, here in the 1850s, the plans of the reforms that would be implemented in the 1860–70s were under discussion. Elena Pavlovna was also known for her philanthropic activities: several hospitals and charity-homes were founded on her generous donations, as well as the community of the Sisters of Mercy during the Crimean War. See, Chicherin, *idem,* 30–3, 73–87; see also B.I. Ivanov, *et al.* (eds), *Istoriia Otechestva s drevneishikh vremen do nashikh dnei. Entsiklopedicheskii slovar'* (Moscow: Bol'shaia Rossiiskaia Entsiklopediia, 1999), 201, 452.
2. On foreign scholars at the St Petersburg Academy of Sciences during the eighteenth century, see A. Vucinich, *Science in Russian Culture* (Stanford: Stanford University Press, 1963), Vol. 1, 145–57, 295–304, 308; on science, Enlightenment, and absolutism in the age of Catherine II, see *ibid.,* Vol. I, 125–35. On the correspondence of Catherine II with Voltaire, d'Alembert, Diderot, and Madame Geoffrin, see S.M. Solov'ev, *Istoriia Rossii s drevneishikh vremen* (Moscow: Mysl', 1994), 18 vols, Vol. 13, 469–75.
3. V.O. Samoilov, *Istoriia Rossiiskoi meditsiny* (Moscow: Epidavr, 1997), 61–2.
4. *Polnoe sobranie zakonov Rossiiskoi imperii s 1649,* Vol. 25, (1798), no. 18474, cited in E. Wieschhöfer (ed.), S.G. Svatikov, *Russische Studenten in Heidelberg* (Heidelberg: Universitätsbibliothek Heidelberg, 1997), 9–10.

5. K. Pollak, 'Josephinum und Pépinière: Der Beitrag des Militärsanitätswesens zur Vereinigung von Chirurgie und Medizin', *Wehrmedizin und Wehrpharmazie,* iii (1985), 134–8.
6. The Baltic provices fell under Russian rule during 1710–21. On Dorpat University, see E.V. Petukhov, *Imperatorskii Iur'evskii, byvshii Derptskii, universitet za sto let ego sushchestvovaniia, 1802–1902* (Iur'ev, 1902), 2 vols, Vol. I, 50–4.
7. *Polnoe sobranie zakonov, op. cit.* (note 4), Vol. 26 (1830), no. 599, cited in Vucinich, *op. cit.* (note 2), Vol. I, 189.
8. On the design of the new educational system by Napoleon Bonaparte in 1806–8, see C. Musselin, *The Long March of French Universities* (New York: Routledge Falmer, 2004), 9–12. On the Napoleonic era and the most extensive changes on the German university scene, see C.E. McClelland, *State, Society, and University in Germany, 1700 –1914* (Cambridge: Cambridge University Press, 1980), 143–4; 151–3.
9. On the foundation of the university system in Russia, see V.O. Kliuchevskii, *A History of Russia.* C.J. Hogarth (trans.), (New York: Russel & Russel, 1960), 5 vols, Vol. V, 161–3.
10. On Göttingen University and its reformation, see McClelland, *op. cit.* (note 8), 58–9.
11. On Vilno University and Czartoryski's activities, see N.K. Schilder, *Imperator Alexander I: Ego zhizn' i tsarstvovanie* (St Petersburg: Izdatel'stvo Suvorina, 1897), 4 vols, Vol. II, 42–5. On Frank's activities as a reformer, see E. Lesky, 'Johann Peter Frank als Organisator des medizinischen Unterrichts', *Sudhoffs Archiv,* xxix (1955), 1–29. Frank's famous *System einer vollständigen medizinischen Polizey* published in 1779 is a multi-volume work on the administration of social welfare including the control of fertility, disease and mortality, and the most prominent example of 'cameralist' public health in Germany. See, J.P. Frank, 'The Civil Administrator: Most Successful Physician', J.C. Sabine (trans.), *Bulletin of the History of Medicine,* xvi (1944), 289–318.
12. On Parrot, see Schilder, *op. cit.* (note 11), Vol. 2, 87–8; see also K. Siilivask (ed.), *History of Tartu University, 1632–1982.* (Tallin: Periodika, 1985), 74–5.
13. On Scherer, see J. Partington, *A History of Chemistry* (London: Macmillan, 1964), 4 vols, Vol. 3, 598.
14. On Struve and his collection of astronomical instruments, see M. Hoskin (ed.), *Cambridge Illustrated History of Astronomy* (Cambridge: Cambridge University Press, 2000), 216–7.
15. P. Miliukov, 'Universitety v Rossii', in F.A. Brockhaus and I.A. Efron (eds), *Entsiklopedicheskii slovar'* (St Petersburg: Izd-vo Brokgauza i Efrona, 1902), Vol. 18, 788–800: 789.

16. F. Engel, 'Introduction' in N.I. Lobatschewsij, *Zwei geometrische Abhandlungen,* F. Engel (trans.), (Leipzig: Vogel Verlag, 1898), 420. On Lobachevskii, see A. Vucinich 'Nikolai Ivanovich Lobachevskii: The Man Behind the First Non-Euclidean Geomentry', *Isis,* 53 (1962), 465–81: 470; see also Vucinich, *op. cit.* (note 2), Vol. I, 314–29.
17. Kliuchevskii, *op. cit.* (note 9), Vol. V, 162–4; see also Vucinich, *op. cit.* (note 2), Vol. I, 233–40.
18. McClelland, *op. cit.* (note 8), 146–7. Heidelberg University, the Grand Duchy of Baden, was founded in 1386; Jena University, Thuringia, in 1558; Leipzig University, the Kingdom of Saxony, in 1409; Giessen University, the Grand Duchy of Hesse, in 1607; Würzburg University, the Kingdom of Bavaria, in 1402 (1582); Göttingen University, the state of Lower Saxony, former Grand Duchy of Hanover, in 1737; and Berlin University, Brandenburg–Prussia, was founded in 1803. See R.A. Müller, *Geschichte der Universität: von der mittelalterlichen Universitas zur deutschen Hochschule* (Munich: Callwey, 1990).
19. On the reaction of the Russian government on the activities of *Burschenschaften,* see Svatikov, *op. cit.* (note 4), 13–7; on the German response to radical nationalism in the 1820s, see M. Kitchen, *Cambridge Illustrated History: Germany* (Cambridge: Cambridge University Press, 1996), 162–4.
20. Vucinich, *op. cit.* (note 2), Vol. I, 198, 221, 242.
21. A classical account on the secret political societies of 1816–25, on the 'catastrophe of 14 December, 1825', its participants, their motives, education, see Kluchevskii, *op. cit.* (note 9), Vol. V, 165–72, see also N.P. Pavlov-Sil'vanskii, *Ocherki po russkoi istorii 18–19 vekov* (St Petersburg, 1910), 56, cited in Vucinich, *op. cit.* (note 2), Vol. I, 239.
22. Muliukov, *op. cit.* (note 15), 791.
23. N.I. Pirogov, *Questions of Life: Diary of an Old Physician,* G. Zarechnak (trans.), (Canton: Watson Publishing International, 1991), 278–9. Written in 1881, Pirogov's *Questions of Life* is a revealing record of social and medical life in nineteenth-century Russia.
24. On foreign scholars at the St Petersburg Academy of Sciences during the first half of the nineteenth century, see Vucinich, *op. cit.* (note 2), Vol. I, 295–304, Vol. II, 66–7. On Pulkovo Observatory, see N.I. Nevskaia, 'Astronomicheskaia stolitsa mira', in Z.I. Alferov (ed.), *Akademicheskaiia nauka v Sankt Peterburge v XVIII i XIX vekakh* (St Petersburg: Nauka, 2003), 237–54: 246–9; see also Hoskin, *op. cit.* (note 14), 272–5.
25. Pirogov, *op. cit.* (note 23), 337.
26. A promising practical chemist, Voskresenskii eventually left his research career entirely for teaching at the St Petersburg Main Pedagogical Institute. A number of notable chemists, D.I. Mendeleev, N.N. Beketov, N.N. Sokolov,

N.A. Menshutkin, and A.A. Alekseev were among his students. On Voskresenskii, see M.N. Mladentsev and V.E. Tishchenko, *D.I. Mendeleev, ego zhizn' i deiatel'nost'* (Moscow: Akademiia Nauk SSSR, 1938), 57–9.

27. Pirogov, *op. cit.* (note 23), 343–4.
28. On the reactionary educational policy of Nicholas I during the 1848 revolution in Europe, see R.G. Eitmontova, *Russkie universitety na grani dvukh epokh: ot Rossii krepostnoi k Rossii kapitalisticheskoi* (Moscow: Nauka, 1985), 58–61.
29. Chicherin, *op. cit.* (note 1), 21.
30. Vucinich, *op. cit.* (note 2), Vol. I, 255, 367.
31. N.A. Belogolovyi, *S.P. Botkin. Ego zhizn' i meditsinskaia deiatel'nost'* (St Petersburg: Tipografiia J.N. Erlikha, 1892), 21.
32. For an example of a recent study on the German laboratory, see L. Otis, *Müller's Lab* (Oxford: Oxford University Press, 2007). For an overview of the historical studies on German chemical laboratory, see A. Rocke, *The Quiet Revolution: Hermann Kolbe and the Science of Organic Chemistry* (Berkeley: University of California Press, 1993), 4.

2

Physiologist-Physicists: Foundation of the Discipline

After their first meeting as young men in Berlin in 1847, Sechenov's two mentors, Emil du Bois-Reymond (1818–96) and Carl Ludwig (1816–95), became the dominant figures in the world of physiology. In counterpoint to the hitherto prevailing influence of the Romantic Naturphilosophie with its central notion of Lebenskraft, or 'vital force', they built the discipline of physiology around the study of function in relation to physical and chemical principles and in separation from structure and morphology. During the first decade of their careers, they developed methodological and theoretical tools that transformed a discipline subservient to anatomy into an independent science, a fusion of physics and chemistry. To achieve this, they successfully applied the method of the controlled experiment from contemporary physics and chemistry to physiological. They designed new physiological apparatus and devices, based on instrumentation used in mechanics, ballistics, and electrical telegraphy that made it possible to trace and measure, with astonishing precision, physical changes in isolated functioning organs.

Sechenov, who studied in in the laboratories of du Bois-Reymond, Ludwig, and Hermann von Helmholtz (1821–94), and was closely acquainted with Brücke, once said that du Bois-Reymond was the driving force and inspiration behind the new trend of physico–chemical physiology introduced by these four men.[1] This is no exaggeration. Sechenov pointed to du Bois-Reymond's tireless encouragement and support of the work and talents of his friends. Du Bois-Reymond's close friendship and co-operation with Brücke, and his careful nurturing of Helmholtz's early career, can be seen in the correspondence of du Bois-Reymond with both Helmholtz and Ludwig. In these letters, du Bois-Reymond is at his most enthusiastic – declaring his belief in a physics-based physiology and the need to sweep away all vitalism. He comes across as devoted to his friends and irreconcilable to his opponents. With a broad education and interests in the sciences, medicine, art, and languages, du Bois-Reymond is a fine representative of the German men of science of his time.

Sechenov also understood the immense debt of Russian physiology to German physiologists, Ludwig in particular. As Ludwig's first Russian student and his life-long friend, Sechenov paved the way for a whole generation of Russian physiologists who studied with Ludwig and went on to make laboratory medicine in Russia. There is no better way to understand the development of mid-nineteenth-century German experimental physiology and its profound influence on the rise of this discipline in Russia, than to examine the early work of du Bois-Reymond and Ludwig in these formative years of the physical trend in physiology, 1843 to 1858.

Emil du Bois-Reymond

Emil du Bois-Reymond was born in 1818 in Berlin, into the family of the Privy Counsellor representing the Swiss canton of Neuenburg in the Prussian Foreign Ministry. Both his parents belonged to the Berlin French community, and Emil attended a French gymnasium from 1830 to 1837. At Berlin University, he studied theology and philosophy, as well as chemistry and mathematics, and in 1839 went on to medicine. He received his medical degree in 1843 and two years later became a founding member of the Berlin Physical Society. At one of the sittings of the Society, he met Lieutenant Werner von Siemens and mechanic Georg Halske, whose electrotelegraphic workshop became indispensable for du Bois-Reymond in the technical matters of his electrophysiological research. In 1848, the first volume of du Bois-Reymond's landmark *Untersuchungen über thierische Elektrizität* appeared in print.[2]

This was the period known as the 'Vormärz', the years which spanned from the Congress of Vienna of 1814–15 to the Berlin March Revolution of 1848. In German art and thought it was the time of various short-lived movements, Classicism followed by a period of Romanticism, which in turn gave way to the realism of the Biedermeier period. The educational reforms of the 1810s in Prussia had helped to create a highly educated and often liberal bourgeoisie élite. The German drive towards unity, which became particularly pronounced after the Napoleonic wars, had economic effects. After the Pan-German customs union, the Zollverein, operating from 1834, a common market of twenty-five million Germans had been brought under the skilful and flexible leadership of the Prussian bureaucracy.[3] Construction of railways and the telegraph lines connecting all the big cities in the German states gave a decisive stimulus for Prussian industrialisation. The first steam-driven railway in Prussia was built in 1837, and a decade later, in 1847, the newly founded firm Siemens & Halske constructed a telegraph line from Berlin to Frankfurt.

Du Bois-Reymond spent his formative years at Berlin University, then one of the most advanced educational centres in Germany, particularly in medical and natural sciences. Founded in 1811 by the Prussian philosophers and reformers Johann Gottlieb Fichte (1762–1814), Wilhelm von Humboldt (1767–1835), and Friedrich Schleiermacher (1768–1834),

Image 2.1

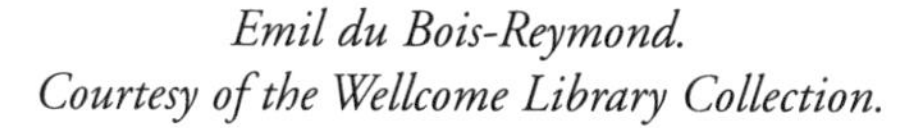

Emil du Bois-Reymond.
Courtesy of the Wellcome Library Collection.

Berlin University represented a new model for the broad education of the human mind and spirit. At the University, humanities were particularly emphasised, for Humboldt preferred to educate free men with an interdisciplinary curriculum rather than produce narrow specialists or professionals. For Humboldt, inclined towards idealist philosophy and the intellectual intuition of Friedrich Wilhelm von Schelling (1775–1854), the pursuit of pure knowledge was the paramount goal of education. In the 1820s, Carl Freiherr Stein von Altenstein (1770–1840) succeeded Humboldt as *Kultusminister.* Von Altenstein, much influenced by Immanuel Kant's (1724–1804) stress on sense intuition and empiricism, began to shift policy gradually towards a greater emphasis on teaching natural sciences. Over the following years, von Altenstein brought in a number of experimental scientists, Gustav Heinrich Magnus (1802–70) and the Rose brothers, Heinrich (1795–1864) and Gustav (1798–1873), all students and

Image 2.2

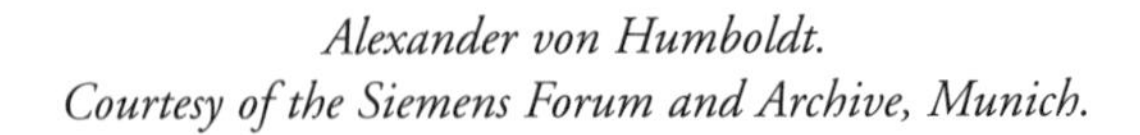

Alexander von Humboldt.
Courtesy of the Siemens Forum and Archive, Munich.

collaborators of the brilliant bench and theoretical chemist Jöns Jacob Berzelius (1779–1848).[4] Von Altenstein agreed to accept Johannes Müller (1801–58) as a successor to Carl Asmund Rudolphi (1771–1832) in the chair of anatomy and physiology.[5] Müller's adherence to the biological tradition of Kant and Goethe and his philosophy that observation and experiment were instruments for understanding the nature of life, was largely responsible for his appointment to both the philosophical and medical faculties.[6] The neo-humanist reforms and the new Wissenschaftsideologie were associated with a strong commitment by the faculty of Berlin University for research and publication, combined with teaching.[7]

The 1830s and 1840s saw the emergence of the first research laboratories at Berlin University. Although not for general education these small laboratories for individual scientists were early innovations in laboratory-

Image 2.3

Ernst Brücke.
Courtesy of the Institute for History of Medicine
at the Humboldt University of Berlin.

based instruction in chemistry and physiology. The wide-ranging researchers Gustav Magnus and Heinrich Rose, in the early 1840s, used their academic laboratories only for personal research with a few advanced students and for the preparation of lecture demonstrations, not for general education.[8] The same was true for Johannes Müller, who pursued investigations ranging from experimental physiology, with the use of vivisectional techniques and chemical methods, to cellular pathology, and from comparative anatomy to embryology. Müller's personal laboratory was also available only for exceptional students.[9]

Among Müller's exceptional students, who were invited to study in his 'Kabinet' during 1840–3, were du Bois-Reymond and his close friend Ernst Brücke (1819–92). Müller's animated anatomy based on observation,

Image 2.4

Heinrich Gustav Magnus.
Courtesy of the Siemens Forum and Archive, Munich.

dissections, and experimentation, and his research interests clearly influenced the choice of dissertation topics of his students. Brücke studied the diffusion of fluids through the living and non-living membranes, a topic of particular interest for Müller in the 1830s. Du Bois-Reymond's thesis was on electric fish, a topic of continuous interest for many natural scientists of the older generation, including Müller and Alexander von Humboldt. The latter was particularly interested in nerve–muscle physiology and, as early as 1797, published research on irritated muscle and nerve fibre, in the two-volume title *Versuche über die gereitzte Muskel- und Nervenfaser.*[10] In 1841, Humboldt brought to Müller, from Paris, the recently published *Essai sur les phénomènes électriques des animaux* (1840) by the Italian physicist Carlo Matteucci (1811–68). Matteucci had demonstrated the electrical negativity of the cross-section of a muscle, and described in his *Essai* the electric oscillation in a tetanised muscle. Aware of du Bois-Reymond's interest in physics, Müller suggested a topic for further research based on the verification of Matteucci's experiments.

Yet, at the same time the early research of both du Bois-Reymond and Brücke, by aim, execution and clear physicalistic bent, were driven by an even stronger influence than that of their primary mentor. 'Brücke and I swore to each other to validate the basic truth that in the organism no other forces have any effect than the common physico–chemical ones.'[11] This commitment to follow the developments in physics and to exploit its methods, technique, and instruments in physiological research was largely due to the influence of Gustav Magnus and his colloquia, which du Bois-Reymond and Brücke attended from 1841. At these informal classes regularly held in Magnus's house at Kupfergraben, the young men became acquainted with other students in physics and chemistry, who formed the Jüngerer Naturforscherverein – the 'association of young natural scientists'.[12]

Gustav Magnus, a collaborator and a close friend of Berzelius, studied in Paris in the 1820–30s as did other French-educated chemists of the time, Justus Liebig (1803–73), Heinrich Rose, Friedrich Wöhler (1800–82), and Robert Bunsen (1811–99). What appealed most to the German chemists and physicists was the more realist and phenomenalist style of explanatory theory, characteristic of such luminaries as Joseph Louis Gay-Lussac (1778–1850) and Louis Jacques Thenard (1777–1857). The emphasis on experimentation, use of mathematical methods, the avoidance of speculation, and the view of nature as ruled by natural laws, were features that German science education lacked at the time. As a true Berzelian, Magnus throughout his career was committed to materialist–mechanist explanation, both in chemistry and physiology, and to categorical rejection of all mystical obscurantism. Berzelius, in his highly influential *Lehrbuch der Chemie* (1827), stated that *Lebenskraft* or 'vital force' was simply a term used to refer to the complexities of animal chemistry and that all the phenomena of life are produced by forces which are absolutely identical to those of inorganic nature.[13] By 1828, a majority of chemists were ready to affirm the possibility, in principle, of the artificial preparation of physiologically familiar substances, and virtually no professional chemist defended transcendent vital forces after the early 1830s.[14] In contrast, in German physiology of that time, the 'vital force' was the dominant notion of Romantic Naturphilosophie. The method followed by German physiologists was less explicitly experimental and more oriented towards the use of analogies and images. Johannes Müller, in particular, was deeply convinced of the absolute distinction between the inorganic and organic realms and of the ultimate action of a vital force in the formation of tissues and organs and in their functioning.[15]

Although du Bois-Reymond's dissertation, *Quae apud veteres de piscibus electricis exstant argumenta* (1843), was more of an historical–literary character, dealing with ancient Greek and Roman literature on the electric fish, it contained a statement that all forces in nature are similar and mechanical: 'We do not need a mysterious servant to sneak in and to perform the job, the challenge is to explain everything by natural forces, matter and energy.'[16] Concurrently with his dissertation, du Bois-Reymond completed a paper presenting a clear description and analysis of the so-called 'frog current', and currents of the resting muscle and nerve, the result of his revision of Matteucci's experiments, and extremely critical treatment of the research done by the Italian physicist. Recommended by Humboldt, the paper was published in J.C. Poggendorff's *Annalen der Physik und Chemie.*[17] In his turn, Brücke, in his dissertation, *De diffusione humorum per septa mortua et viva* (1842), tried to prove that the phenomena of endosmosis are not to be related to any sort of uncertain vital force, but to weighable and measurable, repelling and attracting physico–chemical forces. Also recommended by Humboldt, the abstract, in which Brücke formulated his theory of diffusion of liquid substances through porous membranes, was published in Poggendorff's *Annalen.*[18]

In 1845, du Bois-Reymond and Brücke, together with the other five members of the Jüngerer Naturforscherverein, founded the Physikalische Gesellschaft zu Berlin – the Berlin Physical Society – and from 1849 du Bois-Reymond became its President. The other founding members were four young physicists, all lecturers in physics at the University and Artillery School of Berlin: G. Karsten (1820–1900), W. Heintz (1817–80), H. Knoblauch (1820–95), and W. Beetz (1822–86). Du Bois-Reymond and Brücke, both graduates of the medical faculty, closely associated with their mentor Müller, found themselves most frequently in the company of physicists. The young physiologists intended to separate physiology from medicine and to create the discipline concerned with the study of physics and chemistry of the living organism using the methods, techniques, and instruments of the exact sciences.

Gustav Magnus, the older colleague of the young physiologist-physicists, and a popular professor at both the University and the Artillery School, welcomed his numerous students interested in the natural sciences to his private laboratory at Kupfergraben. At one such colloquia held there, du Bois-Reymond made the acquaintance of a young lieutenant, Werner Siemens (1816–92), Magnus's student at the Artillery school and who also became one of the founding members of the Society.[19] For both du Bois-Reymond and Siemens, the meeting had the most important consequences – life-long friendship and a shared fascination of and fruitful co-operation in the new fields of electrophysiology and electrotelegraphy, which they

Image 2.5

Hermann Helmholtz.
Courtesy of the Wellcome Library Collection.

continued to build up throughout their careers. Their co-operation serves as one of the most telling examples of the nascent unification of the 'science for its own sake' and science-based technology, which eventually brought Germany to the leading position in science and industry by the end of the century.[20] At Magnus's colloquia, du Bois-Reymond and Siemens met the mechanic Georg Halske (1814–90), who, since 1844, had been a partner in the chemical apparatus workshop, Böttcher und Halske, closely associated with such scholar–experimenters as Eilhard Mitscherlich (1794–1863) and Johannes Müller. Halske had developed a particular sense and intuition for the construction of scientific instruments, which proved decisive for the development of ingenious equipment and apparatus for electrotelegraphy as well as for electrophysiology. In 1847, Werner Siemens together with Halske founded the workshop for electrotelegraphy equipment, which soon would develop into one of the most famous German enterprises, Siemens & Halske.[21]

The activity of the Society, as stated in its preliminary report of 1845, was aimed at informing the broader audience of the results of theoretical research, and making them accessible for the use in technology to facilitate technical and industrial improvements.[22] The Society had six sections: general physics, acoustics, optics, heat studies, electricity, and applied physics. The collection of the reports made at the meetings or sent by the members to the Society was published in the annual *Fortschritte der Physik*, intended as a programme statement of the talented and enthusiastic young men with various backgrounds, from military and mechanical engineering to medicine. Judging by the publications in the *Fortschritte*, the major focus of the Society was on measurement technologies and physics applied to various fields, such as the new field of electrical telegraphy. Ferdinand Leonhardt, the skilful watchmaker and mechanic, who worked with Siemens in the Artillery School on the problem of the exact measurement of the velocity of projectiles, was one of the most active contributors on the new electric telegraph, alongside Siemens himself.[23] Telegraphy, acoustics, electromagnetic devices for time measurment, graphic recording of temporal processes related to light, sound, or neurophysiological phenomena were the most enduring interests of the Society's leading members.[24] Demonstration of experiments and presentation of the new theoretical ideas, and discussions on how to utilise scattered and isolated theoretical results became the major attractions of the meetings of the Society. In 1846, an enthusiastic du Bois-Reymond reported to his friend Eduard Hallman about the Society: 'The Physical Society, which is the essential part of my life, begins to flourish. There is no meeting with less than thirty members, and always brilliant experiments.'[25] The Society, which developed into a forum for the practical implementation of the new ideas and findings in physics, gave a strong impetus for the development of science-based technology. Four decades later, in 1887, the *Physikalisch technische Reichsanstalt* – Physico-Technical Imperial Institute – was founded, largely due to the activities of du Bois-Reymond and a generous donation by Siemens. In 1888, Helmholtz, by that time a world-class physicist, became the first President of the newly founded Institute.[26]

Du Bois-Reymond met Helmholtz in 1845, when he was a regimental physician to the Garde du Corps at Potsdam. Apparently, they did not meet each other at Müller's Institute, although Helmholtz, a graduate of the Pépinière, had completed his thesis, *De fabrica systematis nervosi evertebratorum,* under Müller's supervision in 1842.[27] Soon after their first meeting du Bois-Reymond reported to Hallmann his first impressions:

> [I]n our small association of Brücke and I, he [Helmholtz] is the third organic physicist. The fellow has devoured chemistry, physics, and mathematics with a spoon; he entirely shares our viewpoints concerning philosophy of life, and is rich in thought and a new kind of imagination.[28]

In 1846, du Bois-Reymond introduced Helmholtz to the Society, where his exceptional abilities in mathematics and physics, as well as his interest in experimental research, were appreciated by Magnus and other members of the Society. Helmholtz became an active contributor to the *Fortschritte der Physik,* through which his important early works first became known. In 1847, Helmholtz presented his first theoretical work in physics *Über das Prinzip der Erhaltung der Kraft* [*On the Conservation of Energy*] to the Society. Du Bois-Reymond recognised a fundamental universality of the law and its significance for physical physiology. As a framework for all transformations of energy, the law did not allow for such hypothesis as the vital force. He wrote to Ludwig:

> Helmholtz has written a paper that simply cannot be praised enough: the conservation of energy. It is an exposition of the great principle of the constancy of energy and its application to various topics of natural science... it is only because of Helmholtz's profound work (no matter how many objections and changes may be left for the details) that physics, having become a science, has received a goal.[29]

The older physicists, Magnus and Poggendorff, were cautious about publishing the paper in the *Annalen.* Magnus regarded experimental and mathematical physics as separate departments. Poggendorff, acknowledging the theoretical importance of Helmholtz's work, considered its subject somewhat speculative and insufficiently experimental. It was the enthusiastic support of du Bois-Reymond and the younger physicists of the Society that proved decisive in urging Helmholtz to publish the paper as a separate treatise. Du Bois-Reymond recommended it to the publisher G. A. Reimer, and in 1847, Helmholtz's landmark work appeared in print.[30]

Three years later, in 1850, Helmholtz, by that time a professor at Königsberg University, sent du Bois-Reymond a short report *Über die Geschwindigkeit der Nervenwirkung* with a request that he would communicate it to the Society. In measuring the speed of the propagation of the impulse in motor nerves in the frog, Helmholtz applied a method of timing developed by the French physicist Claude Pouillet (1790–1868) in 1844 for ballistic purposes. Apparently, this method became known to

Helmholtz through the research of Werner Siemens, who, at one of the meetings in 1845, presented a historical critical review of the most important methods for measuring short time periods, including Pouillet's method, and his own time-measuring method and device for the use in artillery.[31]

Helmholtz's 1850 paper on the rate of transmission of excitation in nerves provoked questions on the part of the older physicists, as well as Müller. Earlier theories on the rate of the transmission of the nerve impulse had been entirely speculative: the conduction velocity was believed to be infinitely great and immeasurable, analogous to the velocity of light. Johannes Müller, in 1844, stated that the time necessary for a sensation to pass from periphery to the brain and spinal cord and the efferent reflex, which produces a contraction, is too swift to be measured.[32] 'Your work,' du Bois-Reymond wrote to Helmholtz:

> I say with pride and grief, is understood and recognised in Berlin by myself alone.... I had to explain it separately to one after the other – to Dove, to Magnus, to Müller... I brought it forward at the Physical Society, where at any rate we did not have the same difficulty.[33]

By the time Helmholtz conducted a series of experiments that led to the exact measurement of the conduction velocity in nerves, du Bois-Reymond had already developed his theory of the electrical nature of the nerve impulse and was armed to defend Helmholtz's experimental results. Du Bois-Reymond held that propagation of excitation in nerves is essentially conditioned by altered arrangement of the molecules, therefore the rate of propagation is a measurable and even a moderate magnitude since it is a case of molecular action in ponderable bodies. Du Bois-Reymond convinced Müller, Humboldt, and Magnus of the correctness of Helmholtz's experiments and insisted that this important discovery should be reported before the Academies of Paris and Berlin.[34]

Although du Bois-Reymond as President of the Society was most actively engaged in its work, he spent much of his time on his own research, and in 1848–9 published the first volume of his *Untersuchungen über tierische Elektrizität.* He dedicated the first part of the opening volume to Müller and the second to Humboldt, the two scholars who had drawn his attention to the topic, the only one he would pursue throughout his career, and which would bring him renown as the founder of the discipline of electrophysiology.

Du Bois-Reymond understood that his debt to the previous researchers was deep and devoted the first of three sections of the volume to the achievements of his predecessors in the field, on what had been known as 'animal electricity'. Accompanied by the original commentaries and abundant excellent illustrations with clear descriptions, it is a classical specimen of historico–scientific literature, which even contained excerpts from Luigi Galvani's letters to his colleague and friend Lazzaro Spallanzani. It is a masterful and detailed account on the major events in the field: Galvani's *De viribus electricitatis in motu musculari commentarius*, and his animal electricity theory; Alessandro Volta's 'struggle against a supposed animal electricity', and his electromotor theory; the 'contra' research of the Bologna school – Giovanni Aldini's 'exquisite performance' with the arcs of similar metals. The chapter covers Galvani's discovery of the muscle twitch without using metals; Volta's critique and explanation of the phenomenon in terms of mechanical stimulation; Humboldt's defence of Galvani's theory; and the development of the new forms of research without metals by Müller and Bunsen. [35] The second chapter of the historical section continues the account from Hans Christian Oerstedt's discovery of electromagnetism, André-Marie Ampère's invention of the astatic compass in 1820, Leopoldo Nobili's discovery of the electromagnetic action of frog current in 1827, to Carlo Matteucci's work on animal electricity.[36] The second part of the first volume begins with a historical account of the different 'electrical' theories of muscular contraction before Matteucci, for example, the pressure theories of Joseph Priestley and Erasmus Darwin, the electrodynamic theories of Pierre Prévost and Jean-Baptiste Dumas, and Eduard Weber, followed by Matteucci's theory of the 'second or induced contraction'.[37]

Summing up his account, du Bois-Reymond pointed out that Nobili and Matteucci concerned themselves mainly with explanation and measurement. Both used the galvanometer invented by Nobili – basically an improvement of Schweigger's multiplier – which allowed the demonstration of an electrical current on the intact but skinned frog trunk. Matteucci managed to demonstrate the 'injury current' between the uninjured surface and the cross-section of a muscle. In du Bois-Reymond's opinion, Nobili's electromagnetic theory and Matteucci's electrochemical theory lacked a clear physical concept and clear physical methods of deriving and measuring muscle currents.[38] Du Bois-Reymond's own interpretation of the basic molecular processes was analogous to Ampère's interpretation of the magnet. It assumed that the muscles and fibres were composed of strings of so-called peripolarelectric molecules, and allowed a clear – though not correct – explanation of the currents from intact, injured, and contracting muscle as having one cause, the so-called pre-existence theory, which du Bois-Reymond would defend throughout his career.

In short, du Bois-Reymond molded and shaped the separate facts and results accumulated by the previous researchers into a uniform, refined theory, which relied strictly on exact physical methods. It should be noted here that although du Bois-Reymond ascertained the importance of chemistry to physiology, he cast a strong doubt upon whether chemistry could provide a clear concept of the molecular mechanism of muscular contraction. The physical approach of du Bois-Reymond did not give space to any approximated speculations such as Liebig's electrochemical explanation of muscular contraction exposed in his *Die Thierchemie oder die organische Chemie in ihrer Anwendung auf Physiologie und Pathologie.*[39] Du Bois-Reymond felt that Liebig's 'physiological fantasies' were 'useless and pernicious', given that Liebig's erroneous views were almost universally accepted, because of his enormous authority. Indeed, Berzelius had once referred to Liebig's work as 'an easy kind of physiological chemistry produced at the writing desk, the more dangerous the more genius goes into its execution.'[40] This is likely why du Bois-Reymond, in the preface to the first volume of his *Untersuchungen,* called Liebig a 'scourge of God imposed upon physiology in our time'.[41]

The principal novelty in du Bois-Reymond's results presented in his *Untersuchungen* was the demonstration of the injury current in the nerve; he also succeeded in demonstrating the 'negative fluctuation' in tetanized nerves thus confirming one of most ingenious points in his theory – the electrical basis of nervous activity, which made functional connection between organs. This breakthrough became possible due to the improved version of Nobili's multiplier, which du Bois-Reymond managed to make extremely sensitive and which allowed him to avoid numerous difficulties, uncertainties and errors, and obtain accurate and secure results. This unusual sensitivity was achieved by greatly increasing the number of wire coils as compared to the previous multipliers. Du Bois-Reymond started his research with a multiplier which had 6,000 coils, which he then doubled, and by 1848 he reported to Ludwig that his current instrument had 24,160 coils.[42] Many years later in 1882, Halske in a letter to du Bois-Reymond reminisced about their 1847 co-operation in the workshop of the Siemens & Halske:

> I saw in my mind's eye how, with the utmost gentleness and caution, a galvanometer case was fitted with 33,000 coils, only to have no current when at last it was completed. How awful! The wire was wound off and, in its puffed-up condition like a curly wig, must have long continued to be a source of entertainment to the workshop staff.[43]

In the late 1880s, the young physiologists had to make much of their equipment themselves, as necessary devices and apparatus were not commercially available. The need for the advice and help in technical matters drove them into the workshops of the talented craftsmen.[44] It was in the workshop of Siemens & Halske that du Bois-Reymond built his famous sledge induction apparatus using the principles introduced by Oersted and developed by Faraday.[45] The apparatus, which had a secondary coil that could be moved on a sliding carriage, enabled precise graduation and calibration of the intensity of the first and last induction shocks. This piece of equipment, useful both in physics and electrophysiology, seems to have been conceived by all three together, du Bois-Reymond, Siemens, and Halske.

Du Bois-Reymond was well aware of the significance of his new multiplier, as well as his theory of identity of the nervous principle with electricity. This is evidenced in the title-page vignettes for his *Untersuchungen.* The centrepiece of the first volume vignette is the multiplier surrounded by volumes by Galvani, Volta, and Faraday from one side and an electric ray from the other. On the second volume vignette, the horse is struggling with the electric eel in the water surrounded by a fine ornament of leaves, illustrating the statement from Müller's *Handbuch der Physiologie* that 'the electric eel is capable of combating and undermining even a horse, a point that has been nicely exhibited by Alexander von Humboldt in his *Ansichten der Natur.*'[46] Both vignettes are fine representations of the coming era of electricity and the most exciting period in the study of animal electricity. Drawn by du Bois-Reymond himself, the vignettes are worthy of his great-grandfather, a famous artist and director of the Berlin Academy of Arts, Daniel Chodowiecki (1726–1801), whose drawings and engravings were considered the best depiction of the epoch of Friedrick the Great.[47] Artistically gifted, du Bois-Reymond shared a love of art with his closest friend and collaborator of that period Ernst Brücke, who had been a promising painter in his youth and this, in a unique way, determined their challenge to perfect physics to an art and to make art serve physics.[48]

As has already been noted, in 1849 du Bois-Reymond became President of the Berlin Physical Society. He must have already considered himself a respected member of the Berlin physical and physiological élite. The next step was to win Paris and, in the spring of 1850, du Bois-Reymond went there 'to promote his investigations'.[49] Following the advice of Magnus, given to him as early as 1846, du Bois-Reymond planned to counteract the influence of Matteucci's hypothesis and dismiss it as being too vague and lacking rigid experimental proof.[50] Recommended by Humboldt, Bois-Reymond presented his research at the Paris Academy of Sciences before the Commission, which included the physicists Antoine César Becquerel (1788–1878), César Mansuète Despretz (1792–1863), the physiologist François Magendie (1783–1855) and the clinician Pierre Rayer (1793–1867). Du Bois-Reymond's presentation must have been impressive by the novelty of the method and apparatus. In addition, he spoke French 'like a patriarch',[51] with eloquence and splendor so much valued by the

French. The Commission accepted du Bois-Reymond's results regarding the origin of muscular current fluctuation during the voluntary contraction of muscles in the human as valid. At the same time, the Commission admitted that the fluctuation of muscular currents might have been of an external chemical origin, the point du Bois-Reymond had strongly rejected. Nevertheless, du Bois-Reymond saw his presentation as a triumph.[52] There is a curious parallel to his performance at the Paris Academy: his friend Werner Siemens also happened to be in Paris representing German science before the academic committee, headed by the notable physicists Pouillet and Henri Regnault (1810–78). The lecture, entitled *Mémoire sur la télégraphie électrique,*[53] was on his new electrotelegraphic apparatus that would soon win the entire European market.

Du Bois-Reymond had one more mission in Paris besides bringing his own research to attention. Helmholtz wrote to his father:

> Du Bois is in Paris to present his research at the Académie. He will put mine forward too. He is very clever at this, and I have no doubt that he will introduce the German scientists to the French in quite an imposing light.[54]

Ludwig also saw du Bois-Reymond 'struggles in Paris' as the way to gain 'popular reputation necessary to us, after the manner of politicians'.[55] However this was not an easy task. Du Bois-Reymond was disappointed by the cool reception and reluctance of the French scientists to acknowledge the importance of the work of his friends. The French seemed unaware of what had been done by the Germany physiologists. Du Bois-Reymond's report on Helmholtz's method of measuring the propagation of the nervous impulse was not favourably received at the Academy of Sciences.[56] 'The ignorance and limited view of even the best men here is incredible,' du Bois-Reymond wrote to Ludwig. He was astonished that the French Academician, physiologist François Longet (1811–71) 'for instance… does not make any allusion whatsoever to Brücke in the chapter on vision in his new work on physiology.'[57] Du Bois-Reymond referred to Brücke's important works in physiological optics. In his *Anatomical Description of the Human Eye,* Brücke had examined optical media, after-images, stereoscopic vision, and the reflection of light from the retina. Published in 1847, the work had already become the standard anatomo–histological textbook for German ophthalmologists. As for Ludwig, 'You do not exist here any more than Brücke,' continued du Bois-Reymond with indignation, 'they consider Helmholtz to be a madman. Your textbook will be a couple of generations ahead of the French conception.'[58]

In Germany, there can be no doubt the physical physiology introduced by du Bois-Reymond was becoming influential, and by the late 1850s, only a decade later, it was dominant. Du Bois-Reymond's *Untersuchungen* helped greatly in making his research methods readily available to others. Sechenov, who in the late 1850s studied or visited virtually all notable physiological laboratories in Germany, wrote in 1899:

> One can judge on the impact made by this book by the fact that from the end of the fifties up till now there is hardly any single physiologist in Germany who had not tried to study the phenomena touched upon in this work of du Bois-Reymond. Such was the impulse given by his researches.[59]

In the following chapters, we will see how in Russia in the early 1860s the introduction of the physiological laboratory into medical education was to start from the organisation of a small laboratory, centred around du Bois-Reymond's apparatus. This proved the most efficient means for teaching basic physical methods and studying the challenging problems of nerve–muscle physiology. Du Bois-Reymond's electrodes for conducting weak bioelectric currents and his multiplier for detecting and amplifying those currents became the key instruments of a standard physiological laboratory of the late 1850s and early 1860s. These instruments became immediately available first from the Siemens & Halske and later from Sauerwald's workshop in Berlin.[60]

But it was not so in France. Sechenov spent a semester in Claude Bernard's laboratory in 1862 and was astonished to hear only two German names, Gabriel Valentin (1810–83) and Rudolf Virchow (1821–1902), at Bernard's lectures. As for electrophysiological research in Paris, Sechenov noted:

> The different aspects of electrical stimulation of the nerves and muscles had not yet at that time come from Germany to the laboratory at Paris, and Bernard still used a *pince électrique*, compasses with copper and zinc ends for this purpose.[61]

Du Bois-Reymond's methods and apparatus became the major driving force for the rise of electrophysiology in the mid-nineteenth century, and remain in use to this day (in modified variations). They represented the complete transition to physical physiology and were quickly adopted by all physiological laboratories in Germany. Soon practical courses in physiology were offered in Germany, as well as group research in electrophysiology on a scale unthinkable in the medical institutions of France. This transition serves as a particularly telling illustration of the beginning of German dominance in nineteenth-century laboratory medicine.[62]

The decade from 1842 to 1852 was decisive in the foundation of the new discipline of physical physiology. These were also the most productive years for du Bois-Reymond as a researcher and President of the Physical Society. In 1852, Helmholtz, in his essay on the recent achievements of the discipline wrote about du Bois-Reymond's discoveries:

> [T]he fruits of assiduous study, and of ten years' labour consistently concentrated upon one aim, during which the frog and the divisions of the galvanometer were his world – a rare example of methodical observation, of rich knowledge, and of that perspicacity of conception which is learned in the school of mathematics.[63]

Indeed, introduction of mathematical and physical methods into studies of electrical phenomena in the nerve and muscle opened up radically new fields, namely electrophysiology, physiological optics, and physiological acoustics. Effective application of these methods yielded results so impressive and so convincing that the new mechanistic–mathematical approach determined the development of experimental physiology for the next fifty years.

Du Bois-Reymond's comparatively secure financial situation, supported by his father and his wife's inheritance, allowed him to divide his time between research and the activities of the Physical Society during the decade from 1842 to 1852, without the need to seek a university position. In addition, du Bois-Reymond was on particularly good terms with Humboldt: 'I am in touch with him on one point or another almost every two or three days', he wrote to Ludwig in 1849. Baron von Humboldt was highly esteemed by Friedrich Wilhelm IV, King of Prussia, not only for his aristocratic provenance, but also for his exceptional educational and scientific achievements. In 1849, Humboldt presented to the King a splendidly bound copy of du Bois-Reymond's *Untersuchungen*, for which du Bois-Reymond was granted five hundred thalers, and thus, in his own words, he 'was completely safe and secure.'[64]

Independent financially, patronised by Humboldt, patriarch of Prussian science, and having influential connections in the Ministry, du Bois-Reymond could rather easily manoeuvre in a restless sea of academic appointments promoting his friends. When, in 1848, Brücke, who was at the time a lecturer of anatomy at the Academie der Künste and assistant at Müller's Anatomical Museum, was called to Königsberg, du Bois-Reymond arranged that Helmholtz succeeded Brücke at both positions.[65] Helmholtz, at the time at Potsdam in the Garde du Corps, was always backed up by advice and help from du Bois-Reymond, who, he writes, 'tended to me like a mother, to enable me to attain a scientific position'.[66] Brücke did not stay long in Königsberg, and in 1849 moved to Vienna.[67] Du Bois-Reymond was the first candidate at the Ministry to get the position in Königsberg, but since his private means allowed him 'to exist without a position' he negotiated with the Ministry that this position should be given to one or other of his friends, Helmholtz or Ludwig.[68] Support and recommendations of du Bois-Reymond and Brücke proved successful for Helmholtz.[69] For Ludwig, the situation was more complicated. 'Suspected by the ministry of having democratic convictions', he could not possibly think about an appointment to a Prussian university.[70]

Du Bois-Reymond's own university career started in 1854 as a lecturer and as Müller's assistant at the Anatomical Museum. But his teaching did not go at all smoothly, and it was only due to a Ministry official well-disposed to him that he was assigned part of anatomy course where he could deal with physiology. In his letter to Ludwig, du Bois-Reymond wrote: 'my knowledge had become deficient as a result of my protracted one-sided occupation with animal electricity… my sole hope of competing successfully with Müller is based on showing many experiments in the lectures.' Apparently lecture preparation took much time, while the uncertainty regarding his application to a position of extraordinary professor added to his feelings of exasperation. He further lamented to Ludwig: 'And it is with inward wrath that I see my apparatus idly lying about, disarranged and dusty, my work held up and my development interrupted.'[71] Next year, however, things went better, and du Bois-Reymond was promoted to extraordinary professor. Müller's death in 1858 brought du Bois-Reymond the appointment of Ordinary Professor of Physiology. In 1877, he headed the new Institute of Physiology – the best in Prussia – comparable only to the famous Leipzig Institute of his friend Ludwig in Saxony. From now on, his time was shared between teaching and his duties as a permanent secretary of the Prussian Academy of Sciences, as well as his various public and university activities.[72]

Du Bois- Reymond's addresses delivered at the Academy and his speeches on various occasions are of admirable philosophic, historic, and literary value, testifying to his original views and broad interests. But it was not so with his research. Remaining a most faithful savant of physical physiology, he continued along the single path, whereas his friends' interests shifted to other areas of physiology or to other disciplines. Ludwig and Brücke were known as the most wide-ranging physiologists of the time, and Helmholtz abandoned physiology for physics, to become its great representative. Du Bois-Reymond was frequently reproached by fellow physiologists as having a one-sided pre-occupation with a single problem in nerve–muscle physiology. The same was true on the part of physicists. We know from Ludwig's letter that Bunsen did not attach much importance to a physiologist as a physicist. Although recognising his great knowledge in physics, Bunsen nevertheless held that du Bois-Reymond's education was somewhat one-sided.[73] Du Bois-Reymond's reply to Ludwig is quite remarkable and convincing:

> I would unconditionally prefer as a teacher a scholar who works one-sidedly but excels in his subject, to one with an all-round education who has not accomplished anything much in any field. For the former will be far more deeply imbued than the latter with the spirit and method of science, which seems to me to be what matters most in teaching, and will be able to communicate it. One-sidedness presupposes a certain amount of fervour and enthusiasm, and it is this *sacra ignis* that I consider to be a teacher's greatest gift...[74]

Certainly, du Bois-Reymond had such a gift. Many of his students remembered him as an inspiring teacher.[75] It was du Bois-Reymond's enthusiasm and brilliant abilities in physical experimentation that made such an impact on the young Sechenov in 1857. Du Bois-Reymond's laboratory was popular among Russian researchers: at different times, future distinguished physiologists, such as Cyon and Vvedenskii studied there. Nevertheless, du Bois-Reymond's lectures remained unchanged for more that thirty years and moreover, he rejected any new explanatory model such as the alternation theory of his student Ludimar Hermann (1838–1914), which was accepted by the leading figures in the field he had created.[76]

The 'sacred fire' was burning in du Bois-Reymond himself while he was attempting to find a clue to a mystery of nerve–muscle excitation, the 'hundred year dream of physiologists and physicists of the identity of the nerve process and electricity', which he had laid out in the Preface to the first volume of his *Untersuchungen.*[77] He never finished his *Untersuchungen*, the pride of his heart, which remained 'a torso' and which he named a *Monstrum per defectum* in the *Nachwort* to a final part published in 1884.[78] But failure to finish the *Untersuchngen* was not merely due to his staunch adherence to his electromotive theory. The impossibility of understanding and explaining electrophysiological phenomena was inherent in the limits of the physics and chemistry of the late 1840s and the three successive decades. It became possible only after Svante Arrhenius (1859–1927) had introduced the concept of the ion into chemistry, and Walter Nernst and Wilhelm Ostwald suggested the possible connection between ionic distribution and electrophysiological phenomena in 1889–90. Du Bois-Reymond's student, Julius Bernstein (1839–1917), exploring the possible relevance of the works of Nernst and Ostwald to electrophysiology, suggested his membrane theory of bioelectrical potentials in the early 1900. It would be through the work of du Bois-Reymond's talented disciple that the problem of the nature of muscular contraction and nerve impulse was finally unveiled.[79]

Carl Ludwig

Born in 1816 in Witzenhausen into the family of a steward, Ludwig, according to his own account, received an ordinary education in the classical humanistic tradition unrelated to natural sciences.[80] From 1835 to 1840, he studied medicine in Marburg and Erlangen and after graduation he remained at Marburg University to work in Bunsen's laboratory. In 1841, Ludwig was appointed an assistant at the Anatomical Institute under the direction of Hermann Hasse (1807–92), the notable anatomist and physiologist, keen in mircroscopic and physico–chemical research of the blood. In 1846, Ludwig was then promoted to professor.

The Marburg years determined Ludwig's entire academic and research career in two ways. Firstly, during the revolutionary turmoil of 1847–8, Ludwig was actively engaged in university politics. Together with Bunsen, he was a member of the executive committee of the Vaterlandsverein and editor of a liberal newspaper *Neuer Verfassungsfreund*. After the revolution he did not stay long in Marburg, since his reputation as a democrat hampered the prospects of his career there and elsewhere in Germany. Financial exigency, and his somewhat difficult relationship with Hasse, forced Ludwig to accept the chair of anatomy and physiology at the University of Zürich in 1849.[81] Ludwig formulated his political convictions in a letter to du Bois-Reymond before leaving Marburg: 'You know that I only want the best by means of education and discussion.'[82] A reputation as a democrat followed Ludwig. In 1854, when he, by that time an acclaimed physiologist, was called to the Military Medical Academy in Vienna, the Academy's authorities took into consideration a secret police report on him.[83] Ludwig would return to work in Germany only in 1865, after six years in Zürich and ten years in Vienna. In 1869, the Saxon Government would build, equip, and lavishly finance the *Neue Physiologische Anstalt* for him, the best institution for experimental research in Europe.

Secondly, close interactions with Bunsen in Marburg were decisive in shaping Ludwig's physico–chemical outlook. A drastic difference between his obscure and unimportant doctoral dissertation[84] of 1840 and his habilitation thesis of 1842, *Habilitationsschrift*, suggests that Ludwig became intensely interested in the natural sciences during these two years and succeeded in mastering experimental techniques in Bunsen's laboratory.[85] The *Habilitationsschrift*, published in Latin and in German, was original experimental research that demonstrated skilful application of physico–chemical methods to the problem of renal function.[86] In the Preface, Ludwig wrote:

Image 2.6

Carl Ludwig.
Courtesy of the Wellcome Library Collection.

> I have chosen to explain the physical laws governing renal secretion. This treatise of mine follows a new method of inquiry which was first applied to physiology by Magendie and with greater insight by Müller and, most recently by the ingenious Liebig. I am convinced that those who are experienced in these matters will not at all be surprised why, in these few pages, I have little to say about the vital force but rather discuss at length physical and chemical conditions.[87]

Mentioning the physiologists of the classical school, Magendie and Müller, was surely a token of deserved respect. Mentioning Liebig in this context was not only an appreciation of the celebrated chemist interested in physiological problems but also an emphasis on the rigorous rejection of the *vis vitalis* as explanation of physiological phenomena. As I have stressed earlier in this chapter, the vital force had been buried for chemists long before du Bois-Reymond, Brücke and Ludwig started to eradicate its influence in physiology. Ludwig, educated far away from Berlin University with its wide circle of the first-rate natural scientists and experimentalists, became a collaborator of the Berlin physiologist–physicists before they actually met in 1847. What united them at that point was the treatment of physiology as applied physics, and an evident and forceful indictment of vitalism.

There was yet another important implication of Ludwig's Marburg period and his close communication with Bunsen. Bunsen, a keen mathematician and an ingenious inventor of various devices, was intensely engaged in the study of the organic compounds of arsenic, the so-call cacodyl compounds. Berzelius praised Bunsen's results as they confirmed his statement that the organic world is a reproduction of the inorganic.[88] The research on cacodyl compounds remained Bunsen's single work in the field of organic chemistry and it is fair to suggest that it was a sign of the enormous popularity of Liebig's innovations in organic analysis. Bunsen had visited Liebig's laboratory at Giessen as early as 1832, and was one of the first among German chemists to adopt the Liebigian model of group research in his Marburg laboratory.[89]

Like other French-educated German chemists of the time, such as Magnus, the Rose brothers, Liebig and Wöhler, Bunsen was a prolific researcher and a successful teacher whose laboratory first, in Marburg and then in Heidelberg, from the 1840–50s onwards became one of the new centres for science education in Germany, ultimately surpassing their counterparts in Paris.[90] In the 1840s, German organic chemistry, with its spectacular achievements, exerted an important influence on the emerging discipline of physico–chemical physiology. The growing prestige of chemistry as an academic discipline as well as the expanding chemical community set up a tempting example for physiologists such as Ludwig. In any event, the close communication with Bunsen in Marburg was of crucial importance for the future experimental and pedagogical charisma of Ludwig, whose physiological laboratory in Leipzig in the 1860s would become comparable to the best chemical laboratories – Liebig's in Giessen and Bunsen's in Heidelberg.

The importance of Ludwig's habilitation thesis is twofold. Firstly, it was a bold attempt to refute the views of the older generation of physiologists on the function of the kidney. They believed that the unique ability of the kidney to eliminate the poisonous material from the blood is governed by the forces independent of physics and chemistry, a strong argument in support of teleological explanation in organ functioning. Ludwig suggested a model and a simple instrument to show that the kidney acted as a sophisticated hydraulic device, which could mechanically filter the blood. Secondly, and even more importantly, Ludwig initiated and demonstrated an integrative approach to a physiological problem by this work, which would become his life-long interest. His explanatory model of kidney function incorporated a wide variety of methods: histological, microscopic, physical, and chemical.[91] Physical-physiologist *par excellence*, Ludwig would maintain an integrative approach in all his major studies, and in this he would differ markedly from du Bois-Reymond.

Indeed, besides its obvious merits as a physico–physiological research, Ludwig's habilitation thesis was a fine histo–anatomical study. Since glomeruli were of great importance for the physiological part of his research, Ludwig examined them in a hundred kidneys, which had been injected for this purpose: the kidneys of humans, dogs, rabbits, pigs, and horses.[92] He described for the first time two types of glomeruli in the human kidney and settled a number of controversial views regarding the existence of direct contact between glomeruli and tubules.[93] Using the accepted measurements of Krause and his own measurements of the diameters of renal vessels, he showed that the total cross-sectional area of the glomerular vessels was larger than that of either afferent or efferent arterioles.[94] Concluding his microscopic studies of renal anatomy, Ludwig expressed his admiration to the 'previous outstanding morphological findings' of Müller's student Jacob Henle (1809–85), to whom he would become a close friend and collaborator in Zürich.[95] He continued his anatomo–histological studies during his Marburg and Zürich periods, which would prove significant for his later famous works on the structure of the spinal cord, of the heart chambers, and on the innervation of the heart.

In the physiological part of the thesis, Ludwig examined physical processes occurring in the kidneys. He also presented qualitative and quantitative analysis of urine and blood serum, comparing the amount of solid and dissolved substances in both. Comparative studies of urine and blood by Berzelius, Alexander Marcet (1770–1822) – Berzelius's collaborator and friend in London – and by the French pharmacist Louis René Le Canu (1800–71), established that urine is not a product of the kidney's own chemical activity but rather its principal components are already present in the blood. These results also established that 'the composition of serum and urine can be different, and that is not in accordance with simple filtration.'[96] The different composition of blood and urine suggested to Ludwig that only re-absorptive and diffusive processes occurred in the renal tubules, that is, re-absorption of the fluid filtered in the glomeruli and diffusive transport of the other urinary components.[97] Ludwig's acquaintance with the contemporary chemical studies on the blood and urine clearly indicates his intensive communication on the matter with Bunsen and other Marburg chemists, who apparently were aware of the relevant works of Berzelius, Marcet, and Le Canu. In his Preface, Ludwig expressed gratitude to Bunsen and his fellow chemists for supplying him with the books and necessary reagents and preparations.[98]

Ludwig's hypothesis of kidney function refuted the accepted view that the chemical force of endosmosis, or vital, attractive, force was the basic process behind urine secretion. Ludwig also rejected Müller's earlier view that the kidney is a gland and the renal nerves are responsible for urine secretion.[99] He also suggested that 'the third, physical, force of pressure, by which fluids are forced from the blood vessels is responsible for urine secretion.' His explanatory model was essentially mechanistic since it rested on basic principles of hydraulics. Indeed, Ludwig referred to the recent studies in hydraulics, of 1842, by the Prussian military engineer J. Eytelwein. He admitted that, although the application of the laws governing flow resistance allowed some accurate measurements in these studies, calculations of the magnitude of flow resistance were not sufficient.[100] Since 'that part of hydraulics, which deals with pressure exerted by flowing fluids on the walls of tubes, [had] been completely neglected,' he attempted to study pressure development in blood vessels using his own experimental set-up. Soon, however, he became convinced that 'the matter was more difficult and more intricate than any other part of hydrodynamics'.[101]

In his interpretation of the urine formation as a process of ultra-filtration of the blood, Ludwig incorporated his observations of the structure of the renal glomeruli, the relevant laws of hydrodynamics, and certain known similarities in the chemical composition of blood and urine. He believed that the differences in their composition are due to differences between epithelial permeability and 'the second physical force in the kidney, the endosmotic action'. Although he was convinced that 'a true endosmosis does not occur in the kidney',[102] he was not able to prove it experimentally, 'because of the lack of adequate instruments to mimic endosmosis in the kidney'. Therefore, his studies at that point were limited to the quantitative method 'to determine the endosmotic force in both urine and the blood'.[103]

In summary, Ludwig was the first to comprehend that 'the physical process of pressure filtration of fluid is the primary mechanism underlying urine secretion'. He envisaged that certain substances in the blood were pressed out of the vessels at the level of glomeruli due to existing blood pressure, and this process was followed by reabsorption of water in the adjacent tubuli.[104] It was, in essence, his famous theory of renal secretion. Its conclusion was remarkable too: 'Thus we see in these arrangements a simplicity, a harmony, which in its beauty is not surpassed by any natural process.'[105] Indeed, Ludwig's explanation of renal secretion was purely mechanistic and rather simplified, but at the same time it was clear, precise and neat. It powerfully undermined the existing descriptive and vitalistic theories of renal secretion. As a true mechanistic explanation it was supported by an ingenious quantitative methodology, imported from hydraulics, using devices that 'can be used by physiologists and physicists to study phenomena similar to those occurring in the kidneys and to calculate the pressure on the walls of the tube'. A simple apparatus consisting of the metal and glass tubes of diverse diameters and a manometer attached to one of the tubes to measure the pressure against the walls of the expansion tube. Ludwig also devised a simplified method to demonstrate that a flow velocity does not result in filtration or secretion, which could be used for teaching purposes.[106] The device served not just as a useful model for representing the work of the renal vessels, but also as a guiding principle for his future programme of research in renal physiology.

What Ludwig suggested was a radically new approach, a reduction of the complex function of the organ into its active components and a totally new technique for demonstrating the physical essence of the mechanism of organ functioning. This is exactly what Ludwig would later suggest in his seminal textbook on physiology as 'a highest form of physiological investigation':

> To comprehend an activity as a function of the conditions producing it, one combines either theoretically (through mathematical–physical calculation) or practically (through physical–chemical experimentation) a certain number of conditions of known characteristics, which approximate to the organic, and compares the effects they produce with those produced in nature.[107]

Finally, the habilitation thesis dealt quite extensively with renal blood flow, blood pressure, and venous pressure, and it is fair to suggest that these studies can be viewed as an overture to Ludwig's later famous haemodynamic investigations, for which in 1846 he developed an original device, the kymograph (from Greek *kymos* – wave, and *graphein* – to write). This device allowed synchronous registration of blood pressure and intrapleural pressure, and was first described in his article published in Müller's *Archiv* in 1847. The arterial system of a dog was connected to a mercury manometer by fluid-filled tubing, which translated the oscillations of the mercury column to a staff floating on the surface of the manometer. The bobbing stylus was brought against the rapidly moving surface of a smoked drum to obtain a graphic record of blood pressure as it varied with the animal's heartbeat and breathing.[108] The device was an important innovation that opened a graphic registration era in physiological research and teaching. Like du Bois-Reymond's galvanometer, Ludwig's kymograph was instantly adopted to become the key instrument in every physiological laboratory.

The idea of an inscription device has an interesting background dating back to the architect and mathematician Christopher Wren (1632–1723) and the natural scientist and architect Robert Hooke (1635–1703).[109] The most plausible, however, is the suggestion that the report of the French physicist Pouillet published in 1844 in the *Comptes Rendus Hebdomadaires des Séances de l'Académie des Sciences* and the resulting correspondences in France and Germany might have been the immediate stimulus for Ludwig's idea of the kymograph. In his report, Pouillet presented a concept of self-registration of natural phenomena and described some elements of devices which had been in use in various fields such as meteorology and physics, and which began to be used for electro–mechanical purposes. These were the clockwork-driven cylinder for recording various phenomena; the smoked surface, which greatly reduced friction and made possible the greatest faithfulness in reproduction of the recording; the simultaneous registration of tuning forks or other timing devices.[110] Some of these elements, the smoked surface and a stylus in particular, might be the prototypic elements that Ludwig used in his kymograph. There is no direct evidence that Ludwig borrowed his idea from Pouillet's report, at least he never made any reference to it. However, the report was widely circulated among the German physicists and it might be that Ludwig knew about it from his colleagues in Marburg.

We have discussed earlier the influence of Pouillet's ideas on Siemens and through him on du Bois-Reymond and Helmholtz but we do not know which came first: Ludwig's publication of the article in Muller's *Archiv* with the description of the kymograph, or his personal acquaintance with the Berlin physiologist–physicists. Both events occurred in 1847. Interestingly, in his 1852 textbook on physiology, Ludwig mentioned his employment of the float in a mercury manometer when describing his kymograph, which was derived from 'a principle that the famous Watt is supposed to have first introduced' for his steam engine indicator. It means that Ludwig ascribed both his and Helmholtz's use of graphic methods to the principles laid out by James Watt (1736–1819).[111] This reference might be the result of Ludwig's communication with his friends from the Berlin Physical Society, where the precision instruments and graphic methods used in various fields of applied physics had been intensely discussed during 1845–8.[112] What is certain, however, is that Ludwig began to use the kymograph for recording arterial blood pressure as early as December 1846,[113] before his acquaintance with the Berlin physiologists. What is also certain is that leaving aside the question of priority, it can be said that both Ludwig and Helmholtz, independently and simultaneously, in seeking a precision instrument for recording bodily functions relied on the same kind of sources, that is the principles and techniques employed in electrical mechanics, telegraphy, and ballistics. This by itself was a path-breaking advance, which determined the way German physiology was to follow the next half century. It was also an important token of the time: the beginning of the growing co-operation of science and electrical industrial arts, which, in the words of Werner Siemens, was beneficial for both.[114]

In 1847, during his visit to Berlin, Ludwig met du Bois-Reymond, Brücke, and Helmholtz, and became acquainted with Virchow and Traube.[115] Du Bois-Reymond introduced Ludwig to the Physical Society, and judging by the publication in the *Fortschritte der Physik* of 1848, Ludwig took over the section on organic hydraulics. The number of critical reviews published by Ludwig in the *Fortschritte* in 1847–50 was surprisingly small, and he himself admitted that he did not become as active a contributor as his new friends.[116] But his correspondence with du Bois-Reymond, which started immediately after their acquaintance, was a possible way of exchanging ideas, opinions, and information between both friends. Ludwig assigned a particular importance to his communications with the Berlin friends as well as to his earlier co-operation with his Marburg colleagues:

> After working my way through all the medical men's old rubbish for six years as a student, realising at last that it was inherently unfounded, I proceeded... to study natural sciences. When I began this in my twenty-third year, I could not work out any *regula de tri*, I had never yet seen a distilling apparatus, and I had looked at the microscope from afar. Without a teacher, by means of untiring industry, I carried matters to such a point that after a few years I could make organic analysis, manipulate the microscope, and read Fischer's physics [J.K. Fischer, *Anfangsgründe der Physics* (Jena, 1797)], and this is how, step-by-step and finally my means of my contact with Bunsen and our mathematician Stegmann, and by constantly watching Brücke and you, I have worked myself up to my present level.[117]

This passage provides a valuable insight into Ludwig's enormous capacity and talent for self-learning in sciences, and his amazing intuition in the supreme significance of physical, chemical, quantitative, and microscopic methods, which would allow him a total grasp of nearly all areas of physiology.

In du Bois-Reymond, Ludwig found the most devoted friend and colleague. He encouraged Ludwig to write to Humboldt for recommendation to the Ministry when Ludwig found himself in a desperate financial situation in Marburg. 'Every industrious and ambitious man of science, all the more so one who already has achievements like yours behind him, is Humboldt's son; we all are his family.'[118] 'He has always been the guardian angel of all of us', wrote du Bois-Reymond to Ludwig in 1849.[119] Upon Humboldt's request the Ministry agreed to look at the matter, but by that time Ludwig had already received a call to Zürich. It has been mentioned earlier that du Bois-Reymond interceded with the Ministry to secure for Ludwig a position in Prussia, but to no avail.

It was du Bois-Reymond, too, who inspired Ludwig to write and publish a textbook on physiology. Obviously, he felt that Ludwig was the best suited for the task, since the research of Brücke, Helmholtz, and himself was too specialised and more focused on specific physical problems applied to physiology – such as heat, electricity, optics, and acoustics. Ludwig had already contributed to the foundation of physical physiology with his studies on phenomena of diffusion and endosmosis, on haemodynamics, renal function, respiratory processes, and innervation of organs (his discovery of the secretory nerve in the salivary gland, 1850). With his usual perspicacity, du Bois-Reymond pointed out that Ludwig's textbook might become a convenient benchmark for the achievements of the entire period:

> [T]here are epochs in the sciences when it is just as worthwhile to write a textbook as it is to do the most significant work of one's own. I see such an epoch now in physiology. It is due to the fact that our school sees things in a completely new light. Even if there were no new material not yet contained in the old textbooks, a whole new science results simply by compiling the old knowledge according to our way of thinking. And fifty years hence this epoch may be characterised by the publication of your textbook. As a result of it you will be the school's standard-bearer.[120]

In 1852, Ludwig published the first volume of his *Lehrbuch der Physiologie des Menschen,* with a dedication to Brücke, du Bois-Reymond and Helmholtz.[121] The subtitle of the first volume, *Physiologie der Atome, der Aggregatzustände, der Nerven und Muskeln,* [*Physiology of Atoms, Physical State, Nerves, and Muscles*], an area for which all three had laid the ground, is telling. Their explanation of the living processes was from elementary particles of matter governed by physical and chemical laws. Basic principles of osmosis, diffusion, dynamics, electricity, optics, acoustics, and mechanics applied to physiological problems were brought to the fore by the remarkable studies of these four physiologists. Ludwig's *Lehrbuch,* in several later editions, became the most representative for the period. It contained all sections of contemporary physiology to which the author had made valuable contributions and discovered totally new facts, as well as critical analysis of the results of other researchers. Particularly influential was its Introduction, which, together with du Bois-Reymond's Preface to the *Untersuchungen,* could merit as a programme of the new discipline.

In 1858, the year Sechenov came to Vienna, du Bois-Reymond, in his letter to Ludwig, proudly acknowledged that all commanding heights were in the hands of physical physiologists:

> The physical physiology founded by us is now visibly coming out victorious despite all resistance put up by our adversaries. You two [Ludwig and Brücke] in Vienna, Helmholtz in Heidelberg, I in Berlin, and presumably Pflüger and Heidenhein in Bonn and Breslau. I have good young talents here in Rosenthal and von Bezold.[122]

Indeed, a future medical researcher such as Sechenov had a good choice among the laboratories headed by first-rate physical physiologists. Besides, there were a number of well-established laboratories of physiologists of the older generation such as Ernst Heinrich Weber (1795–1878) in Leipzig and Weber's student Alfred Volkmann (1800–77) in Halle. Ernst Weber, and his brothers Eduard (1806–71) and Wilhelm (1804–91), a famous physicist in Göttingen, employing the physico–mathematical approach had produced works of fundamental importance for nerve–muscle physiology as well as for the research in circulation of the blood. The similar measurable and quantitative physical approach was demonstrated in the works of Volkmann on the problems of physiological optics and haemodynamics.[123] By the end of the 1850s, however, it was the laboratories of du Bois-Reymond in Berlin, Ludwig and Brücke in Vienna, and Helmholtz in Heidelberg that offered the most advanced laboratory training in the newest methods and techniques and produced the cutting-edge studies in the newly established physical physiology.

To begin describing the ways these important innovations in physical physiology moved from Germany to Russia, it is appropriate to discuss three of the leading Russian protagonists of the following chapter – Sechenov, Botkin, and Borodin – and their *Wanderjahre* in Europe in the late 1850s. What influenced their choice to stay at one institution or another? What ideas and concepts, experimental skills, techniques, and instruments did they seek to bring back home and in what way did these advantages affect their careers at home? Their studies in German laboratories in the late 1850s are a convenient starting point in delineating the introduction of the laboratory into Russian medicine and Russia's growing independence in science and education. From the time of their studies abroad, the trio maintained strong ties to the physiological, medical, and chemical community in Germany and France, facilitated by personal relationship. German journals became convenient outlets for their publications, since in Russia specialised journals of this kind had only just begun to emerge. France did not possess the same variety of journals on the scale of scientific publications in Germany, so was a lesser outlet. The key professors at the St Petersburg Medico-Surgical Academy, Sechenov, Botkin, and Borodin, had the most decisive influence on the major transformations in Russian medical sciences and education in the second half of the nineteenth century. Their research and pedagogical activities were determined and shaped by the practice and culture of German laboratory medicine. Our journey begins with Sechenov at the physiological laboratories in Berlin and Vienna, then follows Botkin to the clinics in Berlin, Vienna, and Paris, and finishes in Heidelberg with Borodin and a group of young Russian chemists.

Notes

1. I.M. Sechenov, 'O deiatel'nosti Galvani i du Bois-Reymod v oblasti zhivotnogo elektrichestva,' *Trudy Fiziologicheskogo instituta Moskovskogo universiteta,* 5, 3 (1899), 1–8: 4.
2. E. du Bois-Reymond, *Untersuchungen über thierische Elektricität* (Berlin: Reimer, 1848–84: Vol. 1, Part 1, 1848; Vol. 1, Part 2, 1849; Vol. 2, Part 1, 1860; Vol. 2, Part 2, 1884).
3. For an overview of the period, see K. Rothschuh, 'Emil du Bois-Reymond (1818–1896). Werden, Wesen, Wirken', in G. Mann (ed.), *Naturwissen und Erkenntnis im 19. Jahrhundert: Emil du Bois-Reymond* (Hildesheim: Gerstenberg Verlag, 1981), 11–26: 16; see also G. Böhmer, *Die Welt des Biedermeier* (Munich: K. Desch, 1968); and A. Slavin, *The Way of the West: The Formation of Modern Society* (Lexington: Xerox, 1975), 278–80. Political and social events of the time did not affect the young du Bois-Reymond as compared to Rudolf Virchow, who had expressed support of the starving Silesian weavers in June 1844 and who was actually at the barricades during the Berlin revolution of 1848. Du Bois-Reymond, as his friend Ernst Brücke, was at his studies. For a portrayal of the 1848 Berlin Revolution, as seen from the window of his apartment, see du Bois-Reymond to Ludwig, letter dated 22 April 1847, Berlin, in P.F. Cranefield (ed.), *Two Great Scientists of the Nineteenth Century: Correspondence of Emil du Bois-Reymond and Carl Ludwig,* S. Lichtner-Ayéd (trans.), (Baltimore: The John Hopkins University Press, 1982), 9–13. For biographical details, see an excellent biography by H. Boruttau, *Emil du Bois-Reymond* (Wien: Rikola, 1922). For du Bois-Reymond's contributions to neurophysiology, see C. Campenhausen 'Elektrophysiogie und physiologische Modellvorstellungen bei Emil du Bois-Reymond', in Mann, *idem,* 79–104.
4. F. Krafft, 'Jöns Jacob Berzelius und die deutsche Chemie', in F. Krafft (ed.), *Bunsen-Briefe in der Universitätsbibliothek Marburg* (Marburg: Universitätsbibliothek, 1996), 61–71.
5. On Rudolphi and his influence on the development of anatomical and physiological research based on precise methods, and freed from vague and misinterpreted philosophical principles, see J. Müller, 'Gedächtnisrede auf Carl Astmund Rudolphi 1835', *Abhandlungen der Preussischen Akademie der Wissenschaften zu Berlin,* xxiii (1837), 17–34: 27.

6. On von Altenstein, see D. Wendland, 'Preussische Wissenschaftspolitik unter Kultusminister Altenstein (1770–1840) und die Berliner Universität', in H.-U. Lammel and P. Schneck (eds), *Die Medizin an der Berliner Universität und an der Charité zwischen 1810 und 1850* (Husum: Mathiesen Verlag, 1995), 38–43. On Johannes Müller's philosophical commitments and his laboratory in Berlin, see B. Lohff, '[I]n Berlin eine würdige Stätte schaffen', in Lammel and Schneck, *idem*, 55–66; see also T. Lenoir, 'Science for the Clinic: Science Policy and the Formation of Carl Ludwig's Institute in Leipzig', in W. Coleman and L.F. Holmes (eds), *The Investigative Enterprise, Experimental Physiology in Nineteenth-Century Medicine* (Berkeley: University of California Press, 1988), 139–78: 146–8; and K. Rothschuh, *History of Physiology*, G.B. Risse (trans. and ed.), (Huntington: R.E. Krieger Publishing, 1971), 200–3.
7. On the German professoriate in the early nineteenth century, see K. Schwabe (ed.), *Deutsche Hochschullehrer als Elite 1815–1945* (Boppard am Rein:
H. Boldt, 1988).
8. On German university and pedagogical reform, see A. Rocke, *The Quiet Revolution: Hermann Kolbe and the Science of Organic Chemistry* (Berkeley: University of California Press, 1993), 9–14, 28; see also W. Coleman, 'Prussian Pedagogy: Purkynje at Breslau, 1823–1839', in Coleman and Holmes *op. cit.* (note 6), 15–64: 28–38.
9. On Müller's laboratory, see the reminiscences of Heinrich Bidder (1810–94), Professor of Physiology in Dorpat University, who, in 1834, studied with Müller: H.F. Bidder, 'Vor hundert Jahren im Laboratorium Johannes Müllers', *Münchener medizinische Wochenschrift*, vii (1934), 60–4: 61. On Müller's laboratory, see also E. du Bois-Reymond, 'Gedächtnisrede auf Johannes Müller: gehalten in der Leibniz-Sitzung der Akademie der Wissenschaften am 8. Juli 1858', in E. du Bois-Reymond (ed.), *Reden von Emil du Bois-Reymond* (Leipzig: Verlag von Veit, 1912), 2 vols, Vol. 1, 135–317: 200, 212, 263. The most recent study on Müller is L. Otis, *Müller's Lab* (Oxford: Oxford University Press, 2007).
10. On Humboldt's electrophysiological research, see K. Rothschuh, *Alexander von Humboldt et l'histoire de la découverte de l'électricité animale* (Paris: Edition du Palais de la découverte, 1960).
11. Du Bois-Reymond to Hallmann, letter dated May 1842, Berlin, in E. du Bois-Reymond, (ed.), *Jugendbriefe von Emil du Bois-Reymond an Eduard Hallmann* (Berlin: Reimer, 1918), 108.
12. W. Siemens, *Recollections* (Munich: Prestel-Verlag, 1983), 38– 9; see also P. Ruff, *Emil du Bois Reymond* (Leipzig: Teubner, 1981), 50.
13. Cited in A. Rocke, *Nationalizing Science: Adolphe Wurtz and the Battle for French Chemistry* (Cambridge: MIT Press, 2001), 249.

14. On the French-educated German chemists, see *ibid.*, 14–5. Rocke provides a convincing picture of vitalism in chemistry, see *ibid.*, 245–50; see also H. Driesch, *History and Theory of Vitalism*, C. Ogden. (trans.), (London: Macmillan, 1914).
15. Rothschuh, *op. cit.* (note 6), 198. On Naturphilosophie and its concepts in medicine see also *idem*, *Konzepte der Medizin in Vergangenheit und Gegenwart* (Stuttgart: Hippokrates Verlag, 1978), 385–416.
16. Cited in Boruttau, *op. cit.* (note 3), 17.
17. E. du Bois-Reymond, 'Vorläufiger Abriss einer Untersuchung über den sogenannten Froschstrom ubër die electromotorischen Fische', *Annalen der Physik und Chemie,* lviii, (1843), 1–30.
18. E. Brücke, 'Beiträge zur Lehre von der Diffusion tropfbarflüssiger Körper durch poröse Scheidewände', *Annalen der Physik und Chemie,* 58 (1843), 77–94.
19. Siemens, *op. cit.* (note 12), 286–7.
20. This issue is discussed in my forthcoming article '*Drähte und Kabel*: Electrophysiology of du Bois-Reymond and Electrotelegraphy of Werner Siemens'. Du Bois-Reymond and Siemen, both the savants of experimental physics are the most interesting vehicles in delineating the events that led German science and technology to prominence in the second half of the nineteenth century.
21. Siemens, *op. cit.* (note 12), 47–51.
22. G. Karsten (ed.), *Die Fortschritte der Physik im Jahre 1845: dargestellt von der physikalischen Gesellschaft zu Berlin* (Berlin: Druck und Verlag von G. Reimer, 1847), iii.
23. G. Karsten (ed.), *Die Fortschritte der Physik im Jahre 1846* (Berlin: Druck und Verlag von G. Reimer, 1848), 227. See also Siemens, *op. cit.* (note 12), 40–1.
24. T. Lenoir, 'Helmholtz and the Materialities of Communication', *Osiris,* ix (1994), 185–207: 188, 205. Lenoir has argued that the new media technologies that captivated the interest of Helmholtz and his young friends in the Berlin Physical Society, particularly the telegraph and a variety of new audio and visual inscription devices, were crucial in establishing the boundaries for his experiments in physiological optics and acoustics.
25. Du Bois-Reymond to Hallmann, letter dated March 1846, Berlin, cited in Ruff, *op. cit.* (note 12), 53.
26. Siemens was a close friend and relative of Helmholtz: in 1884, Werner's eldest son, Arnold, married Helmholtz's daughter, Ellen, see Siemens, *op. cit.* (note 12), 297. On Siemens and the Institute, see J. Zenneck, 'Werner von Siemens und die Gründung der Physikalisch-Technischen Reichsanstalt', in *Abhandlungen und Berichte* (Munich: Deutsches Museum, 1931), iii, 1–16; see also D. Cahan, 'Werner Siemen and the Origin of the Physikalisch-

technische Reichsanstalt, 1872–1887', *Historical Studies in the Physical and Biological Sciences,* xii, 2 (1982), 252–83.

27. For a comprehensive biography, see L. Koenigsberger, *Hermann von Helmholtz,* F. Welby (trans) (New York: Dover Publications, 1965); see also E. du Bois-Reymond, 'Gedächtnisrede auf Herman von Helmholtz', 1895, in E. du Bois-Reymond, *op. cit.* (note 9), Vol. 2, 516–70. For Helmholtz's correspondence, see R. Kremer (ed.), *Letters of Hermann von Helmholtz to his Wife, 1847–1859* (Stuttgart: Franz Steiner Verlag, 1990); D. Cahan (ed.), *Letters of Hermann von Helmholtz to his Parents: The Medical Education of a German Scientist, 1837–1846* (Stuttgart: Franz Steiner Verlag, 1993).
28. Du Bois-Reymond to Hallmann, letter dated April 1845, Berlin, cited in Ruff, *op. cit.* (note 12), 44.
29. Du Bois-Reymond to Ludwig, letter dated 4 January 1848, Berlin, in Cranefield, *op. cit.* (note 3), 6.
30. Koenigsberger, *op. cit.* (note 27), 38, 42; see also Karsten (ed.), *Die Fortschritte der Physik im Jahre 1848* (Berlin: Reimer, 1850), 10.
31. W. Siemens, 'Über Geschwindigkeitsmessung', in Karsten, *op. cit.* (note 22), 47–72; see also W. Siemens, 'Verfahren zur Messung der Geschwindigkeit von Projectilen mittelst des reibungselektrischen Funkens', *Annalen der Physik,* 66 (1845), 435.
32. J. Müller, *Handbuch der Physiologie des Menschen für Vorlesungen* (Coblenz: Verlag von J. Hölscher, 1833–7), 2 vols, Vol. 1, 16.
33. Cited in Koenigsberger, *op. cit.* (note 27), 64–5; Heinrich Wilhelm Dove (1803–79) was a notable Berlin physicist and meteorologist.
34. H. Helmholtz, 'Über die Fortpflanzungsgeschwindigkeit der Nervenreizung', *Verhandlungen der Königlich Preussischen Akademie der Wissenschaften zu Berlin,* (Berlin, 1850), 14–5. Historians regard this work as the beginning of the era of natural sciences in neurophysiology, see W. Blasius, 'Die Bestimmung der Leitungsgeschwindigkeit im Nerven durch Hermann Helmholtz am Beginn der naturwissenschaftlichen Ära der Neurophysiologie', in K. Rothschuh (ed.), *Von Boerhaave bis Berger: die Entwicklung der kontinentalen Physiologie im 18 und 19. Jahrhundert* (Stuttgart: Gustav Fischer Verlag, 1964), 71–84.
35. Du Bois-Reymond, *op. cit.* (note 2), Vol. 1, Part 1, 31–102.
36. *Ibid.,* 103–26.
37. *Ibid.,* Part 2, 3–30.
38. *Ibid.,* 177.
39. J. Liebig, *Die Thierchemie oder die organische Chemie in ihrer Anwendung auf die Physiologie und Pathologie* (Braunschweig: Vieweg, 1843).
40. Cited in J. Partington, *A History of Chemistry* (London: Macmillan, 1964), Vol. 4, 316.

41. Du Bois-Reymond, *op. cit.* (note 2), Vol. 1, xxxvii. Du Bois-Reymond had great respect for Liebig's talent and the excellence of his earlier, non-physiological works, however, he felt that Liebig 'completely lacks the necessary foundation and critical education' to study physiological problems; see du Bois-Reymond to Ludwig, letter dated 22 April 1848, Berlin, in Cranefield, *op. cit.* (note 3), 14. Helmholtz, too, vehemently criticised Liebig's treatment of the origin of animal heat, published by J. Liebig, 'Über die thierische Wärme,' *Annalen der Chemie und Pharmacie,* 53 (1845), 63–77, see R. Brain and M. Wise, 'Muscles and Engines: Indicator Diagrams and Helmholtz's Graphical Methods', in M. Biogioli (ed.), *The Science Studies Reader* (New York: Routledge, 1999), 51–66: 53–5.
42. Du Bois-Reymond to Ludwig, 4 January 1848, Berlin, in Cranefied, *op. cit.* (note 3), 5.
43. *Ibid.,* 126 (fn. 10). Halske's letters to du Bois-Reymond are located in *Siemens Archives,* Munich.
44. Du Bois-Reymond, *op. cit.* (note 9), Vol. 1, 641.
45. On construction of the sledge induction apparatus, see du Bois-Reymond to Ludwig, letter dated 22 April 1847, Berlin, in Cranefield, *op. cit.* (note 3), 14.
46. Müller, *op. cit.* (note 32), Vol. 1, 65.
47. Emil's mother, Minette Henry, was the grand-daughter of Daniel Chodowiecki, the famous artist, who illustrated nearly the entire contemporary literature. On Chodowiecki, see Staatliche Museen und National-Galerie (ed.), *Grosse Deutsche in Bildnissen ihrer Zeit* (Berlin: Ullstein Drukerei, 1936), 394.
48. Du Bois-Reymond and Brücke experienced a rare harmony in their love for art, philosophy, poetry, and music. In his letter to Ludwig, du Bois-Reymond mentioned what 'long shared life and work' with Brücke meant for him: 'We spent eight years together in a single aspiration, a single passion, a single way of thinking, in which our very different personalities merged every single day. We became men together, and in the otherwise soberly scientific circle of friends around us we guarded a sanctum of art, philosophy, poetry, and blithe, worthy wisdom.' See, du Bois-Reymond to Ludwig, 22 April 1848, Berlin, in Cranefield, *op. cit.* (note 3), 8–9. Born in Berlin into the family of a painter, who later moved to Italy, Brücke from an early age became interested in art and painting. One can judge on his artistic abilities by his self-portraits from his student days, which are now displayed in the boardroom of his great-grandson Franz Theodor von Brücke in the Institute of Pharmacology at the University of Vienna. Ernst was expected to pursue an artistic career but he chose to study medicine and in 1838 entered Berlin University. Throughout his life, Brücke preserved devotion to art that in different ways greatly influenced his research in physiology. In his work on

physiological optics, on the nature of brown colouring, on the sequence of colours in Newton's rings, on subjective colours, one can trace a keen interest of a painter to the classical problems such as light and colour changing by means of natural sciences. See, E. Brücke, *Die Physiologie der Farben für die Zwecke der Kunstgewerbes bearbeitet* (Leipzig: Hirzel, 1866); *idem, Die Beckenlinie antiker Statuen* (Vienna: Gerold, 1888); *idem, Schönheit und Fehler der menschlichen Gestalt* (Vienna and Leipzig: Braumüller, 1891). The sources on Brücke are surprisingly scarce: a biography by Theodor von Brücke, *Ernst Brücke* (Vienna: Springer, 1928) with a complete list of Brücke's works and a detailed account on his interest in art; see also L. Schönbauer, *Das medizinische Wien: Geschichte – Werden – Würdigung* (Berlin: Urban und Schwarzenberg, 1944), 244–6. The English language sourses are: Rothschuh, *op. cit.* (note 6), 234–6; E. Lesky, *The Vienna Medical School of the Nineteenth Century* (Baltimore: Johns Hopkins University Press, 1976). On interest in sensory physiology and in art and aesthetics of du Bois-Reymond, Brücke, and Helmholtz, see P. Cranefield, 'The Philosophical and Cultural Interests of the Biophysics Movement of 1847', *Journal of the History of Medicine,* i (1966), 1–7: 2; see also T. Lenoir, *Instituting Science: Cultural Production of Scientific Disciplines* (Stanford: Stanford University Press, 1997), 127–38.

49. Du Bois-Reymond to Ludwig, letter dated 9 April 1850, Paris, in Cranefield, *op. cit.* (note 3), 57.
50. Du Bois-Reymond to Hallmann, letter dated March 1846, Berlin, cited in Ruff, *op. cit.* (note 12), 26.
51. Du Bois-Reymond to Ludwig, letter dated 9 April 1850, Paris, in Cranefield, *op. cit.* (note 3), 58.
52. Du Bois-Reymond to Ludwig, letter dated 26 August 1850, Berlin, in *ibid.*, 60.
53. Siemens, *op. cit.* (note 12), 84–5.
54. Cited in Koenigsberger, *op. cit.* (note 27), 67.
55. Ludwig to du Bois-Reymond, letter dated 19 September 1850, Zurich, in Cranefield, *op. cit.* (note 3), 63.
56. Koenigsberger, *op. cit.* (note 27), 69.
57. Du Bois-Reymond to Ludwig, letter dated 9 April 1850, Paris, in Cranefield, *op. cit.* (note 3), 57–8. Brücke's research on luminescence in the animal eye and the method of causing luminescence at will in the human eye was continued by Helmholtz, who succeeded Brücke in Königsberg in 1849. Later, Helmholtz asserted that Brücke was 'a hair's breadth off the invention of the ophthalmoscope', which was introduced by Helmholtz in 1851. Cited in Koenigsberger, *op. cit.* (note 27), 74.
58. Du Bois-Reymond to Ludwig, letter dated 9 April 1850, Paris, in Cranefield, *op. cit.* (note 3), 58. Patronised by Alexander von Humboldt, du

Bois-Reymond had contacts in the French scientific community, besides du Bois-Reymond was particularly open to France because of his heritage, however, he preferred to communicate with British physicists rather than with his French colleagues. Du Bois-Reymond travelled to England several times and his talks before the Royal Institution were a great success. In addition to his scientific connections, including such influential men as Michael Faraday (1791–1867), Henry Bence-Jones (1814–73), and John Tyndall (1820–93), du Bois-Reymond had his wife's close relatives in England. He spoke and wrote English fluently. At one point, uncertain about financial support from his father, du Bois-Reymond thought about moving to England as the German chemist August Wilhelm Hoffmann (1818–92) had done. Du Bois-Reymond's financial situation soon improved due to a significant inheritance left for his wife, see, du Bois-Reymond to Ludwig, letter dated 27 December 1854, Berlin, in Cranefield, *op. cit.* (note 3), 87; fn. 196, in *ibid.*, 151.

59. Sechenov, *op. cit.* (note 1), 5.
60. *Auszüge aus Kassenbücher der Firma Siemens und Halske, 1846–66,* in *Siemens Archive, op. cit.* (note 43).
61. Sechenov noted that Bernard was very little acquainted with the German authors, as he did not know German, which was usual with French scientists. It was Bernard's German student, Wilhelm Kühne (1837–1900), a future notable physiological chemist, from whom Bernard learned much about research in German physiology, see I.M. Sechenov, *Autobiographical Notes,* D.B. Lindsley (ed.), K. Hanes (trans.) (Washington: American Institute of Biological Sciences, 1965), 106–7.
62. Allan Rocke convincingly argues that the introduction of the *Kaliapparat* by Liebig in the 1830s can serve as a convenient benchmark for the gradual passing of chemical leadership from France to Germany, see, Rocke, *op. cit.* (note 13), 66.
63. Cited in Koenigsberger, *op. cit.* (note 27), 90.
64. Du Bois-Reymond to Ludwig, letter dated 17 May 1849, Berlin, in Cranefield, *op. cit.* (note 3), 34.
65. Du Bois-Reymond to Ludwig, letter dated 26 June 1849, Berlin, in *ibid.*, 40; see also Koenigsberger, *op. cit.* (note 27), 51.
66. Koenigsberger, *op. cit.* (note 27), 29.
67. In Vienna Brücke's interests shifted from the physiology of sense organs, which he had pursued primarily in Berlin and Königsberg, to various problems of digestive physiology, and also to nerve–muscle physiology. His acclaimed *Vorlesungen über Physiologie* (Vienna: Gerold, 1856), published in several editions, demonstrated the diversity and originality of his own physiological and microscopic studies, in nearly all areas of contemporary physiology and histology. In his *Grundzüge der Physiologie der Sprachlaute*

(Vienna: Gerold, 1856), Brücke introduced the idea of the so-called 'Pasigraphie', the representation of the sounds of different languages with specific signs, an attempt to create a unified phonetic writing method for any foreign language. Interestingly, in his book, Brücke spoke very highly of the research by du Bois-Reymond's father, Félix du Bois-Reymond, on the general alphabet from a physical, physiological, and graphic standpoint, which Félix had undertook in 1811–2: F.H. du Bois-Reymond, *Kadmus oder allgemeine Alphabetik vom physikalischen, physiologischen, und graphischen Standpunkt* (Berlin: Reimer, 1862).

68. Du Bois-Reymond to Ludwig, letter dated 9 February 1849, Berlin, in Cranefield, *op. cit.* (note 3), 22.
69. Koenigsberger, *op. cit.* (note 27), 60.
70. Du Bois-Reymond to Ludwig, letter dated 17 May 1849, Berlin, in Cranefield, *op. cit.* (note 3), 34.
71. Du Bois-Reymond to Ludwig, letter dated 27 December 1854, Berlin, in *ibid.,* 86–7.
72. Recommended by Alexander von Humboldt, du Bois-Reymond was elected to membership at the Prussian Academy of Science in 1851. From 1876, as a permanent secretary of the Academy, he devoted much of his time to the Academy, see I. Jahn, 'Die Anfänge der instrumentellen Elektrophysiologie in den Briefen Humboldts an Emil du Bois-Reymond', *Medizinhistorisches Journal,* ii (1967), 135–56: 142.
73. Ludwig to du Bois-Reymond, letter dated 20 February 1849, Marburg, in Cranefield, *op. cit.* (note 3), 26.
74. Du Bois-Reymond to Ludwig, letter dated 25 February 1849, Berlin, in *ibid.,* 28.
75. The richly decorated album dedicated to du Bois-Reymond's sixty-year anniversary contains forty-nine portraits and autographs of his students. The silver cover of the album is remarkable. In its center there is a relief, which depicts a handsome young man sitting at the table with a famous set of du Bois-Reymond's electrophysiological devices and volumes of du Bois-Reymond's *Untersuchungen* and *Gesammelten Abhandlungen zur allgemeinen Muskel- und Nervenphysik* (Leipzig: Vogel Verlag, 1875–7), 2 vols. The relief is framed by a set of four medallions with the names of Johannes Müller, Alexander von Humboldt, Allessandro Volta, and [Luigi] Aloisio Galvani. In between the medallions on each side there is a nice design of lianas and the experimental animals, a frog, a rabbit, an electric eel, and an electric ray. The first colourful page of the album depicts an imposing building of du Bois-Reymond's Institute at Berlin Friedrich-Wilhelm-Universität on the Dorotheenstrasse and du Bois-Reymond's house on the Carlstrasse 21, where he performed the most part of his breakthrough experiments. The album is preserved in the *Universitätsbibliothek und Archiv* of the Humboldt

University of Berlin. See also P. Ruff und H. Choinowski, 'Eine Festgabe für Emil du Bois-Reymond', *Wissenschaftliche Zeitschrift der Humboldt-Universität zu Berlin,* xvi, 5, (1967), 839–46; and P. Schneck (ed.), *Emil du Bois-Reymond. 1818–96* (Berlin: Gerhard Weinert, 1996).

76. Rothschuh, *op. cit.* (note 6), 223.
77. Du Bois-Reymond, *op. cit.* (note 2), Vol. 1, Part 1, xv.
78. *Ibid.,* Vol. 2, Part 2, 501–2.
79. W. Nernst, 'Elektromotorische Wirksamkeit der Ionen,' *Zeitschrift fur physicalische Chemie,* iv (1889), 129–81; W. Ostwald, 'Elektrische Eigenschaften halbdurchlässiger Scheidewände,' *Zeitschrift für physikalische Chemie,* vi (1890), 71–82. Bernstein's theory appeared in final form in Bernstein's last work, *Elektrophysiologie: die Lehre von den elektrischen Vorgängen im Organismus auf moderner Grundlage dargestellt* (Braunschweig: Vieweg, 1912). For the useful and apprehensive treatment of Bernstein's theory, see T. Lenoir, 'Models and instrumentation in the development of electrophysiology, 1845-1912', *Historical Studies in the Physical and Biological Sciences,* xvii, 1(1986), 1–54: 39–49; see also a classical essay on the topic by P.E. Cranfield 'The Organic Physics of 1847 and the Biophysics of Today', *Journal of the History of Medicine and Allied Sciences,* xii (1957), 407–23: 418.
80. For biographical information, see H. Schröer, *Carl Ludwig: Begründer der messenden Experimentalphysiologie, 1816–95* (Stuttgart: Wissenschaftlische Verlagsgesellschaft, 1967).
81. On Ludwig at the University of Zürich , see P. Zupan, *Der Physiologe Carl Ludwig in Zürich, 1849–55,* Inaugural Dissertation (Zürich: Juris Druck Verlag, 1987).
82. Ludwig to du Bois-Reymond, letter dated 25 May 1849, Marburg, in Cranefield, *op. cit.* (note 3), 36.
83. The police report on Ludwig contains detailed information on Ludwig's appearance, personality, political, and religious views. The report is preserved in the War Records Office in Vienna, and published by P.G. Spieckermann, 'Physiology with Cool Obsession: Carl Ludwig: his Time in Vienna and his Contribution to Isolated Organ Methodology', *Pflügers Arch. Eur. J. Physiol.* 432 (1996), 33–41: 34.
84. C. Ludwig, *De olei jecoris aselli partibus efficacibus* (inaugural dissertation, University of Marburg, 1840). On Ludwig's dissertation, see Schröer, *op. cit.* (note 80), 294, see also Rothschuh, *op. cit.* (note 6), 206.
85. The British physiologist John Burdon-Sanderson pointed out that it was Bunsen from whom Ludwig 'derived training in exact sciences, which was to be of such inestimable value to Ludwig afterwards.' See, J.P. Burdon-Sanderson, 'Carl Ludwig', in B. Jones (ed.), *The Golden Age of Science: Thirty*

Portraits of the Giants of 19th Century Science by their Scientific Contemporaries (New York: Simon and Schuster, 1966), 409–10.

86. C. Ludwig, *De viribus physicis secretionem urinae adiuvantibus* (Marburg: Elwert, 1842); *idem, Beiträge zur Lehre vom Mechanismus der Harnsekretion* (Marburg: Akademische Buchhandlung Hrsg. N.G. Elwert, 1843). I have used the German version, located at the Sudhoffs Institute for the History of Medicine in Leipzig, hereafter referred to as Ludwig, *Habilitationsschrift.* For the English translation, see J. Davis and K. Thurau, 'Contributions to the Theory of the Mechanism of Urine Secretion', *Kidney International,* 46 (1994), 1–23.
87. Ludwig, *Habilitationsschrift, op. cit.* (note 86), iii.
88. On Bunsen's organic research, see Partington, *op. cit.* (note 40), Vol. 4, 283–7.
89. Rocke, *op. cit.* (note 13), 51, 85.
90. On Bunsen in Marburg University, see F. Kraft, 'Robert Wilhelm Bunsen in Marburg', in F. Kraft (ed.), *Bunsen-Briefe in der Universitätsbibliothek, Marburg* (Marburg: Universitätsbibliothek, 1996), 71–83.
91. On Ludwig and the intergrated approach to physiology, see Lenoir, *op. cit.* (note 6), 151–152.
92. Ludwig, *op. cit.* (note 86), 2.
93. *Ibid.,* 14.
94. *Ibid.,* 3.
95. *Ibid.,* 17. Henle, Müller's assistant in Bonn and Berlin, assumed the position of Professor of Anatomy in Zürich in 1840, and in 1844 in Heidelberg. Ludwig had been in intensive correspondence with Henle since 1846, but later strong disagreement on the primacy of functional versus morphologic interpretations of physiological problems eventually interrupted the contact. See A. Dreher (ed.), *Briefe von Carl Ludwig an Jacob Henle aus den Jahren 1846–72* (unpublished thesis, University of Heidelberg, 1980), 68–70, cited in Lenoir, *op. cit.* (note 6), 154; 139–78.
96. Ludwig, *op. cit.* (note 86), 21. Le Canu, professor of the Paris l'Ecole de Pharmacie tried to isolate the red colouring matter of blood and to assess its nature by adding acids, bases, and various salts to it. He coined the word 'hematosine' for it and showed that it contained iron. In 1838, he published his monograph, *Études chimiques sur le sang humain.* Berzelius split the blood into protein, 'globulin', and a coloured component, 'haematin', which contained iron oxide and was able to take far more oxygen then the serum. See, J. Berzelius, *Lehrbuch der Chemie,* F. Wöhler (trans.), (Dresden: In der Arnoldischen Buchhandlung, 1833–41), 10 vols, Vol. 9, 62. Marcet, a lecturer in chemistry at London Guy's Hospital, working in his private laboratory discovered in a urinary calculus a substance he called 'xanthic oxide', since it was soluble in alkalis and on evaporation gave a yellow

substance. He published these results in his *An Essay on the Chemical History and Medical Treatment of Calculus Diseases* in London in 1817, which Berzelius cited in his *Lehrbuch,* see Partington, *op. cit.* (note 40), Vol. 3, 548; 707–8; see also P. Astrup and J. Severinghaus, *The History of Blood Gases, Acids and Bases* (Copenhagen: Munksgaard, 1986), 117.

97. Ludwig, *Habilitationsschrift, op. cit.* (note 86), 13.
98. *Ibid.,* iv.
99. *Ibid.,* 16.
100. *Ibid.,* 16, 18. J. Eytelwein, *Handbuch der Mechanik fester Körper und der Hydraulik* (Leipzig: Fleischer, 1842).
101. Ludwig, *Habilitationsschrift, op. cit.* (note 86), 17.
102. *Ibid.,* 19.
103. *Ibid.,* 21.
104. *Ibid.,* 17. Müller and the British physiologist William Bowman (1816–92), who discovered in 1842 the relationships between glomeruli and urinary tubules, favoured the tubular secretion explanatory model. See, Rothschuh, *op. cit.* (note 6), 247. For a detailed discussion on Ludwig's concepts of renal function see, T.K. Thurau, J. Davis, and D. Häberle, 'Renal Blood Flow and Glomerular Filtration Dynamics: Evolution of a Concept from Carl Ludwig to the Present Day', in C. Gottschalk, R. Berliner and G. Giebisch (eds), *Renal Physiology: People and Ideas* (Bethesda: American Physiological Society, 1987), 31–61. See also K. Hierholzer, 'Carl Ludwig, Jacob Henle, Hermann Helmholtz, Emil du Bois-Reymond and the Scientific Development of Nephrology in Germany', *Am J Nephrol,* xiv (1994), 344–54, 348.
105. Ludwig, *op. cit.* (note 86), 22.
106. *Ibid.,* 18.
107. C. Ludwig, 'Introduction' to the textbook on human physiology, M. Frank and J. Weiss (trans.), *Medical History,* x (1966), 76–86: 85–6.
108. C. Ludwig, 'Beiträge zur Kenntnis des Einflusses der Respirationsbewegungen auf den Blutlauf im Aortensysteme,' *Müllers Archiv für Anatomie, Physiologie und wissenshaftliche Medizin,* iii (1847), 242–302.
109. H. Hoff and L. Geddes, 'Graphic Registration before Ludwig. The Antecedents of the Kymograph', *Isis,* 50 (1959), 5–21: 5–6.
110. On a detailed description of Pouillet's device, see *ibid.,* 14. C.S.M. Pouillet, 'Note sur un moyen de mésurer des intervalles de temps extrémement courts comme la durée du choc des corps élastiques, celles du debandement des resorts, de l'inflammation de la poudre, etc.; et sur un moyen nouveau de comparer les intensités des courants électriques, soit permanents, soit instantanés', *Comptes Rendus Hebdomadaires des Séances de l'Académie des Sciences,* xix (1844), 1384–9.
111. C. Ludwig, *Lehrbuch der Physiologie des Menschen* (Heidelberg: Winter Verlag, 1852), Vol. 1, 330, 333.

112. Brain and Wise argue that precision instruments and graphic methods had been a topic of discussion at the Physical Society long before Ludwig's invention, see Brain and Wise, *op. cit.* (note 41), 52.
113. W.J. Meek, 'Carl Ludwig', *The Gamma Alpha Record,* xxiii (1933), 31–43: 37.
114. Siemens, *op. cit.* (note 12), 246.
115. It is not known where and how du Bois-Reymond first met Ludwig, see Ruff, *op. cit.* (note 12), 46; Schröer, *op. cit.* (note 80), 36.
116. Ludwig to du Bois-Reymond, letter dated 25 September 1848, Marburg, in Cranefield, *op. cit.* (note 3), 19.
117. Ludwig to du Bois-Reymond, letter dated 14 June 1849, Marburg, in *ibid.,* 39.
118. Du Bois-Reymond to Ludwig, letter dated 26 June 1849, Berlin, in *ibid.,* 41.
119. Du Bois-Reymond to Ludwig, letter dated 7 August 1849, Berlin, in *ibid.,* 44.
120. Du Bois-Reymond to Ludwig, letter dated 17 February 1852, Berlin, in *ibid.,* 72.
121. Ludwig, *op. cit.* (note 111), Vol. 1.
122. Du Bois-Reymond to Ludwig, letter dated 7 November 1858, Berlin, in Cranefield, *op. cit.* (note 3), 97.
123. A. Volkmann, *Neue Beiträge zur Physiologie des Gesichtssinnes* (Leipzig: Veit, 1836), and *idem, Die Haemodynamik nach Versuchen* (Leipzig: Veit, 1850). E.H. Weber's collected works were published as *Annotationes anatomicae et physiologicae programmata collecta* (Leipzig: Veit, 1851). On Ernst H. Weber's influence on the development of the exact methods in physiological investigations as well as on the contribution of Eduard F. Weber and Wilhelm E. Weber to the application of physical principles in physiology, see C. Ludwig, *Rede zum Gedächtnis an Ernst Heinrich Weber* (Leipzig: Verlag von Veit & Comp., 1878).

3

A Viennese Prelude: Sechenov's Research at Ludwig's Laboratory

Ivan Mikhailovich Sechenov (1829–1905) came from a family of provincial gentry. His father, an officer in the élite Guards regiment during Catherine II's reign, had property in Kostroma and Simbirsk provinces on the Volga. Sechenov was educated at home and, through his sisters' German governess, became fluent in both German and French. On the advice of his brother, who was an officer, the fourteen-year old Sechenov entered the St Petersburg Military Engineering School. The school was known for its thorough training in technical, mathematical, and physico–chemical disciplines, given by some highly-regarded professors. One was M.V. Ostrogradskii (1801–61), a mathematician of European reputation and a member of the St Petersburg Academy of Sciences with overall responsibility for all mathematics education in the St Petersburg military academies. Ostrogradskii wrote several important texts on mathematical physics, analytical geometry, celestial mechanics, and ballistics.[1] He was a brilliant lecturer and, not surprisingly, Sechenov became increasingly interested in physics and mathematics:

> I had distaste for engineering and everything to do with it, physics was my favourite subject. In the junior officer class I conceived a liking for chemistry. I was good at mathematics, and had I entered the Mathematical Faculty at the University immediately after leaving the Military Engineering School, I think that I might have become rather a good physicist.[2]

After six years of study, Sechenov entered army service as a junior field engineer in Kiev, but soon retired after one-and-a-half years of service. In his decision, Sechenov was influenced by a young lady from a fellow-officer's family. Olga Aleksandrovna – her last name is unknown; we only know that she was a daughter of an exiled Polish physician – eagerly discussed the rights of women and other social problems, and set a high value on science and intellectual work. She spoke to Sechenov about Moscow University as a centre of culture, and about medicine as one of the noblest professions.

In 1851, Sechenov entered the medical faculty of Moscow University. The faculty was one of the best of the Empire, staffed by professors who had studied abroad, mainly in Germany, before starting an academic career.

Image 3.1

Ivan M. Sechenov (right) with Sergei P. Botkin (left). Courtesy of the Wellcome Library Collection.

Nikolai E. Lyaskovskii, Professor of Pharmacognosy, had been a student of Justus Liebig at Giessen; Alexander I. Polunin (1820–88), Professor of Pathological Anatomy, a student of Carl von Rokitanski (1804–78) in Vienna; Fedor I. Inosemtsev (1802–69), Professor of Surgery, had studied at the Charité in Berlin; and Ivan T. Glebov (1806–84), Professor of Anatomy and Physiology, had studied under Johannes Müller and Theodor Schwann in Berlin, and well as in Paris. The professors were knowledgeable, but none were actively engaged in research or publication. Sechenov liked Glebov most of all. A brilliant lecturer and original thinker, Glebov inspired

Sechenov's interest in comparative anatomy. In his physiology course, Glebov ultimately confined himself to the French school. His lectures reflected his admiration for the works of Marie-Jean Flourens, François Magendie, and Claude Bernard. While a student in Paris, Glebov had translated both Magendie's *Précis élémentaire de physiologie* and Julius Budge's *Lehrbuch*[3] into Russian. His lecture demonstrations were focused on some of Flourens's classical experiments: 'pigeons with pinholes in the brain were used to describe disturbances in locomotion and changes in sensitivity caused by the operation'. To Sechenov's surprise, Glebov never discussed the new physico–chemical trend in German physiology in his lectures, and never talked about experiments on electrical stimulation of the nerve and muscle, although 'Germany was long ago full of these experiments'. Even Ernst Weber's famous experiment of 1845 on stopping the heart by stimulation of the vagus nerve was not mentioned.[4] It was Johannes Müller's famous *Handbuch der Physiologie des Menschen* [translated into English by William Baly as *Elements of Physiology*], available from a German bookseller in Moscow[5] that turned Sechenov's interest to German physiology. Sechenov admitted that already, in the third year of his studies, he had lost interest in practical medicine. For him, with his science background and mathematical cast of mind, medicine appeared as 'pure empiricism', and a career as a practising physician seemed not at all suitable.

After graduating, Sechenov decided to use an inheritance from his mother to study in Germany. Berlin University, where 'the celebrated Johannes Müller' taught, attracted him the most. It was a good choice: there was a growing awareness of the vigour of some German universities in the natural sciences and medicine in the late 1850s. When Sechenov arrived in Berlin in 1856, the Russian medical students formed a small colony, including Sergei Botkin, ophthalmologist E.A. Junge (1833–98), and Ludwig A. Bekkers (1832–62), who had been a surgeon under Nikolai I. Pirogov in the Sevastopol campaign in the Crimean War. After studies the young men often gathered together socially, with the effervescent Botkin taking the unofficial lead as life and soul of the party.[6]

Sechenov's acquaintance with Berlin University began with the courses taught by Gustav Magnus in physics and Heinrich Rose in analytical chemistry. For Sechenov, both courses were elementary but he valued the excellent lecture demonstrations:

> Magnus was considered a first-rate lecturer and an extremely skillful experimenter by Helmholtz. [Helmholtz succeeded Magnus in Berlin in 1871.] Magnus always tried to do the experiments in such a way that he could put into action an apparatus shown, or evoke the desired phenomenon, just by means of pulling a string or by some simple

> movement. His course was luxuriously provided with experiments done with such speed that they did not disturb the smoothness of the reading. Carbonic acid was changed in about a quarter of an hour into lumps of loose snow which were thrown among the listeners in the auditorium.[7]

At that time, Sechenov could scarcely have imagined that his first experimental success would be the improvement of Magnus's method and blood-gas pump, and that the state of carbon dioxide in the blood would become his life-long interest.

Sechenov planned to study at Müller's laboratory but, to his disappointment, Müller did not admit physiology students and so Sechenov only attended his lectures:

> In my soul was still hidden the naïve habit, brought from Moscow, of thinking that each famous professor was necessarily a brilliant orator, and I expected to hear in this auditorium an absorbing talk full of wide generalisations, but instead I heard a purely business-like talk with a showing of drawings and preparations in alcohol. This was, however, the last year of Müller's glorious life, and at the lectures he appeared a tired, ill man. In all his movements and in his very speech certain nervousness was felt; he lectured quietly, not raising his voice, and only his eyes continued to burn with the indescribable brilliance which together with the famous name of the scholar became historical.[8]

Sechenov also took du Bois-Reymond's course in electrophysiology. Du Bois-Reymond's lectures 'with many detours into innervation of the heart, intestines, and respiratory movements both by their content and their execution were fascinating.' After lectures, Sechenov and another student worked with the galvanometer in the corridor adjacent to the room, where du Bois-Reymond himself worked. This was basically du Bois-Reymond's laboratory. Sechenov always admitted his indebtedness to 'these very studies in the corridor' that gave him 'means to advance easily into the new area of investigation'.[9] Learning the fine points of electrophysiological techniques at first hand from du Bois-Reymond proved crucial for Sechenov's later research in nerve–muscle physiology.

Another principal place for the study of experimental methods was the laboratory of Felix Hoppe-Seyler (1825–95) in the Virchow Pathological Institute. In Moscow University, medical students were not allowed into the chemical laboratory and Sechenov did not even know how to handle a burner, chemical dishes, etc. He had to gain basic laboratory skills at the private chemical laboratory of dozent Franz Leopold Zonnenschein, at the time one of the best in Berlin, before he could pass on to Hoppe-Seyler's laboratory. Sechenov liked Hoppe-Seyler's laboratory, fine, spacious, and well equipped, and its young director, 'a dear, able and lenient teacher who did not differentiate between Russian and German students'. Hoppe-Seyler welcomed research proposals from his students and if they proved to be reasonable and feasible, gave his encouragement and support.[10] Sechenov's research on the effect of alcohol on humans and animals, a project of his own devising, required experimental study of the influence of alcohol poisoning on body temperature and quantitative analysis of carbon dioxide exhaled by an intoxicated animal. Studying the influence of alcohol on nitrogen metabolism, Sechenov repeated Bernard's experiments on the action of various poisons on the nerves and muscles in the frog, and these results became his first publication in Germany.[11] Next year he moved to Leipzig to study with Otto Funke (1828–79) in the laboratory at the Physiological Institute of Ernst Heinrich Weber. Although Funke's recent research on physico–chemical properties of the blood[12] was of particular interest to him, Sechenov did not stay long in Leipzig and soon moved to Vienna to study with Ludwig.

Sechenov stayed in Ludwig's laboratory at the Kaiserlich-königliche chirurgische Militär-Akademie, the Josephinum, for more than a year during 1858–9. The laboratory at the Josephinum, a closed military institution, was not intended for the students' practical studies and Ludwig accepted only Sechenov and one other researcher to his laboratory. Sechenov continued the experimental part of his thesis and sometimes assisted Ludwig in the experiments on saliva secretion registered by the kymograph. Ludwig used vivisection techniques widely in his investigations but it seems that Sechenov was reluctant to develop in this direction. Studies with Ludwig were instructive as well as entertaining. Ludwig loved to chatter at work, and used to ask Sechenov about Russia. The German professor knew Mikhail Lermontov's poems in translation and Sechenov recited for him *The Gifts of the Terek*, famous for its fascinating descriptions of the wild Caucasian nature.[13] Sechenov was a frequent guest of the Ludwig family, which consisted of his wife, 'a very modest, taciturn woman, and a daughter of fifteen'. His other 'amusements were open air concerts by Johann Strauss in the *Volksgarten*, and trips by steamboat along the Danube'.[14] Sechenov always recognised Ludwig as the decisive influence on his research career, and saw Ludwig as his primary mentor. Sechenov's *Autobiographical Notes* contains many warm and fascinating reminiscences of Ludwig, the 'incomparable teacher':

> Ludwig became an international teacher of physiology known all over the world. To occupy such a position it was not enough to have talent: Helmholtz, while he was a physiologist, and du Bois-Reymond, in all his long activity, had few laboratory students. Besides talent and knowledge,

> Ludwig possessed special abiliies as a teacher, and developed methods of teaching, which made a period spent by a student in the laboratory both useful and pleasant. Invariably friendly and cheerful both in moments of rest and at work, Ludwig took a active part in everything, undertaken under his instructions. He usually worked not by himself, but together with his students, carrying out the most difficult parts of the investigation himself. In publications, his name appeared only now and then beside the name of a student, despite his major contribution to any research pursused in his laboratory.[15]

Ludwig and Sechenov remained close friends throughout their lives. In critical moments of his life, Sechenov always found consolation and comfort with Ludwig, whose goodwill 'did not cease, right up to his death, manifesting itself in all the little turns of my life with warm, compassionate letters.'[16]

Some useful details about the Josephinum laboratory, of the time when Sechenov studied there, can be found in Ludwig's letters to du Bois-Reymond, who was keen to know the expenses and equipment of Ludwig's new institute. It might well be that du Bois-Reymond wanted to use the Viennese example in his fight at the Ministry for increasing their meagre funding for his Berlin laboratory. What we know from Ludwig's letters is that his institute had a lecture auditorium for fifty people, a small chemical laboratory, a mechanical workshop with a lathe, carpenter's bench and vice, and a room where Ludwig performed 'the more delicate jobs'. This room was equipped with kymograph, myograph, multipliers, goniometers, ophthalmometers, scales, air pumps, induction apparatus, spirometer and gasometer, and equipment for analysing air. All the rooms were gas-lit and partly gas-heated or had the stove necessary for chemical work; there was also a heated dog kennel and a rabbit hutch. Ludwig was allowed to use the ice cellar of the Garrison Military Hospital. In his letters to du Bois-Reymond, Ludwig meticulously listed all small expenses, without which laboratory work could not be carried out, such as animals (dogs, rabbits, frogs, tortoises, fish), metal for building apparatus in the Institute's mechanical workshop, glassware, knives, porcelain dishes, reagents, coal, alcohol, mercury, paper, sealing wax, pens, string, and linen. Another important component for the laboratory was its staff. Ludwig had a senior physician-assistant, a mechanical assistant, and an orderly from military service.[17] Although Ludwig admitted that the premises of his institute were limited and could not be compared with 'Brücke's splendid rooms' at Vienna University, he seemed quite happy with the conditions he had been offered at the Josephinum: 'I have hopes of being able to set up a fine and useful institute; they are willing to make available as much money as I desire.'[18]

Ludwig appreciated 'a most friendly welcome in Vienna', and willingness of the Josephinum authorities to support his 'scientific wishes'. He was grateful to Austria for the opportunity to leave his Swiss exile at a time when his political reputation, as he himself admitted, 'was at its worst'.[19] Ludwig was well aware that 'when the Viennese commenced their negotiations, they imposed the strictest secrecy' on him because in Vienna, as in Berlin, 'the nomination was opposed by some serious resistance'. Nevertheless, the appointment went through: the willingness of the Josephinum administration to have one of the best German physiologists outweighed political prejudice. In his letter of acceptance, Ludwig strategically pointed out:

> Extensive experience, which has shown to me not only the attractive, but also the dismal sides of universities makes it appear very likely to me that your Academy will offer the conditions under which a professor can teach efficiently and pursue his own academic development without disturbance.[20]

Ludwig's requests (very high salary of 2,600 florins, pension, etc.) were fulfiled by the order of Emperor Franz-Joseph.[21] We will return to the Josephinum in the following chapter, in connection with the reformation of the St Petersburg Medico-Surgical Academy and the crucial changes in military medical education after the Crimean War.

Before we pass onto Sechenov's blood-gas research in Ludwig's laboratory, it is useful to put into perspective the major developments in respiratory physiology before the late 1850s. From the time of Lavoisier and well into the 1830s, the problem of respiration was of particular interest to chemists. Since Lavoisier's theory of respiratory combustion had been undermined by the data collected over the first third of the nineteenth century, there was no consensus on the gaseous exchange during the respiratory process. Jons Jakob Berzelius held that the blood must take up a considerable amount of oxygen from the air in the lungs and that it was possible that gases existed in the blood.[22] Magnus took up the problem in 1837. In a letter to Berzelius, he mentioned that he had investigated the problem of the possibility of extracting gases from the blood with the help of the air pump alone, without adding the acetic acid to blood.[23] Magnus's experiments with blood gases revealed the existence of carbon dioxide, nitrogen, and oxygen in both venous and arterial blood. Most important, however, was a novel method for the extraction of gases dissolved in the blood. Magnus devised an original apparatus, a 'shortened barometer', which produced a Torricellian vacuum over the blood sample.[24] The apparatus consisted of an inverted 'separator funnel' full of mercury into which the blood sample was introduced from below before closing the stopcock. The level of the mercury was then lowered by evacuating the air from the enclosing bell jar with an air pump. The liberated gas was analysed in Volta's eudiometer according to the well-established practice.[25] In his later investigations of blood gases published in 1845, Magnus pointed to the need for determining the absorption coefficients for gases dissolved in blood. He assumed that the quantities of gases in the blood depended on their absorption coefficients and on their partial pressures, following Henry's and Dalton's laws of absorption.[26] Magnus's explanation of the mechanisms of blood gas exchange in strictly physical terms was responsible for focusing the attention of physiologists onto the problem. In 1844–5, the notable physiologists Gabriel Valentin (1810–83) and Karl von Vierord (1818–84) applied the physical law of diffusion to explain the passage of oxygen and carbon dioxide through the lungs.[27] At the same time, Berzelius and Liebig demonstrated that the quantities of oxygen and carbon dioxide present in the blood were too large to be explained by physical laws alone. The experimental data showed that the formation of bicarbonate salts was the binding agent for carbon dioxide in the serum, and that the red blood corpuscles had a chemical affinity for oxygen. These findings suggested that gases of the blood were in a bound state.[28] Thus, a chemical approach to the respiratory process began gaining ascendancy over the physicalistic explanations.

In 1857, Bunsen published his famous *Gasometrische Methoden,* which proved decisive for the development of the techniques for analysis and extraction of blood gases. Bunsen's methods of gas analysis were basically physical. He applied chemical tests only if they could yield accurate quantitative results. The calculation of the absorption coefficients for nitrogen, hydrogen, carbon dioxide, and oxygen in pure water involved determining physical conditions such as temperature, pressure and volume of the absorbed gas. Bunsen devised the standard absorption coefficients, which allowed the calculation of any of these parameters with accuracy – the major advantage that Magnus's method lacked. Bunsen's apparatus consisted of the straight-tube eudiometer over mercury, and solid absorbents in the form of small spheres on the ends of platinum wires. It enabled one to eliminate a single gas from a mixture of gases in a chemical reaction, and, by measuring the volume, temperature, and pressure before and afterwards, to determine the quantity of the eliminated gas. In the *Gasometrische Methoden,* Bunsen also presented methods of collecting, preserving, and measuring gases, techniques of eudiometric analysis, new processes for determining the specific gravities of gases, and the results of his investigations on the absorption of gases in water and alcohol using the absorptiometer of his own devising. Bunsen's method, based on Dalton's assumption that the partial pressure of a gas controls its absorption by a liquid, was so precise that it became widely used by chemists as a quantitative test for the presence of gases in liquids.[29]

There were two major implications of Bunsen's laws of gaseous absorption for the development of the techniques of blood gas extraction. The first one was the variability of the absorption coefficients with temperature. The two most successful methods, by Lothar Meyer (1830–95) and by Sechenov, included the means of maintaining the blood sample at a constant temperature to obtain accurate results. The second implication was the importance of the relationship between partial pressure and the absorption of gases. Even more significant for the blood gas research done by Meyer and then taken up by Sechenov were Bunsen's standards for absorptive coefficients of water and the method of their determination. Bunsen's approach ultimately suggested an analogue model in which blood reproduced certain relevant features of water, a model Sechenov would exploit in his research on blood gases in 1873.

Bunsen was well aware of the importance of his method for research on the absorption of gases by the blood[30] and suggested that Lothar Meyer undertake a project related to the problem. Meyer was an able candidate for the task. He had studied medicine under Ludwig in Zürich and his interest in chemistry, encouraged by Ludwig, soon led him to Bunsen's laboratory in Heidelberg. Meyer spent two years with Bunsen, from 1854 to 1856, and performed important investigations on blood gases, *Die Gase des Blutes*, which he dedicated to Ludwig.[31] Meyer challenged Magnus's method for determining blood gases using a different principle which involved boiling in a vacuum as suggested by Bunsen. It enabled Meyer to demonstrate, in 1856, that oxygen absorption by the blood had no dependence on a fairly wide range of relatively high partial pressures. It meant that apart from the physical binding of oxygen there must be chemical binding in the blood, but the chemical binding was relatively loose or weak. It also meant that oxygen could be expelled from the blood at low pressures or by shaking the blood with other gases. Meyer suggested two methods of extracting gases from the blood. The first one, '*Auskochung*' (boiling out), eliminated the major defect of Magnus's method, which required prolonged contact with mercury (up to six hours). The Auskochung method reduced the foaming of the blood in the vacuum so that it was possible to extract more carbon dioxide. Meyer's second method involved the evaluation of the effects of pressure on the absorption of gases by the blood using Bunsen's formulae.[32]

Working with Bunsen, Meyer showed special skill in devising instruments. Basically Meyer's apparatus was a modification of the absorption apparatus of his mentor, better adapted to the analysis of blood gases and their extraction from the blood. The blood sample was transferred to the bottom flask *A,* which contained distilled water from which all gases had been boiled out. The content of the flask was brought to the boil so that the air above was expelled to flask *B,* which initially also contained distilled water, and then to tube *C,* where it collected. It was possible to regulate the pressure within the system and there was no prolonged contact of the blood with mercury or water. The apparent disadvantage of Meyer's method was lack of control of the temperature of the apparatus and hence, inaccurate results.[33]

The imperfections of Meyer's first method were improved by Emil Fernet (1829–1905), a professor at the École Polytechnique, Paris. Fernet eliminated condensation and maintained a more constant temperature of the apparatus. In his work published the same year as Meyer's *Die Gase des Blutes,* Fernet gave a lengthy review of the research that had been done on the problem.[34] Meyer's method, and its improved modification by Fernet, provided experimental evidence that the gases associated with respiration were not simply dissolved in the blood, rather they were held by chemical affinities. It was also shown that both pressure effects and chemical effects were involved in the absorption of carbon dioxide by the blood serum and salt solutions.[35] Finally, and most importantly, these findings proved convincingly that both the physical and chemical forces acting upon the same process were not incompatible.

After Heidelberg, Meyer moved to Königsberg in 1856, to study with the mathematical physicist Franz Neumann.[36] In 1858, Meyer accepted the offer of Rudolf Heidenhain to take over the chemical laboratory at Breslau Physiological Institute. In collaboration with Heidenhain, Meyer performed some important studies in the chemistry of the blood.[37] Meyer taught courses in organic, inorganic, and physiological chemistry. Eventually, his interests shifted to theoretical chemistry. In 1864, he published his *Die modernen Theorien der Chemie und ihre Bedeutungen für die chemische Statik,* an extraordinarily clear statement of the fundamental principles of chemistry. Although his most important achievements were in the realm of chemistry, particularly his work on the periodic classification of elements, Meyer's physiological investigations significantly advanced the blood gas research taken up by Ludwig and Sechenov.[38]

Ludwig became interested in the results of Meyer's investigations as the problem of gaseous exchange in the blood moved away from the narrow framework of strictly physicalistic explanations. As early as 1846, Ludwig published a critical review of Karl von Vierord and Gabriel Valentin's physical theory of gaseous transfer in the lungs, the so-called diffusion theory. Ludwig objected to their approach and method of quantitative estimates of red blood cells and lung capacity, particularly to Valentin's reliance upon extensive numerical data as an application of arithmetical arguments to living phenomena.[39] In his studies on the respiratory function of the blood as well as in his research on the renal function, Ludwig increasingly began to perceive that simple application of mechanics and physics could not account for the complex interrelations between physiological functions. Ludwig felt that 'in our field, chemistry in the true sense of the word provides the prospect for the most significant advances.'[40] Interest in chemical transformations in bodily fluids led Ludwig to encourage both of his students, Meyer and later Sechenov, to apply methods of physical chemistry to the problem of gaseous exchange in the blood.

Sechenov's doctoral project required studies of the absorption of gases by the blood. With his background in physics and chemistry, and his technical dexterity, Sechenov was ideally fitted to tackle the problem. He believed, as did his mentor, that accurate measurement of the gas content of the blood would allow for better understanding of the complex phenomena underlying the process of respiration. The apparatus reported by Magnus and Meyer did not meet the requirement of precision.[41] Meyer's results on the content of oxygen in the blood were fairly accurate but his estimates of carbon dioxide were not exact.[42] The commonly employed methods were: heating; formation of a vacuum for a gas in a solution by placing that solution in a closed vessel filled with a different gas, known as the replacement technique; the addition of reagents; the creation of a vacuum with air pumps or by the evacuation of mercury from the filled tubes, known as the Torricellian vacuum. At this point, Sechenov was confronted with the major technical difficulties present in all four methods: uncontrolled heating led to coagulation of blood proteins and entrapment of gases; the added chemical reagents altered the natural structure of the blood; a long contact with mercury resulted in clumping and the precipitation of proteins; and finally, foaming of the blood in the vacuum prevented complete removal of the carbon dioxide.

Sechenov managed to eliminate the inefficiencies of the existing methods and constructed a reliable device, which removed gases almost entirely without adding any chemical reagents. In his improved model, Sechenov combined the best features of the previous methods. In the manner of Emil Fernet, Sechenov's apparatus was combined with a vacuum pump for removing mercury. The apparatus first was completely filled with mercury and then mercury was completely removed to form a Torricellian vacuum. Sechenov evaluated the variations in temperature and vapour pressure on mercury, and the quantity of water present in the blood sample and took into consideration the absorption coefficients of mercury for oxygen and carbon dioxide. He also co-ordinated the size of the blood sample with the size of the collection chamber, adapted the length of the apparatus, and secured constant temperature throughout the experiment. Sechenov's method yielded results of unprecedented accuracy, and Ludwig immediately ordered the new apparatus for his laboratory.[43] In his paper published in 1859, Sechenov pointed out his debt to previous researchers – Magnus, Meyer, and Fernet – and, in a traditional German manner, modestly commented that the only thing that was new about his method was that highly accurate and secure results were achievable. He also pointed to some advantages of his method, namely the easy operation of letting mercury in and out and the possibility to repeat the boiling of the blood in a vacuum until the blood stopped giving off gas.[44] It should be noted that Sechenov's method of gas extraction from the blood, like all previous methods, was a complex and tedious procedure. It took a whole day to extract gases from 100ml of blood. In order to obtain accurate results, the evacuation had to be repeated many times to ensure the absence of gases in the blood. The absorptiometer with blood–gas pump, a prototype of Van Slyke's apparatus,[45] was heavy and bulky, with a large amount of mercury, which should contain no air. There would be many other modifications of the apparatus by other researchers, but Sechenov's highly accurate estimations of blood gases would remain valid for decades to come.

To sum up, the research undertaken in the late 1850s by chemically inclined physiologists such as Meyer, Ludwig, and Sechenov were quite impressive. The exact quantities of gases contained in the blood were estimated, the effects of both physical and chemical forces on the rate of absorption and retention of blood gases were discovered, and the concept of the role of pressure in the exchange of gases between the blood and the lungs was established for the first time. For Sechenov, his Vienna research would become a life-long interest and an overture to his later remarkable investigations on blood gases and salt solutions in St Petersburg. He once remarked, that 'the long fussing with Meyer's apparatus' became the reason 'that I devoted a very significant part of my life to problems of blood gases and the absorption of gases by liquids.'[46]

Notes

1. On M.V. Ostrogradskii's most important mathematical work, see V. Katz, *A History of Mathematics* (Reading: Addison-Wesley, 1998), 748–9.
2. I.M. Sechenov, *Autobiographical Notes,* D.B. Lindsley (ed.), K. Hanes (trans.), (Washington: American Institute of Biological Sciences, 1965), 17.
3. François Magendie (1783–1855) made epoch-making contributions to brain and nerve physiology, pharmacology and pathology, and was notable for the introduction of strychnine and morphine into medical practice, as well as attracting controversy for his groundbreaking vivisection demonstrations. His two-volume *Précis élémentaire de physiologie* (Paris: Méquignon-Marvis, 1816–17) was the first modern physiology book in which doctrine gave way to simple, precise description of experimental facts. He also founded the

Journal de physiologie expérimentale in 1821, and exerted a decisive influence on his successor, Claude Bernard. See, for example, H. Bloch, 'François Magendie, Claude Bernard, and the Interrelation of Science, History, and Philosophy', *Southern Medical Journal*, 82, 10 (1989), 1259–61. Julius Ludwig Budge (1811–80), Professor of Anatomy and Physiology in Bonn and in Greifswald, Prussia, was known for his anatomical and physiological studies on the vegetative nervous system. For the discoveries that sympathetic stimulus produces pupilary dilatation and that oculo-motor nerve stimulation produces the opposite effect, published in his main work *Bewegung der Iris* (Braunschweig: Vieweg, 1855), Budge was awarded the Prix Montyon of the Paris Academy of Sciences. See *Neue Deutsche Biographie* (Berlin: Duncker und Humboldt, 1953–present), Vol. III, 755.

4. Sechenov, *op. cit.* (note 2), 47–8.
5. *Ibid.*, 50.
6. *Ibid.*, 70–1.
7. *Ibid.*, 69.
8. *Ibid.*, 69–70.
9. *Ibid.*, 68.
10. *Ibid.*, 67–8. On Hoppe-Seyler's research group, see J.S. Fruton, *Contrasts in Scientific Style: Research Groups in the Chemical and Biological Sciences* (Philadelphia: American Philosophical Society, 1990), 93–6.
11. I.M. Sechenov, 'Einiges über die Vergiftung mit Schwefelcyankalium', *Virchows Archiv für pathologische Anatomie und Physiologie und für klinische Medizin*, xiv (1858), 356–70.
12. O. Funke, *Lehrbuch der Physiologie für akademische Vorlesungen und zum Selbstudium* (Leipzig: Leopold Foss, 1858–60).
13. Mikhail Iur'evich Lermontov (1814–41), Russia's famous romantic poet and novelist of old aristocratic provenance, served as an officer in the St Petersburg Imperial Guards. In 1837, he was exiled to the army in the Caucasus for his poem *Na smert' poeta* [*On the Death of a Poet*] dedicated to Alexander S. Pushkin (1799–1837), who had been fatally wounded in a duel with a French officer of the St Petersburg Imperial Guards. Four years later, Lermontov's duel with one of the officers of his regimen in the Caucasus was fatal. With their extremes of mood, their aesthetic and moral individualism and above all, with their Caucasian settings, Lermontov's works were a magnet for the Russian composers of the second half of the nineteenth century. Lermontov is also famous for his translations of Lord Byron (1788–1824), Friedrich Schiller (1759– 1805), and Wolfgang Goethe (1749–1832).
14. Sechenov, *op. cit.* (note 2), 82–4.
15. *Ibid.*, 82.

16. *Ibid.*, 87. Ludwig's letters to Sechenov span more than thirty years and reveal a deep and close friendship between the two men who are still remembered as the towering figures of nineteenth-century German and Russian physiology. In his letters, Ludwig used to report, sentimentally, on the achievements of his Russian students, who were particularly welcome to work in his laboratory. Ludwig's letters also offer glimpses of the academic milieu including interesting remarks on his close associates, du Bois-Reymond, Brücke, Helmholtz, Bunsen, and Ostwald, and provide invaluable source of his views on scientific and institutional matters. Located in the Archive of the Russian Academy of Sciences in Moscow (fond 605, opis' 2/1752), Ludwig's letters to Sechenov were first published in M.N. Shaternikov, 'The life of I.M. Sechenov', in I.M. Sechenov, *Selected Works* (Moscow–Leningrad: State Publishing House for Biology and Medical Literature, 1935), ix–xxxvi; see also H. Schröer, *Carl Ludwig: Begründer der messenden Experimentalphysiologie, 1816–95* (Stuttgart: Wissenschaftlische Verlagsgesellschaft, 1967), 248–62. Unfortunately, Sechenov's letters to Ludwig are not preserved. They are believed to be lost, like much of the archival material of Leipzig University, during the bombardment of Leipzig in 1945.
17. Ludwig to du Bois-Reymond, letter dated 14 March 1858, Vienna, in P.F. Cranefield (ed.), *Two Great Scientists of the Nineteenth Century: Correspondence of Emil du Bois-Reymond and Carl Ludwig*, S. Lichtner-Ayéd (trans.), (Baltimore: Johns Hopkins University Press, 1982), 97. On Ludwig's laboratory at the Josephinum, see E. Lesky, *The Vienna Medical School of the Nineteenth Century* (Baltimore: Johns Hopkins University Press, 1976), 238.
18. Ludwig to du Bois-Reymond, letter dated 3 September 1855, Vienna, in Cranefield, *op. cit.* (note 17), 88.
19. Ludwig to du Bois-Reymond, letter dated 10 November 1858, Vienna, in *ibid.*, 99.
20. Ludwig to du Bois-Reymond, letter dated 3 September 1855, Vienna, in *ibid.*, 87.
21. P.G. Spieckermann, 'Physiology with Cool Obsession: Carl Ludwig', *Pflügers Archiv: European Journal of Physiology*, 432 (1996), 33–41: 35.
22. J. Berzelius, *Föreläsninger over Djurkemien* (Stockholm: Marquard, 1806–8), 2 vols, cited it in P. Astrup and J. Severinghaus, *The History of Blood Gases, Acids and Bases* (Copenhagen: Munksgaard, 1986), 75–6.
23. Magnus to Berzelius, letter dated April 1837, Berlin, in E. Hjelt (ed.), *Aus Jacob Berzelius' und Gustav Magnus's Briefwechsel in den Jahren 1828–1847* (Braunschweig: F. Vieweg, 1900), 122–3.
24. Evangelista Torricelli (1608–47) studied under Galileo Galilei (1564–1642) and after Galileo's death succeeded him as mathematician and philosopher to

the Grand Duke of Tuscany. Torricelli's experiments with liquids heavier than water (such as honey and mercury) led him to the discovery of the principle of the barometer, for which he is probably most famous. Torricelli stated that it was the air pressure that sustained the mercury column when the glass tube, after being filled with mercury, was inverted and placed vertically in a bowl full of mercury with its open end at the bottom. See, Katz, *op. cit.* (note 1), 478–80; see also, W.E. Middleton, *The History of the Barometer* (Baltimore: Johns Hopkins University Press, 1964), Ch. 2.

25. H.G. Magnus, 'Über die im Blute enthaltenen Gase, Sauerstoff, Stickstoff und Kohlensäure', *Annalen der Physik und Chemie,* x (1837), 583–606: 594. The eudiometer (Gr. *eudia* fair, clear weather, and *meter*) was invented by Alessandro Volta (1745–1827) for exploding gaseous mixtures by an electric spark and for the study of methane (marsh gas). Finely graduated and calibrated for the volumetric measurements and analysis of gases, the instrument was also used to determine the purity of the air. See J. Partington, *A History of Chemistry* (London: Macmillan, 1964), 4 vols, Vol. 4, 6. On the emergence and development of eudiometric technology in the context of medical and managerial ambitions of physicians and natural philosophers of the late Enlightenment, see S. Schaffer, 'Measuring Virtue: Eudiometry, Enlightenment and Pneumatic Medicine', in A. Cunningham and R. French (eds), *The Medical Enlightenment of the Eighteenth Century* (Cambridge: Cambridge University Press, 1990), 281–318; see also T.H. Levere, 'Measuring Gases and Measuring Goodness', in F.L. Holmes and T.H. Levere (eds), *Instruments and Experimentation in the History of Chemistry* (Cambridge: MIT Press, 2000), 105–36.
26. H.G. Magnus, 'Über das Absorptionsvermögen des Blutes für Sauerstoff', *Annalen der Physik und Chemie,* 66 (1845), 195–6. John Dalton (1766–1844) proved, in 1801, that the absorption, in liquids, of the separate gases in a mixture is dependent on the temperature and partial pressure. Dalton's close friend, William Henry (1774–1836), who held an MD degree from Edinburgh, showed, in 1803, that the amount of a gas dissolved in liquid is proportional to its pressure: $p2 = KN2$, where p is partial pressure, N is mole part in a solvent, K is Henry's constant. Both Dalton and Henry laws allowed to define absorption coefficients for gases dissolved by liquids, an important prerequisite for the development of respiratory physiology in the second half of the nineteenth century. See I.L. Knuniants, *Khimicheskii entsiklopedicheskii slovar'* (Moscow: Sovetskaia Entsiklopediia, 1983), 158–9.
27. K. Vierordt, *Physiologie des Athmens mit besonderer Rücksicht auf die Ausscheidung der Kohlensäure* (Karlsruhe: Groos, 1845), 222–4; G. Valentin, *Lehrbuch der Physiologie des Menschen* (Braunschweig: Vieweg, 1844), 2 vols, Vol. 1, 518–26. On Magnus and his research on blood gases, see Astrup and Severinghaus, *op. cit.* (note 22), 84–5; C.A. Culotta, *A History of Respiratory*

Theory: From Lavoisier to Paul Bert, 1777–1880 (unpublished PhD dissertation: University of Wisconsin, 1968), 89–92, 138–41, 174–6; F. Lieben, *Geschichte der physiologischen Chemie* (1935, repr. New York: Georg Olms Verlag Hildesheim, 1970), 266–7. On other physicalist theories in German physiology, see P.F. Cranefield, 'The Organic Physics of 1847', *Journal of the History of Medicine and Allied Sciences,* xii (1957), 407–23; E. Mendelson, 'Physical Models and Physiological Concepts: Explanations in Nineteenth-Century Biology', *British Journal of the History of Science,* ii (1965), 201–19.

28. Sechenov, *op. cit.* (note 2), 138; see also Lieben, *op. cit.* (note 27), 268–9; Culotta, *op. cit.* (note 27), 176–8.
29. R. Bunsen, *Gasometrische Methoden* (Braunschweig: F. Vieweg, 1857); *idem., Gasometry, Comprising the Leading Physical and Chemical Properties of Gases,* H. Roscoe (trans.), (London: Walton & Maberly, 1857), 128; 138–40. On Bunsen's laboratory and his *Vorlesungen über allgemeine Experimentalchemie,* see T. Curtius and J. Rissom, *Geschichte des chemischen Universitäts-Laboratoriums zu Heidelberg seit der Gründung durch Bunsen* (Heidelberg: Verlag von F.W. Rochow, 1908), 4–26.
30. R. Bunsen, 'Über das Gesetz der Gasabsorption', *Annalen der Chemie und Pharmazie,* 93 (1855), 47–54: 50. On Meyer's research in Bunsen's laboratory, see H. Roscoe, 'Bunsen Memorial Lecture', in B. Jones (ed), *The Golden Age of Science: Thirty Portraits of the Giants of Nineteenth Century Science by Their Scientific Contemporaries* (New York: Simon and Schuster, 1966), 513–54: 546.
31. L. Meyer, *Die Gase des Blutes* (Göttingen: W.F. Kaestner, 1857).
32. *Ibid.,* 55–6.
33. I.M. Sechenov, 'Beiträge zur Pneumatologie des Blutes', *Sitzungsberichte der mathematisch-naturwissenschaftlichen Klasse der kaiserlichen Akademie der Wissenschaften,* 36 (1859), 293–319, reprinted in Sechenov, *Selected Works, op. cit.* (note 16), 3–24: 2; *idem., op. cit.* (note 2), 88.
34. E. Fernet, 'Du rôle des principaux éléments du sang dans l'absorption ou le dégagement des gaz de la respiration', *Annales des sciences naturelles (zoologie),* ser. 4, xiii (1857), 151–2; Plate VII. On Fernet's method, see Sechenov, *op. cit.* (note 33), 3.
35. Meyer, *op. cit.* (note 31), 49. On the importance of Meyer's method in the development of respiratory physiology, see Lieben, *op. cit.* (note 27), 268–9; Astrup and Severinghaus, *op. cit.* (note 22), 88–9; Culotta, *op. cit.* (note 27), 234–5.
36. Franz Ernst Neumann (1798–1895), Professor of Mineralogy and Physics at the University of Königsberg, was a highly influential teacher who made known many of his discoveries in heat, optics, electrodynamics, and capillarity during his lectures. He inaugurated the German 'mathematisch-

physikalische' seminar to introduce his students to research methodology. Many of his students became outstanding scientists, Gustav Kirchhoff among them, see, J. Burke, 'F.E. Neumann', in C.C. Gillispie (ed.), *Dictionary of Scientific Biography* (New York: Charles Scribner, 1970–present), Vol. 10, 26–9.

37. R. Heidenhain and L. Meyer, 'Über das Verhalten der Kohlensäure gegen Lösungen von phosphorsäuren Natron', *Studien des physiologischen Institut zu Breslau,* ii (Leipzig, 1863), 103–24. A student of du Bois-Reymond, Rudolf Heidenhain later developed an unusually versatile investigative career, and often opposed the tendency to reduce vital phenomena to simple physical and chemical processes and their mathematical and physical interpretations. A number of Russian physiologists studied in Heidenhain's laboratory, Pavlov among them. On Heidenhain as a teacher and scientist, see I.P. Pavlov, 'Pamiati Heidenhaina', Rech' na zasedanii Obshchestva russkikh vrachei v St Peterburge, 23 Oct. 1897, in I.V. Natochin, M.A. Pal'tsev, and A.M. Stochek (eds), I.P. Pavlov, *Izbrannye trudy* (Moscow: Meditsina, 1999), 245–56.
38. On Meyer's research in chemistry, see Partington, *op. cit.* (note 25), Vol. 4, 889–91; on his research on blood gases, see P.P. Bedson, 'Lothar Meyer Memorial Lecture', in Jones, *op. cit.* (note 30), 1403–30: 1404–5; on Meyer's system of chemical elements, see J.W. Spronsen, *The Periodic System of Chemical Elements: A History of the First Hundred Years* (New York: Elsevier, 1969), 128–32. On some interesting parallels in the research career of Meyer and Sechenov, see P. Cranefield, 'Robert Bunsen, Carl Ludwig and Scientific Physiology', in F. Kao, K. Koizumi, and M. Vassalle (eds), *Research in Physiology: A Liber Memoralis in Honor of Prof. Chandler McCuskey Brooks* (Bologna: Aula Gaggi, 1971), 743–8: 746.
39. C. Ludwig, 'Erwiderung auf Valentin's Kritik der Bermerkungen zu seinem Lehren vom Athmen und Blutkreislauf', *Zeitschrift für rationelle Medizin,* iv (1846), 183; J.P. Burdon-Sanderson, 'Carl Ludwig', in Jones (ed.), *op. cit.* (note 30), 411; see also Culotta, *op. cit.* (note 27), 162–3.
40. Ludwig to Jacob Henle, letter dated 18 October 1857, Vienna, cited in T. Lenoir, 'Science for the Clinic: Science Policy and the Formation of Carl Ludwig's Institute in Leipzig', in W. Coleman and L.F. Holmes (eds), *The Investigative Enterprise: Experimental Physiology in Nineteenth-Century Medicine* (Berkeley: University of California Press, 1988), 139–78: 155.
41. C. Ludwig, 'Zusammenstellung der Untersuchungen über Blutgase', *Zeitschrift kaiserlich königlich Gesellschaft der Ärzte in Wien,* ix (1865), 145–6.
42. Sechenov, *op. cit.* (note 33), 4.
43. Sechenov, *op. cit.* (note 2), 87.
44. Sechenov, *op. cit.* (note 33), 11, 14–6.
45. Donald Dexter Van Slyke (1883–1971) made his principle contribution to the development of clinical chemistry while in the hospital laboratory at the

Rockefeller Institute in New York. His volumetric apparatus (1917) and his manometric apparatus (1924) laid the basis for the introduction of the gasometric technique in hospitals throughout the world. As in the early blood–gas pumps (including Sechenov's), a Torricellian vacuum was first produced, then the instrument could register the changes in volume or pressure caused by the release and subsequent absorption of the gases. See, Astrup and Severinghaus, *op. cit.* (note 22), 252; see also R.E. Kohler, *From Medical Chemistry to Biochemistry* (Cambridge: Cambridge University Press, 1982), 238–9, 242–3.

46. Sechenov, *op. cit.* (note 2), 87.

4

Berlin Wins over Paris and Vienna: Botkin's View on European Clinics

Sergei Petrovich Botkin (1832–89), eleventh son of the wealthy Moscow tea merchant, was educated at Ennes, one of Moscow's best private boarding schools, known for its excellent instruction in classical and modern languages. But his passion at school was mathematics, which he was taught by Iu. K. Davydov, a young professor from Moscow University. After school, Botkin applied to the mathematical faculty at Moscow but was not accepted. In the early 1850s, the enrolment to any faculty, except medical, was limited. He had to choose medicine, which eventually became his real passion, but he preserved an interest in exact sciences throughout his life.[1] Much later, in 1881, Botkin, in a speech dedicated to Rudolf Virchow, spoke critically about the way medicine had been taught in Moscow in 1850–5, when he was a student:

> The majority of our professors had studied in Germany. They more or less skilfully taught us what they knew; but such mode of teaching, in the form of catechism truths, could not inspire the eager minds of future researchers.[2]

Botkin recalled that Professor Polunin had continued to teach outdated theories. A student of Carl von Rokitanski, Polunin had adhered to the humoral theory of his teacher and was especially keen on Rokitanski's dyscrasia theory, which held that pathological conditions are due to a change in the blood mixture. Of particular importance in such cases were blood plasma and its proteins, and fibrin, ie. the albuminous and fibrinous dyscrasias. It was believed that they might oxidise and thereby disturb the entire balance of the solution. In the 1850s, Rokitanski's theory was overthrown by Rudolf Virchow's cellular pathology. Botkin's friend, Belogolovyi, recalled that at one of the lectures a student asked about Virchow's theory, which was recognised by even Rokitanski himself. Polunin ended the discussion replying: 'Indeed, Rokitanski has admitted that his theory is wrong. It shows only that the genius professor is out of his mind.' Years later, after Botkin and Belogolovyi had graduated from the University, Polunin, *plus royaliste que le roi,* continued to teach Rokitanski's theory as the last word in medical science.[3]

Image 4.1

Sergei P. Botkin.
Courtesy of the Institute for History of Medicine
at the Humboldt University of Berlin.

Professor of Surgery, Fedor I. Inosemtsev, in turn, confined himself to the theory of the French pathologist Marie Francois Xavier Bichat (1771–1802), who focused particularly on the histological changes produced by disease. In his *Traité des membranes en général et de diverses membranes en particulie* (1799), Bichat stated that each organ contained particular 'membranes' or tissues, and described twenty-one tissues including connective, muscle, and nerve tissue. 'So', Botkin wrote, 'we had two catechism truths: the humoral theory of the Vienna school, and the doctrine of tissue pathology of the French.'[4] A skilful surgeon, Inosemtsev was the favourite among medical students for his peculiar-but-kind disposition and honesty. Botkin later admitted that some of his Moscow clinical professors, who stuck mainly to the Paris school, were skilful diagnosticians, able to determine the disease 'at first glance'. Most of all they valued the clinical intuition of a physician and often disregarded so-called scientific diagnostics based on microscopic and chemical analysis.[5]

Botkin was in his fourth year when the news about the desperate situation with the wounded at the Crimean front reached Moscow. From the dispatch of Admiral V.A. Kornilov (1806–54), it became known that there were not enough hospitals, dressing stations or even stretchers in Sevastopol, and that the wounded were being left on the battlefield. Like many of his fellow-students Botkin felt an urge to join the army, but reasoned that he had to finish his studies. After graduation, in September 1855, he left for the Crimea with the second detachment of the Sisters of Mercy, formed by Nikolai I. Pirogov under the patronage of the Duchess Elena Pavlovna. Botkin was appointed to one of the hospitals in Simferopol that became the major place for quartering the sick and wounded after the fall of Sevastopol. Botkin's experience was bitter. His real fight was not at the operating table or at the bedside with wound infection but with poor administration, theft, lack of medical supplies, food, and firewood. Assisting and sometimes performing countless amputations, Botkin realised that he was not able to ligate bleeding vessels because of his poor eyesight and that the career of a surgeon was closed to him.

After three-and-a-half months service in the Crimea, Botkin returned to Moscow. By that time his father had died, so Botkin's older brother took responsibility for family business and the care and education of the younger brothers and sisters. Botkin decided to use the money left to him by his parents to study in Europe. He had no definite plan of where to go, and after a short stay in Königsberg at the clinic of Professor Girsch, he decided to move on to Würzburg to study with Virchow. In 1856, Virchow acquired a position in Berlin, and Botkin followed his mentor.[6] Writing from Berlin, Botkin reported with delight about the new Pathological Institute and its laboratory and described the richness and accessibility of the 'treasures of Berlin University'. Botkin's brother, Vasilii wrote to a friend:

> The other day I got a letter from Sergei. What astonishing work is in full swing now in the European scientific world! And look at the path that modern medicine has chosen: microscopic analysis and chemistry are its foundation; everything is being verified by experiment and observation, absolute theories are being ridiculed. Sergei is awfully disappointed by the state of our medical education: how backward it is in comparison with what is now being done in Germany. On the whole, the surge in the field of natural sciences is quite remarkable in Germany. It has been restricted for so long by the exclusive predominance of philosophy. Imagine, philosophical lecture halls are absolutely empty, only about two or three listeners, truly 'the last of the Mohicans'. The lecture halls for natural sciences, on the contrary, are full.[7]

Image 4.2

The Pathological Institute at the Berlin Charité Hospital.
Courtesy of the Berlin Medical Historical Museum of the Charité.

A broadly educated and cultured man, Vasilii Botkin pinpointed an interesting feature – the decline of the philosophical fame of Berlin University and the growing prestige of its natural sciences.

The new Pathological Institute headed by Virchow was one of the most attractive places in Berlin for medical researchers to work. In the mid-1850s Virchow's great interest was further refinement of his new theoretical formulation, cellular pathology,[8] and an institutional programme in support of scientific medicine. Virchow promoted his theory through an integrative approach: normal histology and physiological chemistry were represented by both courses and the laboratory. Two of his first assistants at the new institute were Felix Hoppe-Seyler, a chemist and histologist with clinical training in Vienna, and Friedrich von Recklingshausen (1833–1910), a clinician, skilful in devising laboratory methods and techniques in tissue pathology.

At the Institute, Botkin continued his studies in histology and microscopic techniques, which he had started under Rudolf A. von Kölliker

Image 4.3

The laboratory at the Pathological Institute at the Berlin Charité Hospital, 1855. Courtesy of the Berlin Medical Historical Museum of the Charité.

(1817–1905) in Würzburg. Virchow's lectures in cellular pathology greatly disappointed Botkin at first: 'Virchow spoke about blood segments [blood cells] with thoroughness, typical for him, and about different morphological types. All these "small details" seemed to me boring and unnecessary.' Botkin soon realised the unique value of these small details in a new pathology that relied heavily on microscopic examinations. Avoiding the speculative assumptions so typical of the humoralists, Virchow based his theory on exact observation and experiment. Botkin, who had his first experience in the military hospital in the Crimea, desperately fighting inflammation and suppuration in post-operative wounds, quickly saw the utility and importance of cellular pathology. Intellectually and institutionally connected with the University and the Charité clinics, Virchow's Institute became, for Botkin, as well as for numerous students and collaborators, a place of the investigation of both 'scientific' and clinical problems, unifying science and medicine.

Image 4.4

Rudolf Virchow.
Courtesy of the Wellcome Library Collection.

Botkin's research topic in Hoppe-Seyler's laboratory at the Pathological Institute was 'clinical and scientific' too: a study of the effect of neutral salts on the circulation of red blood cells that could serve as an explanatory model for venous haemostasis. Botkin improved the method of application of salt solution to the walls of blood capillaries and got new results, which contradicted the conventional physico–chemical explanation of the venous haemostasis. He pointed out that, although the escape of the liquid part of the blood through the walls of vessels due to endosmosis did occur, the main cause of haemostasis was the loss of elasticity of the red blood cells – they became non-resilient and could not pass through the contracted capillary. His explanation was perfectly well in accord with the dictum of the new pathology, *omne cellula e cellula*, and confirmed its rule – altered circumstances make altered cells.[9]

Botkin was interested in practical medicine too and spent much time in the clinic of Ludwig Traube (1818–76), one of the most notable Berlin internists. The clinic was a part of the Charité hospital, situated close to the

Pathological Institute in the Charité garden. In the late 1840s, Ludwig Traube, the best student of the influential Lucas Schönlein (1793–1864) and his assistant at the department for chest diseases became the first civilian to be employed at the Charité. Traube's clinical approach and views had been formed during his postgraduate training in Vienna under Rokitanski and Josef Skoda (1805–81). His early interest in experimental research in the Breslau physiological laboratory of Jan Purkinje (1787–1869) during his student years, and his later keen interest in pathology shaped his experimental thinking. In 1846, together with Reinhardt and Virchow, he began to publish the periodical of the new school, the *Beiträge zur experimentellen Pathologie und Physiologie.* It ran for only one year and was superseded by *Virchows Archiv für pathologische Anatomie und pathologische Physiologie und klinische Medizin* [hereafter *Virchows Archiv*]. Influenced by their teacher Johannes Müller, Traube and Virchow demanded exactitude and consistency not only in physiology but also pathology. In his first published experimental work of 1847, on the causes and nature of pathological changes in the lung parenchyma as a consequence of cutting the vagus nerve, Traube stated that only experiment combined with observation could make pathology what it should be, an exact science.[10] Virchow's ideal of clinical research was that practical medicine would become applied theoretical medicine and theoretical medicine would become pathological physiology. His concept was close to Carl Wunderlich's (1815–77) 'physiological medicine', whose intent was to replace anatomical research with physiological investigations. Wunderlich's *Archiv für physiologische Heilkunde,* founded in 1842, gave its name to the new trend in German clinical medicine. In his *Geschichte der Medizin* (1869), Wunderlich pointed out that pathology was merely the physiology of the diseased man, and it required the same means, methods, and logical argument that were used in the science of the healthy man.[11] The new concept opposed what Wunderlich called the ontological personification of diseases in *Naturphilosophie.* On the other hand, in contrast to the French pathological anatomical doctrine of specificity, based on the purely descriptive method, physiological medicine began to be seen as strictly scientific – it relied on the new methods of investigation, graphic, microscopic, and chemical.[12]

Traube, one of the first clinicians of this new stamp in Germany, strove to reconstruct clinical medicine along strictly physiological lines. In his 1856 work on the connection of heart and kidney diseases, he developed the idea of variation in blood pressure, specifically in hypertension as a cause of the cardiac hypertrophy, which had been observed in chronic nephritis. Traube's explanation of clinical phenomena in heart and kidney diseases was based on the results of Carl Ludwig's findings and on his own careful observation and experimental research on animals with the improved Ludwig's haematodynamometer. Traube's synthesis of clinical and laboratory medicine found its finest expression in his three volumes of *Gesammelte Beiträge,* with fine pictures of Ludwig's kymograph and its application, and with the core chapters on experiments on animals, pathological–physiological–clinical investigations and studies in physico–clinical diagnostics.[13]

Traube's younger brother, Moritz (1826–94), was also engaged in laboratory research. Forced by family obligations to maintain the family's large wine business, Moritz set up a private laboratory first in Breslau, where he used the opportunity to work at Rudolf Heidenhain's laboratory, and then in Berlin. Trained in medicine and chemistry under Liebig in Giessen, Moritz investigated the biochemical processes of diabetes mellitus and distinguished two forms of the disease. His experimental studies extended over much of physiological chemistry and his original ideas and experiments proved of importance for general chemistry – oxygen-carrying ferments, 1857, and semipermeable membranes, 1867, are among his most interesting discoveries.[14] Moritz's studies on the source of the energy for muscle contraction and on respiration were of particular interest and help for his brother's clinical research. Ludwig Traube re-examined the relationship of the blood gases to dyspnoea. He attributed dyspnoea to carbon dioxide excess but not lack of oxygen. Moritz Traube held a similar view on the problem in his earlier studies on the influence of carbon dioxide on the respiratory and circulatory system.[15]

Ludwig Traube's strong background in physiology and experimental methods, combined with his clinical experience, distinguished him among contemporary clinicians. Theodor Billroth, one of Traube's numerous students, pointed to Traube's exceptional skill as a clinical teacher.[16] Another of Traube's student, Russian physician N.A. Belogolovyi, noted that Traube's extraordinary power of observation and keen clinical intuition helped him to grasp individual peculiarities in an ordinary clinical patient. Traube's analysis of the patient's condition was even more interesting and instructive for the physicians who were more experienced in the nuances of diagnostics than for the students. His lectures always contained something new and interesting that one could not find in any textbooks – a thought or observation, as well as fresh ideas or comments on the problems that were not yet understood. Not surprisingly, Botkin preferred Traube's lectures to any of the other notable clinicians in Vienna and Paris. Botkin became a close friend of Traube's, and frequently visited him while in Europe.[17]

In the autumn of 1858, Botkin and his friend Ludwig A. Bekkers came to Vienna for the winter semester. On Sechenov's request, Ludwig gave a series of lectures on the circulation of the blood and the innervation of the blood vessels to a group of Russian students. Ludwig enjoyed lecturing. His lectures were splendidly arranged and were accompanied by skilful vivisection experiments. After the course, the grateful participants invited the professor to a dinner in his honour. Ludwig knew from Sechenov that Botkin had recently married in Vienna, and was very friendly towards the young couple.[18] Botkin was greatly impressed:

> Ludwig's lectures surpassed all my expectations for clarity and completeness. Ludwig is the best physiologist I have ever heard; his personality is nice, his simplicity and courtesy are astonishing.[19]

At the same time, Botkin was quite disappointed with his clinical studies in Vienna: 'I am displeased with Vienna... you cannot learn much here. It is a complete waste for a decent person to stay in Vienna more than three months.'[20] His acquaintance with the traditionally famous Viennese medical instruction began at the clinic of Professor Johann von Oppolzer (1808–71), the notable clinical teacher and a consulting physician famous for his rapid 'blitz' diagnoses. Botkin, however, liked neither the professor, nor his clinic:

> Oppolzer is an excellent observer, a sharp diagnostician, and in general a kind of good practical physician.... But how often he transgresses science: he cannot be considered a good clinician in a full sense of this word. Frequently he disregards chemistry, pathological anatomy, even physiology.[21]

Botkin's criticism of Oppolzer accords well with Billroth's fine remark on the representatives of the Prague and Vienna medical schools:

> In the case of Skoda and Oppolzer there was always a gap between medical art and modern physiology – between practice and theory – that was artificially and ineffectively bridged over. Appreciation of the big things in natural phenomena as well as in social life was almost entirely absent in Skoda and Oppolzer.[22]

Billroth contrasted the Vienna school to the Berlin school of Lucas Schönlein and Johannes Müller and pointed to Schönlein and Müller's extraordinary encyclopaedic knowledge of the natural sciences and complete command of the physiology taught in those days. There is no question that Botkin favoured Traube, the most talented student of Schönlein, and Virchow, the student of Müller.

It is also fair to conclude that Botkin, an ardent proponent of Virchow's ideas, felt that such revolutionary new teaching was not accepted in Vienna, where pathological anatomy was still ruled by Carl Rokitanski. Rokitanski's humoral pathology of the dyscrasias was one of the targets of Virchow's sharp criticism at the time when Virchow attempted to establish his new cellular pathology based on microscopic and experimental method.[23] The young Botkin was extremely sensitive towards any criticism of Virchow's theory.

Many years later he wrote: 'In those days, Virchow was accessible to few, his teaching was far from being common knowledge, as now, and his method of research and mode of thinking was open only to exceptional people.'[24] Sechenov recalled an argument he had with Botkin in Vienna, about the role of the cellular principle in physiology and pathology. Sechenov could not accept Virchow's static and localised anatomical idea of cellular pathophysiology. It took Ludwig's interference via a letter to Sechenov to reconcile the two friends.[25] The echo of this dispute can be traced in some of the statements in Sechenov's doctoral dissertation:

> The principles of cellular pathology are erroneous, because they are based on the assumption of the physiological independence of the cell, or at least of its domination over the surrounding environment. The theory of cellular pathology is an extreme expression of the purely anatomical trend in physiology. The only correct approach to pathology in our time is molecular [physico–chemical].[26]

After Vienna, Botkin with his wife settled in Paris. The experimental part of his doctoral thesis *O vsasyvanii zhira v kishkakh* [*On the Absorption of Fat in the Intestine*] had been already completed in Hoppe-Seyler's laboratory in Berlin, and he was planning to acquaint himself with Bernard's research on digestion. Bernard had devised a series of experiments to test his theory that neutral fats were rendered absorbable through the action of the pancreas and its secretion. He had established that pancreatic juice had a saponifying action on fat, ie. breaking it down into fatty acid and glycerin.[27] Botkin's own experiments confirmed that neutral fats before the absorption through the walls of the intestine had undergone a process of saponification in order to be soluble in water. He demonstrated that the impairment of the elastic epithelial layer of the intestine led to the impairment of the process of the absorption of fats.[28] Botkin attended Bernard's course on the liquids of the organism at the Collège de France.[29] During the late 1850s, Bernard, working in a dark, cold, damp laboratory of the Collège de France, completed and consolidated a number of important investigations. A very skilful and capable experimenter, Bernard might have inspired the young Botkin, who seemed to be carried away with his own experiments at that time:

> I was not satisfied with reading, attending lectures and visiting clinics, so I arranged a small laboratory at home and started working like mad. I finished the work with the blood and had got new data that suggested to me an interpretation of the results that I had obtained during my experimental research in Vienna.... I wrote a short paper on the diffusion of haematin and ferric pigment. Bernard became interested in this work and published it.[30]

By that time, the experiments of Bernard and Kölliker on the action of curare on nerves and muscles, 'had caused a lot of excitement, and experiments with the influence of various poisons on the muscles and the nervous system were very much in vogue.'[31] Bernard, fascinated by curare as 'an instrument, which dissociates and analyses the most delicate phenomena of the living organism', had carried out numerous experiments on dosage and progressive action of curare on the nervous system, which revealed the process of death by curare. Botkin was also captivated by the experiments with curare:

> I started with frogs, and sitting at the experiments I came across a new curare, atropine sulphate; I had to repeat all the experiments which had been done with curare before. I was so fascinated by novelty of the methods (I had never applied them in my research), the successful results... that I was sitting with the frogs from morning till night; I would have sat with my experiments longer if my wife had not turned me out of my home laboratory. She had no patience with me during my madness, as she called it.... Anyway, I've learnt much from this work.[32]

Botkin's interest in the traditionally famous Paris school of medicine was no less than in Bernard's experimental work. One of the leaders of Parisian clinical medicine at that time was Armand Trousseau (1801–67). A disciple and admirer of Pierre Bretonneau (1771–1862) and his doctrine on specific inflammations and specific diseases, Trousseau continued Bretonneau's studies on diphtheria, croup, and typhoid fever and described their various forms and stages. Trousseau published much, and his most recent work, *Traité élémentaire de thérapeutique et de la matière médicale,* of 1858, was a success. In opposition to Wunderlich and his group, Trousseau stressed the specific nature of disease, which, he held, dominated all pathology, all therapy, and all medical science. He believed that the natural history of diseases resembles that of animals or plants; it deals, in the same way, with specific properties, which differentiate the species. Not surprisingly, Trousseau was an antagonist of cellular pathology:

> It [cellular theory] regards the living organism as a small world consisting of heterogeneous and independent elements, therefore it rejects general treatment that cannot affect the elements, which are dissimilar and to some extent counteracting one another. It forgets about a human being and thinks only about the cells and in that way disappears in a huge number of extremely small values.[33]

Though quite familiar with the work of, for example, Pasteur, Virchow, and Bernard – that is, chemistry, microscopy, and experimental physiology

– Trousseau refused to develop in this direction. He was against the intrusion of these methods into the domain of clinical medicine; in his eloquent lectures he lamented that with the emergence of new sciences in medicine, therapeutics had become neglected and no one thought about how to relieve the sufferings of patients or how to cure them. The last of the great classic clinicians, Trousseau, believed that 'from the day a young man wants to be a doctor, he must visit the hospital'. Likewise he would 'profoundly regret the time the student would lose in acquiring too extensive a chemical knowledge... Science renders the mind lazy. Please, a little less science and a little more art, gentlemen'.[34] Though some of the Parisian clinicians, the so-called 'eclectics' saw the limitations of pathological anatomy and advocated the use of physics, chemistry, animal experimentation, and microscopy, the radical empiricism of the previous generation of the classic school of Laennec and Louis and skepticism of Corvisart and Andral remained strong enough among the Parisian clinicians. For scientifically educated and oriented clinicians, especially those adhering to the Virchow–Traube approach, it became obvious that the techniques used by Parisian clinicians – statistics, physical examination, and macroscopic pathological anatomy – were not enough to explain the cause of the disease. New powerful tools came to the fore – animal experimentation, medical chemistry, and microscopy applied in clinical investigations had already yielded valuable results. In Parisian clinics, as Botkin became convinced, scientific methods were prejudiced by the majority of hospital physicians and largely ignored by them:

> Trousseau holds his clinic in a rather routine way; being satisfied with hospital diagnostics [that is, based only on practical observations but not on experimental data], he prescribes an absolutely empirical treatment.... Trousseau is considered here one of the best therapeutists; his lecture halls are always full. I think one of the main reasons of his success is that his oratory always wins over the French.... In Codemont's urology clinic microscopic investigations were disregarded: in two cases of kidney stones and resulting bladder disease, despite positive evidence in the urine, the diagnosis and treatment were wrong. The blunder was made before my eyes by the best medical authorities of the city.... In the infant's clinic the mortality is severe. That partly is due to poor care, badly heated wards, etc. There is an epidemic of croup here, and nearly all children die....[35]

Botkin might have sympathised with some of Trousseau's views, in particular, the great powers of observation at the bedside, but he realised that there was a gap between the advances of pathological anatomy and diagnostics, which had become exact sciences, and therapeutics, which still remained an art. However, he did not accept 'therapeutic nihilism' with its dictum that scientifically educated physicians attached no importance to the art of healing. He felt that the intuition of the physician was still a significant factor, especially in the cases where 'exact knowledge' was still powerless. On the other hand, he also believed that a clinician should apply to the patient all means available in modern scientific medicine:

> Physicians must be scientifically educated. The clinic should not be entrusted to even a very good physician without science background. The diagnoses of such a physician would be always of a hospital character and his treatment of the patient without experimental basis would inevitably be of an empirical kind.[36]

For Botkin, the Viennese and Parisian clinicians reluctant to embrace the radically new cellular pathology and experimental physiology were well behind the German clinical school. Trousseau, who then dominated the Parisian medical scene, and Oppolzer, of the same stature in Vienna, were, in Botkin's opinion, representatives of a once-famous but now outdated hospital medicine. The clinically oriented laboratory research and systematised, well-planned training available in Virchow's institute and in Traube's clinic embodied for Botkin an important link between science and the clinic. The microscopic, chemical, and physiological methods were new tools rendering the art of medicine quantitative, bringing it closer to exact science.

Botkin's images of medicine in late 1850s Paris, Vienna, and Berlin in their own way confirm the historical assumption that, after the mid-nineteenth century, Germany became dominant because of the integration of clinical and laboratory medicine. Indeed, although as historian Erwin Ackerknecht pointed out, a new medicine, 'laboratory medicine', made its appearance in Paris under the leadership of Louis Pasteur, Claude Bernard, and the Société de Biologie as early as 1848, for many decades it was unable to conquer French medicine.[37] Botkin's images also reflect the intellectual vitality of the French school famous for its realism and virtuosi in diagnostics, as medical Paris continued to attract Russian physicians even during the age of the ascendancy of German medicine. Paris and Vienna, with their rich medical traditions, gave a final touch to the formation of the research and clinical experience of Botkin, who would soon initiate the re-orientation of Russian medicine towards the laboratory.

Notes

1. N.A. Belogolovyi,. *S.P. Botkin: ego zhizn' i meditsinskaia deiatel'nost'* (St Petersburg: Tipografiia J.N. Erlikha, 1892), 10; see also *idem., Vospominaniia i drugie stat'i* (Moscow: Tipografiia Aleksandrova, 1897), 251–375: 263–4.

These are the most widely used sources on Botkin, written by Nikolai A. Belogolovyi, Bokin's close friend and colleague. Their correspondence is published in N. Sadovskaia, *Perepiska S.P. Botkina s N.A. Belogolovym* (Moscow: Izd-vo Akademii Nauk, 1939). Botkin's biography, a volume in a popular Russian series *Zhizn' zamechatel'nykh liudei* [*The Life of Remarkable People*] by E. Nilov, *Botkin* (Moscow: Molodaia Gvardiia, 1966), although written in a fictional style, contains some useful information. The most notable works that attempted to define Botkin's clinico–physiological thinking and his influence on the development of the Russian therapeutic school is D.D. Pletnev, *Russkie terapevticheskie shkoly* (Moscow–Petrograd: n.p., 1923). A comprehensive analysis of Botkin's neurogenic theory is found in F.R. Borodulin, *S.P. Botkin i nevrogennaia teoriia meditsiny* (Moscow: Medgiz, 1949); and in V.O. Samoilov, *Istoriia Rossiiskoi meditsiny* (Moscow: Epidavr, 1997), 116–21. For Pavlov's work in Botkin's laboratory, see D. Todes, *Pavlov's Physiological Factory: Experiment, Interpretation, Laboratory Expertise* (Baltimore: Johns Hopkins University Press, 2002), 297–302.

2. S.P. Botkin, 'Rech, proiznesennaia v obshchestve russkikh vrachei po povody iubileia 25 letnei professorskoi deiatel'nosti Rudolfa Virkhofa, 1881', *Ezhenedel'naia klinicheskaia gazeta,* cited in Belogolovyi, *Botkin, op. cit.* (note 1), 29–30. On Rokitanski's anti-localisation views, see E. Lesky, *The Vienna Medical School of the Nineteenth Century* (Baltimore: Johns Hopkins University Press, 1976), 106–15.
3. Belogolovyi, *Vospominaniia, op. cit.* (note 1), 284–5.
4. Belogolovyi, *Botkin, op. cit.* (note 1), 30.
5. Belogolovyi, *Vospominaniia, op. cit.* (note 1), 284.
6. Belogolovyi, *Botkin, op. cit.* (note 1), 21.
7. V.P. Botkin to P.V. Annenkov, letter dated 1856, Paris, cited in Nilov, *op. cit.* (note 1), 42. Vasilii Petrovich Botkin (1811–69), the oldest son in the family, was a well-known publicist and literary, music and art critic, See V.P. Botkin, *Literaturnaia kritika: Publitsistika: Pis'ma* (Moscow: Sovetskaia Rossiia, 1984).
8. R. Virchow, *Die Cellularpathologie in ihrer Begründung auf physiologische und pathologische Gewebelehre* (Berlin: Hirschwald, 1858). On Virchow's institutional programme of the late 1850s, see R. Maulitz, 'Rudolf Virchow, Julius Cohnheim and the Program of Pathology', *Bulletin of the History of Medicine,* 52 (1978) 167–72; on reception of Virchow's theory in 1855–65 in Russia see L. Shumeiko, *Die Rezeption der Zellularpathologie Rudolf Virchow in der Medizin Russlands und der Sowjetunion* (inaugural disseration, University of Marburg, 2000), 46–89.
9. S. Botkin, 'Über die Wirkung der Salze auf die circulirenden rothen Blutkörperchen', *Virchows Archiv,* xv (1858), 173–6; *idem.,* 'Zur Frage von

dem Stoffwechsel der Fette im thierischen Organismus', *Virchows Archiv,* xv (1858), 380–2.

10. L. Traube, *Über Ursachen und Beschaffenheit derjenigen Veränderungen, welche das Lungenparenchym nach Durchschneidung der Nv. Vagi erleidet* (Berlin: A. Hirschwald Verlag, 1846), iii; see also *idem., Gesammelte Beiträge zur Pathologie und Physiologie* (Berlin: Hirschwald , 1871), 2 vols, Vol. 1, v–vi.
11. C. Wunderlich, *Geschichte der Medicin* (Berlin: Hirschwald, 1869), 13.
12. On the new clinical methods, see V. Hess, 'Klinische Experimentalstrategien im Kontext: Ludwig Traube, Carl August Wunderlich und das Fieberthermometer', in C. Meinel (ed.), *Instrument: Experiment historische Studien* (Berlin: Diepholz, 2000), 316–24. On the German clinical school see, K. Faber, *Nosography: The Evolution of Clinical Medicine in Modern Times* (New York: AMS Press, 1930), 59–94.
13. L. Traube, *Über den Zusammenhang von Herz- und Nierenkrankheiten* (Berlin: Hirschwald, 1856), 6–7. On Traube's biography, see H. Stangier, *Ludwig Traube: sein Leben und Werk* (Freiburg: Buchdrukerei Theodor Kehrer, 1935). On Traube's relationship with Virchow, see R. Virchow, 'Errinerung an Ludwig Traube', *Berliner klinische Wochenschrift,* xvi (1876), 200–9.
14. On Moritz Traube's research in physiological chemistry, see F. Lieben, *Geschichte der physiologischen Chemie* (1935, repr., New York: Georg Olms Verlag Hildesheim, 1970), 235– 41; on his research in chemistry, see J. Partington, *A History of Chemistry* (London: Macmillan, 1964), Vol. 4, 307; see also G. Rudolph, 'M. Traube', *Dictionary of Scientific Biography* (New York: Charles Scribner, 1970–present), Vol. 13, 451–3.
15. L. Traube, *Gesammelte Beiträge zur Pathologie und Physiologie, op. cit.* (note 211), Vol. 1, 288–9; M. Traube, 'Über die Wirkungen des Kohlenoxyd-Gases auf den Respirations- und Circulations-Apparat', *Verhandlungen der Berliner medizinischen Gesellschaft* (Berlin: Hirschwald, 1867), 67.
16. T. Billroth, *Über das Lehren und Lernen der medizinischen Wissenschaften an den Universitäten der deutschen Nation* (Vienna: Gerold, 1876), 102, 103. A student of Rudolf Wagner, Müller, Schönlein, von Graefe, and Bernard von Langenbeck, Theodor Billroth (1829–94) had, for a quarter of a century, been an active and successful teacher, investigator and surgeon in three important universities, in Berlin, Zürich, and Vienna. Billroth was a close friend and admirer of Nikolai Pirogov. On Pirogov and Billroth, see V.I. Razumovskii, 'N.I. Pirogov, ego zhizn', nauchno-obshchestvennaia deiatel'nost' i mirovozzrenie', *Russkii khirurgicheskii arkhiv,* xxiii (1907), 11–37. On Billroth's Vienna period, see Lesky, *op. cit.* (note 2), 293–404.
17. Belogolovyi, *Botkin, op. cit.* (note 1), 23–4; *idem, Vospominaniia, op. cit.* (note 1), 301–2.

18. I.M. Sechenov, *Autobiographical Notes,* D.B. Lindsley (ed.), K. Hanes (trans.), (Washington: Americam Institute of Biological Sciences, 1965), 85. Botkin's marriage to A.A. Krylova took place in Vienna in May 1859.
19. Botkin to Belogolovyi, letter dated 2 January 1859, Vienna, cited in Belogolovyi, *Vospominaniia, op. cit.* (note 1), 309.
20. Botkin to Belogolovyi, letter dated 2 February 1859, Vienna, in Belogolovyi, *Vospominaniia, op. cit.* (note 1), 310.
21. *Ibid.*, 309.
22. T. Billroth, *The Medical Sciences in the German Universities* (New York: Macmillan, 1924), 229–30. Oppolzer's most notable contribution was his *Klinische Vorlesungen über specielle Pathologie und Therapie,* published in 1866. On Oppolzer and Josef Skoda (1805–81) and their impact on the development of Viennese medicine, see Lesky, *op. cit.* (note 2), 118–28.
23. On Virchow's critique of the humoral theory published in Rokitanski's *Handbuch der pathologischen Anatomie,* see L.J. Rather, 'Virchow's review of Rokitanski's *Handbuch* in the *Preussische medizinische Zeitung*', *Clio Medica,* iv (1969), 127–40.
24. Belogolovyi, *Vospominaniia, op. cit.* (note 1), 320.
25. Ludwig to Sechenov, letter dated 14 May 1859, Vienna, in M.N. Shaternikov, 'The life of I.M. Sechenov', in I.M. Sechenov, *Selected Works* (Moscow-Leningrad: State Publishing House for Biology and Medical Literature, 1935), xii.
26. *Ibid.,* xiii; see also Sechenov, *op. cit.* (note 18), 87.
27. C. Bernard, *Mémoire sur le pancréas* (Paris: Baillière, 1856), 73; see also M.D. Grmek, *Le Legs de Claude Bernard* (Paris: Fayard, 1997), 30–1; J.M.D. Olmsted and E.H. Olmsted, *Claude Bernard and the Experimental Method in Medicine* (New York: H. Schuman, 1952), 53–5; and F.L. Holmes, *Claude Bernard and Animal Chemistry: The Emergence of a Scientist* (Cambridge: Harvard University Press, 1974), 1–32.
28. S.P. Botkin, *O vsasyvanii zhira v kishkakh: Dissertatsiia,* in *Voenno-meditsinskii zhurnal* (St Petersburg: Voennoe Ministerstvo, 1860).
29. C. Bernard, *Leçons sur les propriétés physiologiques et les alterations pathologiques des liquids de l'organisme* (Paris: J.B. Baillière, 1859).
30. Botkin to Belogolovyi, letter dated March 1859, Paris, in Sadovskaia, *op. cit.* (note 1), 23.
31. Sechenov, *op. cit.* (note 18), 80. See also C. Bernard, *Introduction à l'étude de la médicine expérimentale* (1865), H. Green (trans.), (New York: Dover Publications, 1957), 88.
32. Belogolovyi, *Vospominaniia, op. cit.* (note 1), 327–8.
33. Cited in Faber, *op. cit.* (note 12), 92; on Bretonneau's doctrine of specificity, see *idem,* 44–5.

34. A. Trousseau, *Clinique médicale de l'Hôtel-Dieu de Paris* (Paris: J.-B. Baillière, 1873), 3 vols, Vol. III, 139, cited in E. Ackerknecht, *Medicine at the Paris Hospital, 1794–1848* (Baltimore: Johns Hopkins University Press, 1967), 10–11; see also Faber, *op. cit.* (note 12), 28–58.
35. Botkin's report to the President of the Medico-Surgical Academy, Dubovitskii, about studies abroad, 1860, in *Rossiiskii gosudarstvennyi voenno-istoricheskii arkhiv*, Moscow, fond 316, opis' 32, delo 12. On Trousseau and the debate on tracheotomy versus intubation in the French Academy of Medicine, see G. Weisz, *The Medical Mandarins: The French Academy of Medicine in the Nineteenth and Early Twentieth Centuries* (Oxford: Oxford University Press, 1995), 169–73.
36. Botkin's report, *ibid.*
37. Ackerknecht, *op. cit.* (note 34), xiii. The literature on the Paris school is extensive, see G. Weisz's review, 'Reconstructing Paris Medicine', *Bulletin of the History of Medicine*, 75 (2000), 105–19; on the French school of experimental science, see J. Lesch, *Science and Medicine in France: The Emergence of Experimental Physiology, 1790–1855* (Cambridge: Harvard University Press, 1984).

5

'Alt Heidelberg, du feine...'

Heidelberg and its old university, *Academia Ruperto-Carola,* occupies a particular place in the history of nineteenth-century Russian science – nearly all Russian intellectuals of the 1860s studied there or at least visited the famous university town. The historian and jurist Sergei Svatikov, in his unpublished paper *Russische Studenten in Heidelberg* (1906), wrote:

> For a hundred years Heidelberg was home to Russian youth searching for knowledge. More than a thousand Russian students studied at the old University during that period, and having returned to Russia brought back with them precepts of the great representatives of German science.[1]

Indeed, Heidelberg University had been a centre of scientific and philosophical thought in Europe for its nearly five-hundred-year existence. The small town in southern Germany, located in the picturesque valley of the river Nekkar, had been a favourite place for Russian students since the late eighteenth century. The reforms of Alexander I in the early nineteenth century brought in their wake a need for professionals in jurisprudence, and an interest in political science among Russian intellectuals. The Heidelberg law faculty, traditionally strong in public law and political economy, attracted the most Russian students at that time. In the late 1850s, with the rapid growth of experimental sciences, the Heidelberg laboratories, chemical and physiological in particular, acquired fame for research and training.[2] The notable Russian jurist and writer Boris Chicherin described the Heidelberg of 1858:

> [I]n this small corner of Germany, there were many outstanding scholars that any university would envy. It was a scientific centre in every sense of the word. Besides Mohl, here taught the famous Mittermeier, one of the most learned criminologists of Germany, who was already in his old age; a classicist Vangerow, the best expert in pandects, who trained a whole generation of jurists; Geisser, a historian who taught at that time the history of the French revolution. In other branches of science the University was famous through other even more celebrated names. Often one could see three scholars, luminaries in natural sciences of the time, walking together: Helmholtz, Bunsen, and Kirchhoff.[3]

In the late 1850s, a large group of young Russian scientists, most of them chemists and naturalists, worked there. The core of the so-called 'Geidel'bergskii kruzhok' [the Heidelberg circle] was Mendeleev, Borodin, and Sechenov. The 'Heidelberg period' is well documented in their letters, memoirs, diaries, and official reports. They convey a rich scientific life, which was evident especially among the younger chemists, Friedrich August Kekulé (1829–96), Carl Emil Erlenmeyer (1825–1909), Julius Lothar Meyer, the Briton Henry Roscoe (1833–1915), and the Russians Dmitrii Mendeleev (1834–1907), Leonid Shishkov (1830–1908), and Alexander Borodin. Without doubt, the Heidelberg period had a particular importance for Russian chemistry, as well as for the introduction and development of certain ideas of German laboratory culture in Russian science. We know from Borodin's correspondence that the idea of creating a Russian Chemical Society first materialised in Heidelberg. The 'Heidelberg circle' by itself already represented a small society of young chemists and naturalists. In the autumn of 1860, Nikolai Zinin, Russia's leading organic chemist, visited Heidelberg. Together with younger colleagues Borodin, Mendeleev, and Shishkov, Zinin represented Russian chemistry at the epoch-making 1860 International Congress of Chemists in Karlsruhe. Mendeleev's letter from Heidelberg to his teacher, Professor of Chemistry A.A. Voskresenskii, contains a detailed account of the sessions of the Congress and what it had accomplished. Upon returning home from Heidelberg, the young Russian chemists continued their meetings in St Petersburg. This group, known as 'khimicheskii kruzhok', remained active until the foundation of the Russian Chemical Society in 1868.[4]

The towering figure in Heidelberg chemistry was the Director of the Chemical Institute, Robert Bunsen (1811–99). It is interesting that the impressions of the French chemist Pierre Marcellin Berthelot (1827–1907), who was in Heidelberg at that time, on the research conditions in Bunsen's Institute were similar to those of Mendeleev. Neither of them was impressed by the Bunsen Institute. Nevertheless, both Berthelot and Mendeleev happened to be in Germany at an important transitional point in its intellectual history. In the 1850s, a kind of competitive entrepreneurial fever began to catch hold in the German states with regard to academic laboratory sciences, especially chemistry. The completion of Bunsen's substantial institute in 1854 may be seen as a starting point in the chain of events, which led to the emergence of fabulous new chemical and physiological institutes in Berlin, Leipzig, Munich, and Göttingen with lavish operating budgets and matching salaries, within twenty-five years.[5]

For Russian chemists, Heidelberg of the late 1850s had comfortable and, in a way unique, conditions for research. Writing to the Trustee of the St Petersburg education district from Heidelberg, Mendeleev pointed to the

Image 5.1

Robert Bunsen.
Courtesy of the Wellcome Library Collection.

major reasons why it was so difficult to pursue experimental research in the Russian universities – a lack of means, adequate facilities, and time. Scientists usually had a heavy teaching load to earn their living and, at best, were part-time researchers.

> In Russia the lack of means occurs first, because even in St Petersburg one cannot find a good mechanic, or a good druggist, therefore we have to do a lot of unskilled work that takes so much energy and time; second... we do not have enough laboratories at hand. Here [in Heidelberg] anyone can have his own laboratory, and a professor has a university laboratory. Therefore in little Heidelberg there are five private laboratories.... Thirdly, the majority of our institutional laboratories have no assistants; as a rule in any laboratory abroad, assistants introduce students to experimental manipulations, supervise the analyses, arrange everything for lecture demonstrations, etc. – all these in our laboratories is the responsibility of a professor.[6]

Indeed, there were a few chemical laboratories in St Petersburg: at the University, in some specialised institutions such as the Medico-Surgical Academy, the Artillery Academy and Technological Institute, and some small private laboratories. These laboratories, however, frequently lacked space, and had little means and equipment for experimental work, not to mention training for the students. Additionally, they were usually understaffed by technicians, who kept the laboratory in working order, and by assistants, who relieved the professors of their teaching duties, a usual practice in Germany. In Heidelberg, every possible opportunity was available for acquiring experimental skills in the university laboratory or for pursuing research in a private laboratory. It was quite easy even to set up a small private lab for a particular project. Russian scientists in Heidelberg had time and flexibility for individual studies; they were comparatively well funded by the Russian government, so they could afford instruments and reagents for their work and enjoyed freedom from formal duties. Borodin and Mendeleev often mentioned the low cost of living in Heidelberg, and the availability of instruments and reagents at a relatively low price. Their letters contain detailed accounts of chemical and glassware workshops in Paris and Bonn and pharmaceutical shops in Darmstadt and Paris, as well as comments on quality and prices of chemical instruments and reagents.[7]

For an experienced chemist such as Mendeleev, studying in Bunsen's university laboratory with a lot of students, all of them beginners, was of little value. Mendeleev had already received training in the natural sciences at the St Petersburg Main Pedagogical Institute – with A.A. Voskresenskii in chemistry, Heinrich Lenz in physics, and M.V. Ostrogradskii in mathematics, all luminaries of Russian science. His teachers recognised his talents – in 1859, after two years of work as a Privatdozent at St Petersburg University, Mendeleev was sent abroad for advanced studies in chemistry. Sechenov recalled, 'Mendeleev, of course, became the leader of our group, despite his young age [twenty-five-years old] he was already an established chemist and we were just students.'[8] In Heidelberg, Mendeleev was investigating capillary phenomena and the deviations of gases and vapours from the laws of ideal gases. According to Mendeleev, in Bunsen's laboratory even the weights were rather bad, and there was no tidy quiet place to work with the delicate equipment that he used for his research. Therefore he set up a small laboratory in his apartment with his own equipment and even furnished it with a gas supply. But what Mendeleev did value in Bunsen's laboratory was 'a school itself, with a lot of workers', and the spirit of science and freedom of the old university.[9]

Borodin, a graduate of the Medico-Surgical Academy, had quite different goals to Mendeleev. He had been sent to study abroad by the Conference of the Medico-Surgical Academy before embarking upon his academic career.

Nikolai Zinin, Borodin's mentor and the Academy's leading reformer and principal advocate of basic science training for physicians, instructed Borodin to study special research methods, which are of particular importance for both pure and applied chemistry – for example methods of gas analysis, investigation techniques by titration of solutions, chemical reactions in sealed tubes under high pressure – at the laboratories of Robert Bunsen in Heidelberg, of Adolph Wurtz, Marcellin Berthelot, and Henry Sainte-Claire Deville in Paris, and of August Hofmann in London.

Zinin also recommended Borodin 'to study the application of chemistry in physiological research in the laboratories of Theodor Scherer in Würzburg and Justus Liebig in Munich.' Borodin was to report on achievements of European laboratories to justify Zinin's ambitious plan for creating the Natural History and Sciences Institute at the Medico-Surgical Academy before the military–medical authorities.[10]

From Borodin's reports to the President of the Academy, we know that he was keen to study pure organic chemistry. Obviously, Borodin was reluctant to work at the so-called chemico–medical or chemico–physiological laboratories. The first were laboratories for medical students not familiar with basic chemical methods. Usually such laboratories were equipped just for elementary analytical work, such as analysis of urine by means of titration. Borodin pointed out that no trained chemist, like himself, could be possibly satisfied with the degree of purity, accuracy, and precision of the operations produced there. Studies in the chemico–physiological laboratories required thorough physiological training, where chemists had to devote much time to physiology, which was by no means Borodin's intention.[11]

Borodin thought of himself as a trained chemist, not a beginner, and he preferred to work in the private laboratory of Emil Erlenmeyer, a Privatdozent at the University, which had adequate equipment for gasometry. For Borodin, as for Mendeleev, studies in Bunsen's university laboratory proved to be of no interest or convenience – the equipment was set up for study using the analytical methods developed by Bunsen himself. There were too many students, and too much time was wasted waiting to use ovens or apparatus. From his report to the Academy's Conference, one gets an impression that Borodin wanted to justify his preference to study at a private laboratory paying twice as much as for the university's laboratory, and to use his own equipment, reagents and vessels, which he had bought in Paris and Darmstadt.[12] Borodin presented a rather critical account of the university laboratories and their leaders:

> Bunsen, limiting himself to a narrow frame of development of a few analytical methods for physico–chemical research, lost any interest in

> chemistry as a science (especially organic chemistry) and long ago fell behind in it.... I do not attend any lectures.... [L]ectures of Bunsen and of Kirchhoff are too elementary, and Helmholtz gives, according to duty, a very elementary course on the history of development instead of lectures on his physiological research.[13]

Borodin, however, attended their lectures to get acquainted with the manner of instruction and to see demonstrations of experiments. He also visited Bunsen and Kirchhoff's laboratory unofficially and observed their work in the application of a gas flame to the qualitative and quantitative determination of potassium, sodium, lithium, etc., in various minerals that laid the foundation of spectral analysis. In Erlenmeyer's laboratory, Borodin had favourable conditions for his research – a convenient schedule, and excellent equipment. Apparently Erlenmeyer and Borodin had much in common: both had studied medicine and both were converted into chemists by their teachers, Liebig and Zinin respectively, the leading organic chemists of the time in Germany and Russia. Finally, Erlenmeyer's personality, and his close association with the leading structural chemists August Kekulé and Hermann Kolbe, were of great interest to Borodin. While in Heidelberg, both Borodin and Mendeleev contributed to Kekulé's new journal *Kritische Zeitschrift für Chemie, Physik und Mathematik,* which Erlenmeyer edited. Half of its hundred-and-fifty subscribers in the 1860s were Russian.[14]

Among members of the 'Heidelberg circle', it was Sechenov who most appreciated training in Bunsen's laboratory. For a physiologist engaged in the research on blood gases, studies in gasometry at Bunsen's laboratory were indispensable. As mentioned earlier, the methods of quantitative gas measurements became immensely important for the respiratory physiology research of the time. Sechenov went to Heidelberg after he had completed his research in Ludwig's laboratory, and Ludwig certainly influenced Sechenov's decision to study with Bunsen. There can be no doubt that Sechenov's attraction to chemistry received its first impetus from Bunsen's famous course *Allgemeine Experimentalchemie.*[15] It was in Bunsen's laboratory that Sechenov acquired an introduction to precise chemical investigation, which he later perfected in Mendeleev's laboratory, and which was to dominate his research on blood gases and salt solutions. Of Bunsen himself, Sechenov said with particular warmth:

> Knowing that I was a medical researcher, Bunsen suggested that first of all I study alkalimetry and analysis of mixtures of atmospheric air with CO_2. I knew of Bunsen's perfect goodness and simplicity, and talked to him without being embarrassed. Bunsen lectured excellently; he had the unconquerable habit at lectures to smell the odorous substances he described, however

> harmful and bad the odours were. Someone recounted that once Bunsen had smelled something until he had fainted. Long ago he had paid for his love for explosive substances with his eye, but at any opportunity he produced explosions during his lectures. He was everyone's favourite, and students used to call him Papa Bunsen, although he was not yet an old man.[16]

Sechenov could also appreciate Bunsen's excellent knowledge of mathematics and physics evident in his numerous investigations. In his research and teaching, Bunsen emphasised the experimental side of science. He was an expert glass-blower and devised many pieces of equipment which became famous, the gas burner used in every laboratory for one. Sechenov learnt much from Bunsen in designing various kinds of devices, which he later applied in improving his absorptiometer, and in constructing a stationary gas analyser for the investigation of expired air, and a portable gas analyser. Later Sechenov took lessons from the noted Berlin glass-blower Geissler, and began to make the glass parts for his devices himself.[17]

In Heidelberg, Sechenov also studied in Helmholtz's laboratory and even performed minor research on permeability of the transparent media of the eye to ultraviolet rays. Ludwig, in his letters to Sechenov in Heidelberg, always sent 'heartfelt regards to Bunsen and Helmholtz.'[18] Later, Helmholtz mentioned in a letter to Brücke about Sechenov's research on the fluorescence of the crystalline lens. Ludwig kindly encouraged Sechenov to maintain instruction by Helmholtz, and mentioned that indeed Helmholtz was fond of him. This feeling was mutual, for even Sechenov's brief contact with Helmholtz impressed him so much that later he compared it to 'seeing for the first time the Sistine Madonna in Dresden.... From his quiet figure with thoughtful eyes breathed peace as if he were not from this world'.[19]

Sechenov, Borodin, and Mendeleev were in their twenties when they studied in Germany. According to Borodin, the Russians in Heidelberg were divided into two groups, 'those who do nothing, the aristocrats, the Golitsins, the Olsuf'evs, etc., and those who study. These stick by one another and meet for dinners and for evenings together.'[20] Sechenov recalled that at the meetings, usually held at someone's apartment, they used to read aloud favourite passages from Alexander Pushkin and Alexander Herzen, and when a discussion turned to Russian affairs, hot and noisy arguments occurred, interrupted by stories and recollections. Sometimes the young people criticised German professors and ridiculed German students for their duels and schoolboy tricks, and the local community for its tendency to gossip.[21] At these meetings, Borodin used to entertain the audience with piano music. Knowing that Sechenov was a passionate lover of Italian operas, Borodin played all the principal arias from Rossini's *Il Barbiere di Siviglia.* Sechenov never missed an opportunity to attend Italian operas to hear

Ermina Frezzolini and Adelaide Borghi, famous singers, whom he so admired. Later, he recalled that after performances students used to arrange wild ovations at the theatre-door, and even unharnessed horses to pull the carriages with their favourite actresses themselves.[22] In Berlin, Botkin attempted to 'improve Sechenov's musical taste by listening to German music' and they often attended concerts of F. Liebig's orchestra, famous for its masterful performance of German composers, Beethoven, Mozart, Haydn, and Glück.[23]

It was in Heidelberg, during 1860 and 1861, that Borodin composed and first performed his three chamber-music ensembles, which bore the influence of Felix Mendelssohn's music. Interestingly, according to Sechenov, Borodin 'carefully concealed from his friends that he was a serious musician'. Nevertheless, he acquired the reputation of a musician and played a great deal of chamber music there, and it was at these musical parties that he first met his future wife, Ekaterina Sergeevna Protopopova (1832–87), who was in Heidelberg for treatment. An accomplished pianist, she astonished Borodin with her perfect pitch and easy understanding of modulations, and played for him pieces from Robert Schumann, 'music he did not know, music that kept him awake'.[24] They travelled to Baden-Baden for concerts and to Mannheim where they heard for the first time Richard Wagner's *Tannhäuser, Lohengrin,* and *Der fliegende Holländer.*

Sechenov, Borodin, and Mendeleev enjoyed travelling together across Europe during vacation, without worries or constraint. Beyond the world of Russia for the first time they were fascinated by the picturesque mountains of Lower Saxony, the so-called 'Saxon Switzerland' and by Tyrolean villages in Austria. They marvelled at Alpine peaks, glaciers, waterfalls, and lakes in Switzerland, the wonders of nature recalled in their letters.[25] They rarely missed an opportunity to visit art museums or architectural monuments. Their letters, diaries, and recollections are full of images of European art – the galleries and museums of the Vatican, Milan, and Florence. Sechenov was greatly impressed by the cathedrals, frescoes and Renaissance sculpture, the graceful architecture of Venice, and the beautiful surroundings of Naples. In Rome, Sechenov communicated with young Russian artists in their usual place of gathering, Caffe Greco, and spent much time with the artist Alexander Ivanov who, at that time, was working on a series of pictures illustrating David Strauss's *Das Leben Jesu.*[26] Mendeleev admired the rich collections of Egyptian and Greek art in the Berlin Neues Museum.

The passion for travelling, widespread among the European educated middle classes was largely due to the extension of railway networks throughout Europe in the 1840–50s.[27] The letters of our travellers are full of interesting remarks and comparisons of the cities they visited, modes of life, and national peculiarities in cuisine and entertainment. As if with an ironic smile, the young men shared impressions of their Christmas vacations in Paris. They attended dance classes in Closerie de Lilas, theatres and dinners after performances, and masquerade parties and balls. It was an unforgettable time for all of them. For all the delights of travel, however, Sechenov would always remember the 'romantic country' of his youth:

> [O]ne could not help loving the Germany of that time with its simple, good, open-hearted people. Germany then returns time even now as a peaceful and quiet landscape, with the lilac, apple and cherry trees in bloom, showing as white spots against the green background of the clearing cut up by vistas of poplars.[28]

It seems that Heidelberg had a particular charm for the young people. Mendeleev's *Geidelbergskii dnevnik* conveys a very special emotional atmosphere. The last entry before his departure from the town where he spent two years runs:

> Farewell to Heidelberg.... A mist covered the valley, it was a cold morning, clear mountains, and a haze behind the sun... I haven't noticed anything around me on my way to Darmstadt – my eyes were full of tears... all were dreams of the past.[29]

Some twenty years later, in 1877, Borodin visited Heidelberg and wrote to his wife, Ekaterina, a most poignant letter full of reminiscences that reveal how young-at-heart he had remained:

> As we came nearer to Heidelberg I became so excited that I did not notice that it was already dusk... I was taking in everything, every hillock, every lane, every house, every village – everything brought back at once memories of those happy times... I pressed my face against the window to hide my tears, and gripped the handle of my umbrella to prevent myself from crying like a child.[30]

Borodin, Sechenov, and Mendeleev's memories of Heidelberg are best expressed by Svatikov, who belonged to a later generation of Russian Heidelberg students:

> [A]ll of them without exception, looking back at their youth, remember the university in the same way, as a stronghold of free science, and the old town between the mountains in white bloom, sweet-smelling, a symbol of bright youth, so beautiful and so irrevocably passing... *Alt Heidelberg, du feine...*[31]

Though a touch romantic, the quote reflects a feature relevant to our story: Heidelberg of the late 1850s represented for the young Russian scientists the rigour of research scholarship and the academic freedom of

German science, ideals they hoped to bring back home. For Sechenov, and Borodin in particular, it was the teaching and research laboratory, the principal innovation they introduced to the St Petersburg Medico-Surgical Academy.

Now we shall see how the laboratory was adapted to the Russian setting. It was no novelty for Russia to look to the West in search of ideas for the educational reforms. In its organisational plan and its goals, Moscow University, founded in 1755, was a true embodiment of the spirit of the Western academic tradition. The new universities in Kazan, Kharkov, and St Petersburg, founded at the beginning of the nineteenth century, in their conception and organisation, were modelled after Göttingen University, which, as Napoleon is reported to have said, belonged not to Germany but to all Europe.[32] The reformation of the St Petersburg Medico-Surgical Academy in the late 1850s followed the example of the Prussian Friedrich-Wilhelms-Institut, which was institutionally and intellectually connected to Berlin University. The essential part of the modernisation of the St Petersburg Medico-Surgical Academy was the emulation of the German laboratory, which was instrumental in putting the emphasis on practical training in natural sciences and setting the teaching–research standards for the faculty. The next chapter will examine how it came that the Medico-Surgical Academy happened to be in the right place at the right time for the successful introduction of these innovative practices.

Notes

1. S.G. Svatikov, *Russische Studenten in Heidelberg,* E. Wieschhöfer (ed.), (Heidelberg: Universitätsbibliothek Heidelberg, 1997), 78.
2. On the institutionalisation of experimental physiology in Heidelberg University, see A. Tuchman, *Science, Medicine, and the State in Germany: The Case of Baden, 1815–1871* (New York: Oxford University Press, 1993), 113–51.
3. B.N. Chicherin, *Vospominaniia: puteshestvie za granitsu* (Moscow: Sever, 1935), 88. On Carl Joseph A. Mittermaier (1787–1867), see *Neue Deutsche Biographie* (Berlin: Duncker und Humboldt, 1953–present), Vol. 17, 584–5; on Carl Adolf Vangerow (1808–70), see *Allgemeine Deutsche Biographie,* herausgeben von der historischen Kommission bei der Akademie der Wissenschaften (Leipzig: Duncker, 1875–1912), Vol. 39, 479–80. Pandects (from L. *pandecta,* Gr. *pandektes* all-receiving) are the digest on decisions, writings, and opinions of Roman jurists, the major compilation of the Roman civil law.
4. M.N. Mladentsev and V. E. Tishchenko, *D.I. Mendeleev: ego zhizn' i deiatel'nost'* (Moscow: Akademiia Nauk SSSR, 1938), 171, 180–2, 194, 250–8; on the Congress in Karlsruhe, see also B. Bensaude-Vincent,

'Karlsruhe, septembre 1860: l'atome en congrès,' *Relations Internationales,* 62 (1990), 149–69.

5. A. Rocke, *Nationalizing Science: Adolphe Wurtz and the Battle for French Chemistry* (Cambridge: MIT Press, 2001), 240.
6. Mendeleev to the Trustee of St Petersburg Educational District, letter dated September 1860, Heidelberg, cited in Mladentsev and Tishchenko, *op. cit.* (note 4), 224–6.
7. Mendeleev to L.N. Shishkov, letter dated December 1859, Heidelberg, cited in Mladentsev and Tishchenko, *ibid.,* 159–62; see also Borodin's letters to Mendeleev from Paris, 1860, published in S.A. Dianin (ed.), *Pis'ma A.P. Borodina* (Moscow: Gos. Muz. izd-vo, 1950), Vol. 4, 241–8.
8. I.M. Sechenov, *Autobiographical Notes,* D.B. Lindsley (ed.), K. Hanes (trans.), (Washington: American Institute of Biological Sciences, 1965), 91.
9. Mendeleev to Shishkov, letter dated December 1859, Heidelberg, in Mladentsev and Tishchenko, *op. cit.* (note 4), 116. On Mendeleev's laboratory in Heidelberg, see Borodin to his mother, letter dated November 1859, Heidelberg, in S.A. Dianin, *Pis'ma A.P. Borodina, 1857–1871* (Moscow: Gos. muz. izd-vo, 1927), Vol. I, 34; see also Sechenov, *op. cit.* (note 8), 91.
10. N.N. Zinin, 'Letter of Instruction', cited in N.A. Figurovskii and I.I. Solov'ev, *Aleksandr Porfir'evich Borodin: A Chemist's Biography,* C. Steinberg and G.B. Kauffman (trans.), (Berlin: Springer-Verlag, 1988), 136–7.
11. A.P. Borodin, 'Otchet o zagranichnoi komandirovke, 1859–62', in Dianin, *op. cit.* (note 7), Vol. 4, 254–60: 259.
12. Borodin to Dubovitskii, letter dated 10 February 1860, Heidelberg, in Dianin, *ibid.,* Vol. 4, 239–41: 240. Borodin bought an apparatus for the work in sealed tubes from Marcellin Berthelot in Paris. While in Heidelberg he usually bought chemical reagents from E. Merck, a chemical pharmaceutical shop in Darmstadt. On E. Merck, see *Deutsches Apotheken-Museum im Heidelberger Schloss* (Heidelberg: Schnell und Steiner Verlag, 2000). E. Merck was formed in 1827 from the company Engel-Apotheke which had been founded in 1654 and had been in the possession of the Merck family since 1668.
13. Borodin, *op. cit.* (note 11), 255.
14. Borodin published three articles in 1860 and 1862 on the derivatives of benzidine in Erlenmeyer's *Zeitschrift,* see Borodin, *op. cit.* (note 11), 259–60. On Mendeleev's relationship with Erlenmeyer, see D.I. Mendeleev, 'Dnevniki 1861 i 1862 godov', in S.I. Vavilov, *et al., Nauchnoe nasledstvo* (Moscow: Akademiia Nauk SSSR, 1951), 3 vols, Vol. 2, 95–259: 112–26. In 1868, Erlenmeyer received a call to the Munich Polytechnic School. He was co-author of the three-volume *Lehrbuch der organischen Chemie* (1867–94) and editor of the *Zeitschrift für Chemie und Pharmazie* and of Liebig's

Annalen der Chemie. On Erlenmeyer's research and editorial activities, see A. Rocke, *The Quiet Revolution: Hermann Kolbe and the Science of Organic Chemistry* (Berkeley: University of California Press, 1993), 250–7; see also A. Costa, 'August Carl Emil Erlenmeyer', in *Dictionary of Scientific Biography* (New York: Charles Scribner, 1970–present), Vol. 4, 399–400. On Erlenmeyer's Heidelberg time, see 'Personalakten 1857–63', in Heidelberg University Archive, PA 1524.

15. Bunsen's lectures included much of his own work. His lectures, however, had very little alteration with time. Bunsen never mentioned the periodic law discovered by his students Mendeleev and Lothar Meyer in 1869–70 who were jointly awarded the Davy Gold Medal for their achievements by the Royal Society of London in 1887, see J. Partington, *A History of Chemistry* (London: Macmillan, 1964), 4 vols, Vol. 4, 283.
16. Sechenov, *op. cit.* (note 8), 88, 90–1. Former students of Bunsen always spoke of him with admiration. Henry E. Roscoe was reported to say: 'As an investigator he was great, as a teacher he was greater, as a man and friend he was greatest,' see Partington, *op. cit.* (note 15), Vol. 4, 282. Mendeleev mentioned Bunsen in his *Dnevniki* as 'very nice as usual', see Mendeleev, *op. cit.* (note 14), 153.
17. The Memorial Museum of I.M. Sechenov at the Moscow Medical Academy presents the gas pump constructed by Sechenov at Ludwig's laboratory in Vienna, some apparatus made by Sechenov in collaboration with M.N. Shaternikov in the 1890s, a glass-blowing table that Sechenov used in his work, see, G.G. Grigoryan (ed.), *Pamiatniki nauki i tekhniki v muzeiakh Rossii* (Moscow: Znanie, 1996), 56–7.
18. Sechenov, *op. cit.* (note 8), 88.
19. *Ibid*, 89–90; see also I.M. Sechenov, 'German fon Gel'mgolts kak fiziolog', *Russkaia Mysl'*, xii (1894), 28–37.
20. Borodin to his mother, letter dated 25 November 1859, Heidelberg, in Dianin, *op. cit.* (note 9), Vol. I, 36.
21. *Ibid.*, and Sechenov, *op. cit.* (note 8), 91–2.
22. Sechenov, *ibid.*, 70–1. Adelaide Borghi (1826–1901), Italian mezzo-soprano, sang in Vienna, Paris, and Italy. She was famous for her full-toned, vibrant voice and passionate temperament. Ermina Frezzolini (1818–84), Italian soprano, was admired for her smooth and expressive legato singing (exploited by Verdi). She was noted too for her power and brilliance, in modern manner, and she excelled in dramatic roles. See S. Sadie (ed.), *The New Grove Dictionary of Opera* (London: MacMillan, 1992), 302, 551. On the debates between adherents of Italian operas and German music, see V.P. Botkin, 'Ital'ianskaia i germanskaia muzyka', in V.P. Botkin, *Literaturnaia kritika. Publitsistika. Pis'ma* (Moscow: Sovetskaia Rossiia, 1984), 30–9. On Sechenov's love of music, see A.V. Nezhdanova, *Materialy i issledovaniia*

(Moscow: Iskusstvo, 1967), 59–65. The young Antonina Vasil'evna Nezhdanova (1873–1950) was a close friend of the Sechenovs during the 1890s in Moscow. Nezhdanova represented the famous Russian vocal school of the late nineteenth and early twentieth century, she sang at the Bolshoi Theatre (lyrico-coloratura soprano).

23. N.A. Belogolovyi, *Vospominaniia i drugie stat'i* (Moscow: Tipografiia Aleksandrova, 1897), 306.
24. A.P. Borodin, Sonata for 'cello and piano in B Minor, String sextet for two violins, two violas, and two violoncellos in D Minor, and Trio for violin, 'cello and piano in D Major. For a list of Borodin's musical compositions and his music during the 'Heidelberg period,' see S.A. Dianin, *A.P. Borodin: Zhizneopisanie, materialy i dokumenty* (Moscow: Gos. muz. izd-vo, 1960), 35, 44–9; see also Borodin to his mother, letter dated 31 March 1860, Heidelberg, in Dianin, *op. cit.* (note 9), Vol. 1, 38.
25. Dianin, *A.P. Borodin: Zhizneopisanie, ibid.*, 44–61; see also Borodin to his mother, letter dated 28 October 1860, steamboat Capitole, Italy, in Dianin, *op. cit.* (note 9), Vol. 1, 54; Sechenov, *op. cit.* (note 8), 74–6; Mendeleev, *op. cit.* (note 14), 126–7.
26. Alexander A. Ivanov (1806–58) is famous for his monumental painting *The Appearance of Christ to the People* (1837–57), *The State Tret'iakovskaia Gallery*, Moscow) which might have an influence of Strauss's interpretation of the personality of the Christ. David Friedrich Strauss (1808–74), the German theologiest and philosopher, in his *Das Leben Jesu* published in 1835–6 refuted the authenticity of the Gospels and presented Christ as an historical person.
27. In his discussion of Helmholtz's journies in the early 1850s, Kremer has pointed to several new types of journey which became widespread among the European educated classes, such as 'Wanderjahre', in which a student spent several years at some foreign university; shorter 'Ausbildungsreisen' for specialised knowledge of particular medical institutions; 'Forschungsreise' for specialised research in some particular place; 'Congressreisen' to attend international conferences; and 'Erholungsreisen' for pleasure to cultural and scenic centres of Europe. See R. Kremer, 'Introduction', R. Kremer (ed.), *Letters of Hermann von Helmholtz to his Wife, 1847–1859* (Stuttgart: Franz Steiner Verlag, 1990), xi–xxvi: xix–xx.
28. Sechenov, *op. cit.* (note 8), 95. Mendeleev once remarked ironically: 'Sechenov likes Germans a little bit more than me, defends Germans and even German women, apparently following a Russian habit, he wants to show his gratitude for German hospitality.' See, Mendeleev to the Protopopovs, letter dated 30 January 1860, Heidelberg, cited in Mladentsev and Tishchenko, *op. cit.* (note 4), 190–1.
29. Mendeleev, *op. cit.* (note 14), 126.

30. Borodin to E.S. Borodina, letter dated 30 July 1877, Heidelberg, in S.A. Dianin, *Pis'ma A.P. Borodina, 1872–1877* (Moscow: Gos. muz. izd-vo, 1947), Vol. 2,, 167.
31. Svatikov, *op. cit.* (note 1), 78.
32. P. Miliukov, 'Universitety v Rossii', in F.A. Brokgaus and I.A. Efron (eds), *Entsiklopedicheskii slovar'* (St Petersburg: Izd-vo Brokgauza i Efrona, 1902), 789.

PART II

The St Petersburg Medico-Surgical Academy and Experimental Science

6

Military Medical Education: The Aftermath of the Crimean War

The tremendous loses from the diseases among all the belligerents during the Crimean War made it clear that military medicine needed serious changes to meet the demands of modern armies.[1] After the Crimea, the major emphasis in British, French and Russian military medicine was given to hygiene and sanitary measures to combat the common enemy, epidemics. Russian reforms in military medicine and training, prompted by her painful defeat, were furthered by adopting the major innovation of German university-based medical teaching and research, namely the laboratory. In the late 1850s, new chemical, physiological and clinical laboratories were established at the St Petersburg Medico-Surgical Academy that made it a propitious institution for the young medical scientists to pursue their research and to introduce new laboratory-based training for the students. The interests of the military in modernising the Empire's pre-eminent military institution and the plans of the Academy's reformists, in institutionalising experimental science converged, to the benefit of Russian science

Parallel developments in European and Russian military medicine and training took place prior the Crimean War. By the 1780s, a number of centres for military and naval medical teaching had emerged in France, Russia, and the German states in response to the increasing need of the European powers for military surgeons and medical personnel. By the close of the eighteenth century, the system of military medical education was well established. Through such institutions as the Medizinisch-chirurgische Akademie in Vienna, the Medico-Surgical Academy in St Petersburg, the Medizinisch-chirurgisches Institut in Berlin, and the Val-de-Grâce in Paris, military medical schools offered some of the best surgical and medical instruction of the eighteenth century. Military medicine emphasised surgery, and the army hospitals stressed these skills in teaching. In French military hospitals, surgical demonstrations in anatomical 'amphitheatres' became part of military medical instruction as early as the 1770s. Surgical service in military hospitals provided the best means to receive clinical experience at the bedside, which was largely neglected in the education of medical students at universities.[2] Established specifically for military health needs, which intensified the demand for training 'physician–surgeons', the military medical schools often became in practise, and even in name, schools for both medicine and surgery. By the very nature of military service, distinctions on the battlefield between surgical and medical practices were often nonexistent, and the movement to draw surgery and medicine more closely together, evident everywhere after 1750, was especially pronounced in the military medical schools. Training was eminently practical, and had its merits in supplying useful technical knowledge on a comparatively high level. Clinical teaching was heavily stressed in both parts of the curriculum, medical and surgical – the students usually spent two years of internship in practical instruction and supervised care of patients at the hospital.[3]

Military medical schools were set up in large well-endowed military hospitals, usually controlled by their leading surgeons. The Viennese Medizinisch-chirurgische Akademie, the Josephinum, built in 1783–5 in response to the results of the War of the Bavarian Succession, used the Garrison Hospital for clinical and surgical practice. Directed by Emperor Joseph II's chief surgeon Johann Alexander Brambilla, the Josephinum was lavishly endowed. In 1785, Joseph II ordered from Tuscany the unique and very expensive anatomical collection of beautifully crafted, life-sized wax models, which served as a useful tool in anatomical studies.[4] In St Petersburg, the Medico-Surgical School was established in 1786 at the Land Forces and the Kronstadt Admiralty hospitals, the two largest hospitals set up during Peter I's wars of the 1720s. In 1798, the School was transformed into the Medico-Surgical Academy. Its Prussian counterpart, the Pépinière, or 'seed-bed', a school for royal medical officers, was set up in 1795 in the Charité, the large military garrison hospital founded by Friedrich Wilhelm I in 1726. The Hôpital Val-de-Grâce, formally a monastery founded by Anna of Austria in 1645, became a large military hospital in 1793. Two years later, the French government set up a military medical school there, headed by the outstanding army medical inspector Jean François Coste (1741–1819), a protégé of Voltaire, and the chief physician of the French forces and later of Napoleon's Grande Armée.[5]

In the first quarter of the nineteenth century, governments of the constantly battling European powers continued to pour large sums of money into military hospitals and their schools, achieving strong reputations by offering the best conditions and facilities for training military doctors. The most famous among the military medical schools at that time was the Val-de-Grâce in Paris. Two decades of war brought French military surgeons to the fore, notably Dominique-Jean Larrey (1766–1842), a highly skilful battlefield amputator, who developed the first effective mobile ambulance. Larrey, as well as two other brilliant military surgeons, René Nicolas Desgenettes (1762–1837) and P.F. Percy (1754–1825) taught and worked at the Val-de- Grâce.[6] In his discussion of the antecedents of the Paris school,

Erwin Ackerknecht pointed to military service as one of the most important factors in strengthening the surgical point of view in French medicine. The leading French practitioners were trained as surgeons and often served for lengthy periods in the armies of the Republic and the Empire. The leaders of the new internal medicine, Marie François Xavier Bichat (1771–1802), René Théophile Hyacinthe Laennec (1781–1826), and François Broussais (1772–1838), all expressly stated that they handled the data of internal medicine like surgeons. Surgeons are, by definition, localists and anatomists, and anatomists turn easily to the study of pathological anatomy, in which the Paris school became particularly prominent in the fisrt half of the century.[7]

Service in Napoleon's armies threw together such influential figures as Larrey and Desgenettes with Broussais who became, in 1815, a second professor at the Val-de-Grâce and, in 1820, its chief physician. Solidly entrenched in this hospital and school, he started his conquest of Parisian medicine and made Val-de Grâce one of the most illustrious teaching institutions. Broussais's teaching and his skills as orator attracted a large number of students. His new doctrine, 'physiological medicine', published in *Examen de la doctrine médicale généralement adoptée*, ascribed more importance to the functional disorders of the diseased organs than to their anatomical alterations. He vehemently opposed 'ontology', the 'creation of disease entities from symptoms'. In most cases, a disease was a lesion of the organ, due to inflammation, especially of the gastrointestinal tract. Treatment, therefore, should be local bleeding by leeches and a restricted diet. Broussais's simple explanation and methods were widely accepted by professional army doctors. In the course of time, the most able young military physicians and surgeons, who studied under Broussais, filled leading positions in the French army's medical corps.[8]

There were no comparable developments in Britain as far as military-medical education was concerned. Britain did not possess specialised military medical schools like those in continental Europe. The British army depended solely on civilian practitioners, educated mostly at university medical faculties and attracted to the military service by sufficient rewards and honours. The military hospitals in Fort Pitt and in Plymouth were overburdened in wartime but did not instruct military doctors in peacetime. The Chatham military hospital in Fort Pitt was under excellent regulation, with a good library and a constantly increasing museum of anatomy and natural history.[9] The German military doctor, Adolf Mühry (1810–88), reported in 1835 that English military surgeons had greatly distinguished themselves by their learning and original works on geographical distribution of disease, and their zealous cultivation of natural history.[10] Efforts of some medical authorities, such as Sir George Ballingall (1780–55), to expand specialised military medical teaching at the civilian colleges met with stiff opposition, and his lengthy campaign to establish a chair of military medicine in London was without success. Ballingall, a holder of a controversial chair in military surgery at Edinburgh, was greatly impressed by the continental schools, especially by 'the truly imperial establishment' in Vienna.[11] Indeed, the Vienna Josephinum, despite disruptive closures in 1822–4 and 1848–54 due to unsuccessful military campaigns and political unrest in the Habsburg Empire, remained one of the leading specialised institution for training military surgeons and physicians in Europe.[12]

Its German counterpart, the Berlin Pépinière at the Charité hospital began to acquire prominence by the early 1830s. After the Wars of Liberation, in 1818 the Pépinière changed its French title to the Prussian-styled Medicinisch-chirurgisches Friedrich-Wilhelms-Institut. The Charité's leading surgeons Carl Ferdinand von Graefe (1787–1840) and Johann Nepomuk Rust (1775–1840), who both had teaching responsibilities both at the Institut and at Berlin University, tried to put the Charité under the control of Ministry of Education. They believed that the military medical authorities blocked the efforts of the University to make greater use of the Charité for teaching purposes.[13] Nevertheless, the Charité with its excellent teaching resources remained, as far as administration and clinical education were concerned, in the hands of the military, and the Institut's graduates traditionally had priority in getting appointments in the Charité among other privileges. The Charité, together with the Institut and the University, constituted Berlin's principal institution for training military and civilian doctors. Some important improvements in the Institut's operation, which followed medical legislation in 1825, proved particularly beneficial for military medical teaching in Berlin. One of the most important was the introduction of the theoretical and practical courses in microscopy, chemistry, and pharmacology to the Institut's curriculum.[14] By the late 1830s, courses in the natural sciences at the Institute were given by some distinguished University professors: Heinrich Wilhelm Dove in physics, Eilhard Mitscherlich in chemistry, Johannes Müller in anatomy and physiology. Helmholtz, writing about his education at the Institut, said that he had received a far broader knowledge of all of the natural sciences that was usually given to students of mathematics and physics at Berlin University.[15]

The clinical teaching at the Charité began to acquire a new shape with the appointment of Lucas Schönlein (1793–1864) in 1839. Schonlein lectured in German, not Latin, and represented the type of modern clinician who believed that the stethoscope, the microscope, and the test-tube were prerequisites for efficient clinical diagnosis. He introduced the new diagnostic methods of percussion and auscultation, and accurate pathological, microscopical and chemical examination of discharges, blood, and tissues.[16] Nikolai Pirogov, who studied in the Charité in the 1830s, wrote that it was a period of rapid transition in German medicine, 'the beginning of its ceremonial entry into the exact sciences', best represented by Lucas Schönlein and Johannes Müller. Pirogov left a lively depiction of the leading Charité professors, Rust, von Graefe, and Johann Friedrich Dieffenbach (1792–1847), and of the way they practised and taught surgery.[17]

In St Petersburg, the Medico-Surgical Academy between the early nineteenth century and the 1820s underwent a series of administrative changes connected with the educational reforms of Alexander I. In 1802, the Medico-Surgical Academy was passed over to the newly founded Ministry of the Interior. Its influential official, Count P.A. Stroganoff (1774–1817), suggested a plan for the re-organisation of the Academy and, in 1805, invited Johann Peter Frank to implement this project. During his short stay as President of the Academy, Frank set up therapeutic and surgical teaching clinics (for thirty and thirteen beds respectively), pharmacological and veterinary departments, introduced lectures in Latin, and a new curriculum. The new curriculum, in particular, earned the opprobrium of the Academy's leading professors. In their view, the old curriculum, which had been in operation since 1795, provided well-rounded general science education and adequate practical instruction in the hospital. In the first year, courses in botany, mathematics, physics, chemistry, anatomy, and physiology were given; in the second year, the same courses, except mathematics, and courses in *materia medica*, pharmacy, pathology, and therapy. The third year program included courses in botany, pathology, therapy, surgery, and practical anatomy (on corpses). French, Latin, logic, natural history, and drawing classes were also given. The drawing was practical – military doctors were expected to be able to draw military topography as well as wounds.[18] The fourth year was devoted to clinical practice in obstetrics, therapy, and surgery. After the examinations, the candidates to doctoral degree in medicine and surgery worked the final, fifth year at the Land Forces and Admiralty Hospitals, before entering the state service in the army and fleet. Frank's reforms at the Medico-Surgical Academy, although aimed directly at narrowing the gap between academic and practical teaching, seriously undermined the established basis of military medical instruction at the Academy – a course in surgical diseases was introduced for the junior students, while the senior students were freed from the work at the hospitals, and natural science courses became auxiliary.

Frank's career at the Medico-Surgical Academy came to an end in 1806, due to the interference of Iakov V. Vil'e, Chief Medical Inspector of the Russian Army and also known as Sir James Wylie (1768–1854), being a Scotsman in origin. Using his influence with Alexander I, Vil'e managed to have Frank dismissed, and have himself appointed President of the Medico-Surgical Academy in 1808. An acting surgeon during the Napoleonic campaign of 1804–14, Vil'e ruled the Academy from 'the battlefield'. An excellent administrator and gifted surgeon, he did much for the re-organisation and improvement of medical services in the army that proved crucial during the Napoleonic campagn and Russo–Turkish war of 1828–9.[19] The twenty years of his presidency, however, were marked by total lack of improvement in educational matters and by two major administrative changes. In 1810, the Medico-Surgical Academy was transferred to Ministry of Education, and in 1822, again to the Ministry of the Interior. Inconsistency and controversy over the educational reforms of Alexander I, then followed by the reactionary policy of Nicholas I, were in great measure responsible for the poor state of the Medico-Surgical Academy, which lacked resources, adequate teaching facilities and qualified teaching personnel. According to G.G. Skorichenko, historian of the Academy, the 1820–30s were a period of decline.[20]

In 1838, the Medico-Surgical Academy was taken over by the War Ministry and Count P.A. Kleinmichel (1793–1869), a close friend of Nicholas I, became its curator. According to contemporaries, Kleinmichel's influence at the court was limitless. As Chief of the Office of Public Roads and Communications at War Ministry, Kleinmichel headed the construction of the first railway between St Petersburg and Moscow. At the outbreak of the Crimean War, Kleinmichel initiated the construction of the first telegraph lines, connecting St Petersburg and Warsaw, and St Petersburg and Sevastopol in the Crimea by Siemens & Halske.[21] Fortunately for the Academy, Kleinmichel was keen on educational matters too, and his activities as curator had some important consequences for military medical teaching in St Petersburg. At Kleinmichel's request, the War Ministry allotted a considerable sum of money for the construction of the new Land Forces Hospital for 1,340 beds, which opened in 1840. The War Ministry could afford it: forty per cent of state revenue was assigned to military expenditures. In 1842, Kleinmichel facilitated the transfer of Nikolai Pirogov, an acclaimed surgeon and anatomist from Dorpat University to the newly founded department of hospital surgery and anatomy at the Medico-Surgical Academy. Pirogov's monumental five-volume *Anatomia Topographica* (1843) was recognised as a useful tool for anatomical studies throughout Europe.[22] With Kleinmichel's support Pirogov set up hospital clinics, therapeutic and surgical, in the old Land Forces Hospital for two thousand beds. The largest was the surgical clinic, for one thousand beds, which Pirogov headed himself. The hospital clinics were intended to:

Image 6.1

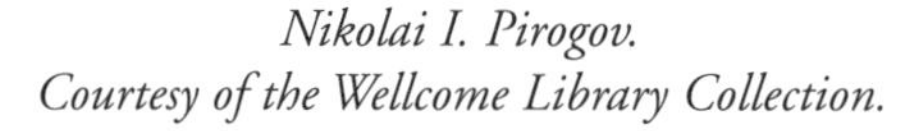

Nikolai I. Pirogov.
Courtesy of the Wellcome Library Collection.

> [R]emove the important lacuna in medical training by providing the students with hospital practise, since the departmental clinics only introduced the students to the basic concepts of disease and its symptoms, diagnosis, and cure.[23]

One of Kleinmichel's most important contributions to the Medico-Surgical Academy, according to Pirogov, was the introduction of 'a new and previously unknown element in the policy of appointment. All vacant and newly opened departments were to be filled by professors who held a university degree.' Since the time of Vil'e, the Medico-Surgical Academy's authorities had been reluctant 'to admit "aliens" into the ranks of the staff, which led to nepotism among the professors.'[24] Apparently, Kleinmichel decided to break the established order. He realised the benefits of attracting the best scholars for teaching science courses, in order to raise the prestige of the Medico-Surgical Academy and its science instruction. In 1841, a distinguished scholar from the St Petersburg Academy of Sciences, Karl

Eduard Eichwald (1795–1876), was invited to teach zoology. Another notable academician, Karl Ernst von Baer (1792–1876), was awarded the chair of comparative anatomy and physiology at the Medico-Surgical Academy, a position specially created for him. In Baer, the Academy acquired an embryologist with a Europe-wide reputation and an excellent microscopist and histologist. The anatomy department possessed a rich collection of microscopic preparations, a gift from Vienna University, and the best available achromatic microscope by Plössel. Baer taught the embryology course *Consectus historiae evolutionis hominis et reliquorum animalium*, and a practical course in microscopy and histology. The course in microscopy and histology was quite a novelty. Clinicians were apathetic to microscopic examination, seeing it as totally unrelated to their daily work.[25] Both Pirogov and Baer ascribed great educational value to practical studies in microscopic anatomy and histology. They emphatically reported to the authorities of the Medico-Surgical Academy that 'microscopy and histology should be taught at the Academy not only for the sake of prestige of the institution, but in order to acquaint the students with the latest developments in medical science.'[26] Pirogov promoted the creation of a histology chair to separate it from zoology and comparative anatomy, but his idea materialised only in 1857.

Clinical medicine at the Medico-Surgical Academy was led by Carl Seidlitz (1798–1885). After graduating from Dorpat University and studies abroad, Seidlitz had served as a senior army doctor during the Russo–Turkish War of 1828–9, and then as a chief physician at the St Petersburg naval hospital. A broadly educated scientist and experienced clinician, Seidlitz had a deserved reputation among both his foreign and St Petersburg colleagues for his studies of cholera and a monograph on scorbutic inflammation of the heart sac. His *Klinischer Bericht* (1846) was a unique example of scientific and practical work. Seidlitz was the first in Russia to use percussion and auscultation both at the hospital and his private practice. His therapy was highly scientific, based on rational symptoms, considerable experience and a profound knowledge of the effect of medicines.[27] Professor N.F. Zdekauer, Seidlitz's student and assistant, later wrote about his mentor:

> Professor Seidlitz was the first at the Academy to deliver lectures on semeiology, to teach us objective methods of percussion, auscultation, measurements and chemical studies of discharges and excreta of the organism. He showed us how to use the microscope to see blood corpuscles and urea crystals, and explained the objective symptoms of diseases. He taught us differential diagnosis, explained the anatomic character of a disease during post-mortem examinations.... He taught general therapy, based on

> conclusion of specific therapy, and acquainted students with embryology.... A new spirit arose in our Academy. [28]

Together with Seidlitz and Baer, Pirogov implemented his project in creating the Institute for Practical Anatomy, which was opened in 1846. Although the Institute was housed in wooden barracks, and the conditions there left much to be desired, the new facility allowed Pirogov to give practical instruction in pathological anatomy and in surgical anatomy for the students. The Institute also had rooms for microscopy studies and for experimental work on animals 'in the search for rational methods of treatment'. Although studies at the Institute for the students were optional, in the evenings the Institute room, where Pirogov and some of his colleagues performed autopsies, would be full of students. During autopsies, Pirogov taught the students the basic approach in identifying pathological processes as well as 'rational access to various organs and surgical techniques.' One of the students reminiscenced about studies at the Institute:

> In the evenings, the huge room, with lots of corpses surrounded by students, dressed in long black aprons, was dimly lit by oil lamps and shrouded with tobacco smoke. This view produced a strange impression, resembling more a cave in Dante's hell than a place of study.[29]

At the Institute, Pirogov had an able prosector, Wencheslav Leopol'dovich Gruber (1814–90), former assistant for the famous Viennese anatomist Joseph Hyrtl (1810–94).[30] Gruber had served the Medico-Surgical Academy for forty years and deserved a reputation as one of the best anatomists of his time. Combining strenuous work at the hospital clinic and private practice, Pirogov continued his anatomical and pathological studies. He supervised or performed autopsies on all patients who died in hospital clinics in the Academy. During fifteen years of his service at the Academy, he managed to perform some 12,000 autopsies and recorded a detailed analysis of his observations.

In 1847, during the Caucasian war, Pirogov went to the acting army in the Caucasus 'to investigate the application of the ether vapours during the operation in the field hospitals'. Due to his efforts all the large military hospitals in the Caucasus were supplied with the apparatus for narcosis of his own construction.[31] In the preface of *Otchet o puteshestvii po Kavkazu* [*Report on Travel in the Caucasus*], published in 1847, Pirogov wrote:

> Russia... during the siege of Saltov in the Caucasus showed not only the possibilities but also the indisputably beneficial effect on the wounded of using ether during surgical operations on the very battlefield. We hope that

> from now on, the ether apparatus, like the surgical knife, will constitute the necessary equipment for every doctor on the battlefield.[32]

The work was immediately translated and published in France and Germany.

Kleinmichel's campaign to restore 'glory' to the Medico-Surgical Academy was not without success. Indeed, under the War Ministry, the Academy received further material means for expansion and improvement of its teaching facilities, including a new hospital, hospital clinics, an Institute for Practical Anatomy, an improved chemical department, and an expanded and enriched library and natural science museum. The introduction of new diagnostic methods in the clinic, and the practical courses in microscopy, histology, semeiology, and anatomy, were important advances in Russian military medical teaching of the 1840s. Pirogov, Baer, Seidlitz, and Eichwald, all active in research and publication, constituted a very close, cohesive, and eminent team to push ahead a programme of medical instruction with a strong practical emphasis.

Despite certain important improvements, the Medico-Surgical Academy still lacked adequate research facilities and instructors capable to carry out independent investigations and to train future medical scientists. In the late 1840s, towards the end of the reign of Nicholaus I, Russia was confronted not only with ineffectiveness of her educational system but also with her major problem - a slow and outdated economy based on serfdom, which could not compete with the rapidly growing capitalist economies of the West. It took defeat in the Crimea, one of her most humiliating and bloodiest wars, for the new Tsar and new Government to acknowledge the urgent need for crucial reform.

The record of Russian military medicine during the Crimean campaign was a bitter one. Military medical administration appeared disastrous, and the complete absence of communication between the army and its hospitals and their supplies aggravated the horrifying state of medical services at the Crimean front. The selfless work of Russian physicians and medical personnel alone could not save the desperate situation and improve the delivery of medical services in war. In a series of letters to his wife from Sevastopol, in November 1854, Pirogov summarised his experience in the Crimea:

> War is a traumatic epidemic. The priority in the treatment of patients in war should not be assigned to medicine, or to surgery, but to an efficient administration. [...] The battle of Inkerman cost the Russians 11,000 dead, and over 6,000 wounded. I found over 2,000 wounded huddled in filthy quilts soaked in blood. We worked for ten days to sort them and separate those who needed urgent operations.[33]

In another letter from December 1854, he reported that the sick were badly treated. The transport was abysmal; the wounded were transported in open carts. They lacked warm clothes, and yet spent nights in the fields or in unheated peasants' huts. Apparently, some of these letters were written in the hope that the Grand Duchess Elena Pavlovna, a close friend of Pirogov's second wife, would read them and make attempts to change the situation.

The hell and heroism of the Crimean War were best related in *Sevastopol'skie rasskazy,* written in 1855 by Leo Tolstoy, who served as an artillery officer for eleven months in the besieged Sevastopol. His depiction of the military hospitals in the city, in terms faithful to the actual experience of the men's dreadful sufferings while awaiting amputation, became a portrayal of war:

> Now if your nerves are strong, go through the door on the left: in this room dressings and operations are performed. You will see doctors with their arms blood-stained to the elbows and their faces pale and gloomy, acting at the bed on which a wounded man is lying under chloroform with open eyes as in delirium, saying meaningless, sometimes simple and touching words. The doctors are doing the disgusting but beneficial work of amputation. You will see how a sharp curved knife pierces the white healthy body; you will see how suddenly the wounded man comes to himself with a horrible, heart-rending cry and curses; you will see how a feldscher throws the cut arm into the corner; you will see how another wounded man who lies on a stretcher in the same room, looking at the operation on his comrade, and writhes and moans not from the physical pain but from moral sufferings of expectation, you will see awful soul-shaking scenes; you will see war not as a splendid array of troops in accurate formation, with music and beating the drums, flying standards and prancing generals but war in its true expression, in blood, suffering and death.[34]

Poor medical arrangements, insufficient preparations for the proper care of the sick and wounded, especially for the common surgical operations, lack of surgeons, dressers, and experienced nurses, constant want for the basic appliances of a workhouse sick-ward and other urgent supplies such as lint became universal at the military hospitals in Sevastopol, Constantinople, Scutari, and Balaclava.[35] The opposing British army experienced the same desperate situation as the Russians. The dispatches from the battlefield published in *The Times* in October 1854 were shocking:

> Our victory has been glorious... but there has been a great want of proper medical attention.... The number of lives which has been sacrificed by the want of proper arrangements and neglect must be considerable.... Here the French are greatly our superiors. Their arrangements are extremely good,

> their surgeons more numerous, and they have also the help of the sisters of Charity who have accompanied the expedition in incredible numbers.[36]

The harrowing conditions at the British hospitals in Scutari and severe losses among the Russians are related in Florence Nightingale's letters from the Crimea:

> [T]he mortality of the operations is frightful. We have Erysipelas, Fever and Gangrene. Russian wounded are the worst. It appears certain that she [Russia] has been drained of every man she can afford. It is thought that the estimate of 500,000 losses is not at all too large. She was losing 3,000 men per day at the time of the bombardment.[37]

This was all the more tragic since the belligerent armies suffered severely from diarrhoea and dysentery and a full-scale cholera epidemic, and these took more deaths than did the actual fighting. In the siege of Sevastopol, the Russian army lost in sickness alone more than 183,000 men as compared with 128,669 lost as casualties. Total French losses reached 95,000, of these only 20,000 died in action or from wounds received in action; the remaining, 75,000 were accounted for by sickness. British losses given by Lord Panmure, Secretary of State for War in 1856 were 20,000; out of these no more than 4,000 were killed in action.[38] In the words of Pirogov, the Crimea was 'a military trauma for all participants and a national calamity for Russia.'

The defeat in the Crimea had a profound and long-lasting impact on the development of Russia's economic and social resources. To quote one of Russia's diplomatic circulars to Europe in 1856 'Russia is collecting her strength.'[39] This meant, first and foremost, the modernisation and reformation of all institutions of society, from military, economic, and educational to governmental, juridical, and social. Despite his obvious will to follow the course of reforms, Alexander II realised that the reforms should not undermine the autocratic system, the security of which rested on its most important pillar, the army. Military and economic reforms therefore became the main concern for the government. Russia's military strength would henceforth depend upon industrialisation to equip the army with the latest weaponry, and to construct a railway network to sustain its mobility. This, in turn, necessitated the development of a network of modern research and educational institutions, where science could serve military and economic objectives.

Bringing the military reform of 1861–74 to a successful conclusion was essentially the result of the energy and perseverance of the War Minister, Count Dmitrii A. Miliutin, and the trust that Alexander II placed in him.[40] Miliutin, and his brother, Nikolai, author of the Peasant Reform plan of

Image 6.2

Crimean War, Russia: appalling conditions as seen in the interior of the Russian hospital in Sevastopol. Wood engravingafter E.A. Goodall, 1855. Courtesy of the Wellcome Library Collection

1861, alongside their close associates, insisted that the reforms must be produced only by the autocratic power aided by progressive and able advisers, who could rise above the outdated privileges of the Russian nobility for the welfare of the Empire. A man of culture with extremely wide-ranging knowledge, author of numerous works on mathematical and military subjects, Miliutin understood the necessity of giving officers a decent education and a thorough professional training. Miliutin fostered improvements in the system of military education to provide Russia with a new nucleus of officers with scientific training adapted to the various specialities of the army. Miliutin expressed the attitude widespread among Russian liberal state ministers and academics that science-based education was essential for the economic and military strength of the country.

The Crimean War once again showed that a mass army required the conquest of camp diseases, dysentery, cholera, and typhus to ensure its efficiency, and this in turn depended upon the development of medical knowledge and practices, as well as the creation of modern facilities for teaching and research. In 1861, P.A. Dubovitskii (1815–68), President of the Medico-Surgical Academy, found a powerful supporter for the reformation of the Academy in Miliutin. They closely collaborated in a campaign to convince the Government that Russian military medical education had fallen behind that of Europe, and required a massive and continuous infusion of funds. In an 1865 report to the War Ministry on the necessity for the reformation of the Medico-Surgical Academy, Dubovitskii pointed to the leading military medical schools – the Val-de-Grâce in Paris, the Chatham Hospital in Great Britain, the Friedrich-Wilhelms-Institut in Berlin, and the Josephinum in Vienna – as examples to follow. To secure the Medico-Surgical Academy's position under the military authority, Dubovitskii deliberately stressed that these schools, specialising in training army doctors and surgeons, remained subordinate to the War Ministry.[41] Unsurprisingly, Dubovitskii first referred to the British and French, to whom Russia had lost the war, although the example of the Chatham army school was out of date – the Chatham army medical school was opened after the Crimean War, in 1860. Its primary function was to give five-month courses of instruction to the newly joined medical officers. In 1862, the school was transferred to the Royal Victoria Hospital at Netley.[42] Both the Netley and Val-de-Grâce schools after the Crimean War emphasised military hygiene in their curriculum. Dubovitskii's reference to the Prussian and Austrian schools seems more appropriate to the St Petersburg reformers with their interest in introducing laboratory-based training for military doctors. Indeed, in Berlin by the 1850s, the reputation and significance of basic medical sciences, physiology in particular, had already dramatically increased. The major figure in the field, Johannes Müller, also taught at the Friedrich-Wilhelms-Institut (Pépinière), and Müller's Anatomical Institute and laboratory at the university was available for selected senior students of the Pépinière, such as Helmholtz and Virchow who received thorough training in physiology and microscopic anatomy under Müller.

The Josephinum example was telling too. The Vienna Josephinum re-opened in 1854, in large part as a reaction to the Crimean War. Emperor Franz Joseph realised that even though Austria had managed to stay out of the war, she had suffered from its consequences.[43] In view of the perceived military medical needs, the reformers of the Josephinum seemed to ascribe a particular place to laboratory training, nominating Carl Ludwig to the chair of physiology and zoology in 1855. Modernisation of the Josephinum was directly associated with the setting up of the physiological institute for

Ludwig. Furthermore, the involvement of an acclaimed scientist would help to re-establish the prestige of the once-famous Josephinum.

Similar developments in Russian military medical education came into prominence after the Crimean War when, in 1860, the reformist administration of the St Petersburg Medico-Surgical Academy invited Ivan Sechenov and Sergei Botkin, two brilliant proponents of German medical science, to set up laboratories for physiological and clinical research and teaching. Three years later, with the completion of the Natural History Science Institute, which housed chemical and physical laboratories, Nikolai Zinin instituted practical classes in chemistry. These laboratories were the new forms of scientific organisation supported by the state, which allowed systematic investigative activities for an independent researcher and training of future investigators and a growing number of students. Sechenov's laboratory marked the beginning of the independent and productive Russian school of physiology, Zinin's laboratory inaugurated the move to scientisation of medical curriculum, and Botkin's laboratory set the foundation for bridging the gap between science and the clinic. Another pronounced aspect of the first laboratories was the formation of connections to an educated part of society through the dissemination of ideals and practices of science and scientific medicine which, in Russia of the 1860s, came to be directly associated with progressive social values. During the liberal upsurge of the 1860s, changes within science and medicine were also believed to call forth changes in the social and political structures of society. For many young intellectuals, such as Sechenov, Botkin, and Borodin, their duty to society converged with their duty to science, and that was evidenced by the unprecedented prestige of science and its advancement in the 1860s, the years which witnessed the great reforms and great hopes and disappointments.

Notes

1. Historians of medicine have pointed to the crucial changes in military medicine during the period followed the Crimean War. There has been little discussion on the impact of war on the aims and concerns of the nineteenth-century military–medical education as well as on how scientific medicine influenced military–medical instruction. For a comprehensive overview on the military–medical teaching and hospitals in 1780– 1830, see T.N. Bonner, *Becoming a Physician: Medical Education in Britain, France, Germany, and the United States, 1750–1945* (Oxford: Oxford University Press, 1995), 52–8. On the importance of the interaction between war and medicine, see R. Cooter, 'War and Modern Medicine', in W.F. Bynum and R. Porter (eds) *Companion Encyclopedia of the History of Medicine* (London: Routledge, 1993), 2 vols, Vol. II, 1536–74; see also,

R. Cooter, M. Harrison and S. Sturdy (eds) *War, Medicine and Modernity* (Stroud: Sutton, 1998).

2. D. Weiner, *The Citizen–Patient in Revolutionary and Imperial Paris* (Baltimore: John Hopkins University Press, 1993), 42–3.
3. On the unification of medicine and surgery in eighteenth-century France, see T. Gelfand, *Professionalizing Modern Medicine: Paris Surgeons and Medical Science and Institutions in the 18th Century* (Westport: Greenwood Press, 1980), 136–7: 154; see also Bonner, *op. cit.* (note 1), 54–6. On practical training at the Berlin Charité, see Dr Schickert, *Die militärärztlichen Bildungsanstalten von ihrer Gründung bis zur Gegenwart* (Berlin: Ernst Siegfried Mittler, 1895), 33–8; and Medizinhistorisches Museum Charité. *Die Berliner Charité: vom Collegium medico-chirurgicum zum Universitätsklinicum* (Berlin: Humboldt-Universität zu Berlin, 1997), 13.
4. M. Skopec and H. Gröger (eds), *Anatomie als Kunst: anatomische Wachsmodelle des 18. Jahrhunderts im Josephinum in Wien* (Vienna: Brandstätter, 2002), 64–6; see also H. Wiklicky, *Das Josephinum: Biographie eines Hauses: die medizinisch-chirurgische Josephs-Akademie seit 1785: das Institut für Geschichte der Medizin seit 1920* (Vienna: C. Brandstätter, 1985). The beautiful collection is displayed in the Museum for the History of Medicine in the Vienna Josephinum. The second existing collection of wax models is in the Imperiale e Reale Museo di Fisica e Storia Naturale in Florence.
5. J.F. Coste, *Du genre de philosophie propre à l'étude et à la pratique de la médecine* (Nancy, 1775). Coste adhered to Hippocratism and was sceptical about physics, chemistry, anatomy, and botany. He insisted on a two-year internship in teaching hospitals, and general education for the physician. See also, E. Ackerknecht, *Medicine at the Paris Hospital* (Baltimore: Johns Hopkins University Press, 1967), 27.
6. On military–medical experiences of Coste, Larrey, Desgenettes, and Percy, and plentiful bibliography, see D. Weiner, 'French Doctors Face War: 1792–1815,' in C. Warner (ed.), *From the Ancien Régime to the Popular Front: Essays in the History of Modern France in Honor of Sheppard B. Clough* (New York: Columbia University Press, 1969), 51–73; see also Weiner, *op. cit.* (note 2), 282–92, and Ackerknecht, *op. cit.* (note 5), 21. On the Val-de-Grâce hospital and school, see D. Larrey, *Memoirs of Military Surgery, and Campaigns of the French Armies on the Rhine, in Corsica, Catalonia, Egypt, and Syria; at Boulogne, Ulm, and Austerlitz; in Saxony, Prussia, Poland, Spain and Austria.* R.W. Hall (trans.),(Baltimore: Joseph Cushing, 1814), 4 vols, Vol. 1, 52, 60. On Percy and Larrey, see J. Rieux and J. Hassenforder, *Histoire du service de santé militaire et du Val-de-Grâce* (Paris: Charles-Lavauzelle, 1951), 33–7.
7. Ackerknecht, *op. cit.* (note 5), 25.

8. On Broussais at Val-de-Grâce, see Ackerknecht, *op. cit.* (note 5), 61–81: 62–4; see also *idem*, 'Broussais, or a Forgotten Medical Revolution', *Bulletin of the History of Medicine*, xxvii (1953), 320–43. The influence of Broussais's doctrine declined in the early 1830s, and gave way to eclectics with their anatomo–pathological orientation. The most famous of the Parisian clinicians of the second half of the nineteenth century, Jean-Martin Charcot (1825–93), in 1867, recognised the passage from symptom to lesion and the appreciation of disease, not as an independent unit, but as a change in function, the two cornerstones of modern medicine. He attributed both advances to Broussais. Claude Bernard was critical of Broussais, yet he was dependent on Broussais armamentarium in his own fight against Hippocratism and scepticism, see C. Bernard, *Introduction à l'étude de la médecine expérimentale* (Paris, 1865): English translation, H.C. Green (trans.), *Introduction to the Study of Experimental Medicine* (New York: Dover Publications, 1957), 19, 95, 119, 150. Four years after his death, in 1838, Broussais's body was brought back to Val-de-Grâce and buried beneath the monument erected in his honour by public subscription in 1841. On the epidemiological laboratory at Val-de-Grâce in the 1870s, see M. Osborne, 'French Military Epidemiology and the Limits of the Laboratory', in A. Cunningham and P. Williams (eds), *The Laboratory Revolution in Medicine* (Cambridge: Cambridge University Press, 1992), 189–208.
9. G. Williamson (ed.), *Catalogue of Preparations, etc., in Morbid Anatomy and Experimental Physiology, Contained in the Museum of the Army Medical Department, Fort Pitt, Chatham* (London: Smith, Elder & Co, 1845).
10. A. Mühry, *Observations on the Comparative State of Medicine in France, England, and Germany, during a Journey into these Countries in the year of 1835,* E.G. Davis (trans.), (Philadelphia: Waldie, 1838), 99–101.
11. G. Ballingall, *Remarks on Schools of Instruction for Military and Naval Surgeons* (Edinburgh: Balfour and Jack, 1843), 3–15, cited in Bonner, *op. cit.* (note 1), 125; see also G. Ballingall, *Introductory Lecture to a Course of Military Surgery, Delivered in the University of Edinburgh* (Edinburgh: Hugh Paton, 1846); M.H. Kaufman, *The Regius Chair of Military Surgery in the University of Edinburgh, 1806–55* (Amsterdam: Rodopi, 2003).
12. K. Pollak, 'Josephinum und Pépinière', *Wehrmedizin und Wehrpharmazie,* iii (1985), 134–8: 136.
13. C.F. Graefe, 'Das medizinische und chirurgische Klinikum bei der Universität zu Berlin,' letter to the Ministry, dated 4 February 1811, Berlin, cited in Bonner, *op. cit.* (note 1), 123; see also Schickert, *op. cit.* (note 3), 18–20; and *Die Berliner Charité, op. cit.* (note 3), 19.
14. On the Institut's professors, who later constituted the first faculty of the Berlin University, see R.L. Kremer, 'Between Wissenschaft and Praxis: Experimental Medicine and the Prussian State, 1807–1848', in G. Schubring

(ed.), *Einsamkeit und Freiheit' neu besichtigt* (Stuttgart: Franz Steiner, 1991), 156. On the Institut's curriculum and professors, see also Schickert, *op. cit.* (note 3), 12–16, 132–3; and *Die Berliner Charité, op. cit.* (note 3), 9–11; and D. Cahan, 'Introduction', in D. Cahan (ed.), *Letters of Helmholtz to his Parents: the Medical Education of a German Scientist, 1837–1846* (Stuttgart: Franz Steiner Verlag, 1993), 1–30: 11–12.

15. H. Helmholtz, 'Das Denken in der Medizin' cited in Cahan, *op. cit.* (note 14), 9.
16. On Schönlein and the Institut, see *Die Berliner Charité, op. cit.* (note 3), 23–4; see also J. Bleker, *Die naturhistorische Schule, 1825–1845: Ein Beitrag zur Geschichte der klinischen Medizin in Deutschland* (Stuttgart: G. Fischer, 1981).
17. N. Pirogov, *Questions of Life: Diary of an Old Physician* (Canton: Science History Publications, 1991), 341–69.
18. N.I. Ivanovskii (ed.), *Istoriia Imperatorskoi voenno-meditsinskoi (byvshei Medico-khirurgicheskoi akademii za 100 let. 1798-1898* (St Petersburg: Tipografiia Ministerstva vnutrennikh del, 1898), 15–42; see also V.O. Samoilov, *Istoriia Rossiiskoi meditsiny* (Moscow: Epidavr, 1997), 68–73.
19. Iakov Vasil'evich Vil'e (James Wylie, 1768–1854) studied medicine in Edinburgh and Aberdeen. In 1790, he came to Russia as a physician to Prince Golitsin. Vil'e took part in more than fifty battles during the Napoleonic wars and Russo-Turkish campaign of 1827–9. He published a number of works on field surgery, pharmacopoeia, contagious diseases, cholera, and plague. In 1823, he started the *Voenno-meditsinskii zhurnal,* Russia's major military medical journal till now. For Vil'e's biography, see Ivanovskii, *op. cit.* (note 18), 240–54.
20. G. Skorichenko, 'Imperatorskaia voenno-meditsinskaia akademiia. Istoricheskii ocherk,' *Stoletie voennogo ministerstva* (St Petersburg: Izd-vo voennogo ministerstva, 1902), Vol. IX, 27–77: 57.
21. In his *Recollections,* Werner Siemens presented a fascinating story of his experiences in Russia during the Crimean War in 1853–5, and his personal impressions on Count Kleinmichel, see W. Siemens, *Recollections* (Munich: Prestel Verlag, 1983), 96–118.
22. Pirogov's rich experience as a military and hospital surgeon is reflected in his numerous works on surgery, anaesthesia, and on Asiatic cholera, most notably in N. Pirogov, *Grundzüge der allgemeinen Kriegschirurgie: nach Reminiscenzen aus den Kriegen in der Krim und im Kaukasus und aus der Hospitalpraxis* (Leipzig: Verlag von F. C. Vogel, 1864). In the first chapter, Pirogov related his experience in both Crimean and Caucasian wars in field hospitals and transportation of the wounded, the two major challenges the Russian army encounted during these campaigns. The literature on Pirogov is extensive, especially in Russian and German. For a bibliography on

Pirogov, see G. Zarechnak, 'Introduction' in Pirogov, *op. cit.* (note 17), xi–xxx. On Pirogov and the medical profession in Russia, see N.M. Frieden, *Russian Physicians in an Era of Reform and Revolution, 1856–1905* (Princeton: Princeton University Press, 1981), 5–11.

23. Pirogov, *op. cit.* (note 17), 417, 424.
24. *Ibid.,* 415–16.
25. On the introduction of microscopy into Parisian medical education in the late 1830s and early 1840s, see A. La Berge 'Medical Microscopy in Paris, 1830–1855', in A. La Berge and M. Feingold (eds) *French Medical Culture in the Nineteenth Century* (Amsterdam: Rodopi, 1994), 296–326; see also *idem,* 'Dichotomy or Integration? Medical Microscopy and the Paris Clinical Tradition', in A. La Berge and C. Hannaway (eds), *Constructing Paris Medicine* (Amsterdam: Rodopi, 1998), 275–312: 280.
26. M. Lavdovskii, *Istoricheskii ocherk kafedry gistologii i embriologii v Imperatorskoi voenno-meditsinskoi akademii* (St Petersburg: Tipografiia tovarishchestva 'Narodnaia Pol'za', 1898), 1, 4–5, 8–9.
27. Pirogov, *op. cit.* (note 17), 415.
28. N.F. Zdekauer, 'Zametki o professore K.K. Siedlitze,' *Russkaia starina,* i (1891), 46.
29. Cited in Samoilov, *op. cit.* (note 18), 98.
30. *Rossiiskii gosudarstvennyi voenno-istoricheskii arkhiv,* fond. 1, opis' 1, delo 2714.
31. N.I. Pirogov, 'Puteshestvie po Kavkazu v 1847', in P. A. Dubovitskii (ed.), *Zapiski po chasti vrachebnykh nauk* (St Petersburg: Izd-vo Jungmeister, 1849), 10.
32. N.I. Pirogov, *Otchet o puteshestvii po Kavkazu* (Moscow: Meditsinskaia literatura, 1952), i.
33. N.I. Pirogov, *Sevastopol'skie pis'ma i vospominaniia* (Moscow: Meditsinskaia literatura, 1950), 28–30, 35.
34. L.N. Tolstoy, 'Sevastopol' v dekabre mesiatse,' *Povesti i rasskazy* (Moscow: Khudozhestvennaia literatura, 1966), 2 vols, Vol. 1, 64.
35. On the hospital administration of the British army, see J. Shepherd, *The Crimean Doctors: A History of the British Medical Services in the Crimean War,* 2 vols (Liverpool: Liverpool University Press, 1991). For the French side there is no equivalent modern research. The same applies to the Russian side, *cf.* J. Curtiss, *Russia's Crimean War* (Durham: Duke University Press, 1979), 461–71. The classical account of the siege of Sevestopol by the chief Russian engineer Eduard I. Totleleben, *Opisanie oborony goroda Sevastopolia,* 2 vols (St Petersburg: Tipografiia N. Tiblena, 1863–74) has an appendix about Russian hospital organisation, translated into French: E.I. Totleben, *Service sanitaire des hôpitaux russes pendant la guerre de Crimée, dans les années 1854–1856* (St Petersburg: Tip. Tiblena, 1870).

36. Cited in S. Goldie (ed.), *Florence Nightingale: Letters from the Crimea, 1854–1856* (Manchester: Mandolin, 1997), 17–18.
37. *Ibid.,* 38. For Pirov quotation, see *op cit* (note 33).
38. W. Baumgart, *The Crimean War: 1853–1856* (London: Arnold, 1999), 215–17, see also L.G. Beskrovny, *The Russian Army and Fleet in the Nineteenth Century: Handbook of Armaments, Personnel and Policy* (Gulf Breeze: Academic International Press, 1996), 51; and G. Parker (ed.), *Cambridge Illustrated History: Warfare* (Cambridge: Cambridge University Press, 1995), 216–20.
39. Cited in W. Baumgart, *Peace of Paris 1856: Studies in War, Diplomacy, and Peacemaking* (Oxford: Clio Press, 1981), 201.
40. D.A. Miliutin, War Minister (1861–81), General-Field Marshal (1898), Corresponding (1853) and Honorary (1866) Member of the St Petersburg Academy of Sciences, professor of the Military Academy; on his views on the reforms and the basic principles of Russia's inner and foreign politics, see P.A. Zaionchkovskii, 'Biograficheskii ocherk', in P.A. Zaionchkovskii (ed.), *D.A. Miliutin, Dnevnik* (Moscow: Publichnaia Biblioteka, Otdel rukopisei, 1947–50); see also D.A. Miliutin, *Vospominaniia general-feldmarshala grafa D.A. Miliutina* with the Introduction by L. Zakharova, 'D.A. Miliutin, ego vremia i ego memuary' (Moscow: ROSSPEN, 2003). The classical treatment of Miliutin's military reforms is in P.A. Zaionchkovskii, *Voennye reformy 1860-1870 godov v Rossii* (Moscow: Izd-vo Moskovskogo universiteta, 1952); see also P. Miliukov, C. Seignobos and L. Eisenmann, *History of Russia* (New York: Funk & Wagnalls, 1968–9), 3 vols, Vol. 3, 45–7; and N. Riasanovsky, *A History of Russia* (New York: Oxford University Press, 2000), 374–8.
41. *Rossiiskii gosudarstvennyi voenno-istoricheskii arkhiv*, Moscow, fond 1, opis' 1, 1865, no 26984
42. Archives, Wellcome Library, London, notes on history and records: Fort Pitt, Chatham, Kent, and Netley.
43. Baumgart, *op. cit.* (note 38), 34–42.

7

The Winds of Change: Reformation of the Medico-Surgical Academy

The reforms of the 1860s and 1870s, the result of a peculiar compromise between the liberal reformists and conservatives in the government of Alexander II, drastically changed the existing economic and social structure of the country. Ultimately, the abolition of serfdom and the agrarian reform of 1861, aimed at modernising the country's economy, facilitated Russia's transition to an industrial society. The government began to invest huge amounts of money in the development of heavy and war industries, equipped by steam-driven engines, and in the expansion of the railway and telegraph networks. The industrial banking movement, and foreign investment, then spread to Russia, where the government needed capital for army reform and economic development. Contrary to the scale of transformation in the economic sector, the socio–political structure of the country remained virtually the same. Autocratic ruling was as firm as ever, and the country was governed through edicts, police action, and the power of the army. The uneven application of the emancipation of serfs, who were forced to pay their former masters for their land, and the estate division of society, not only hindered the economic development of the country, but also produced the growing social tension, peasant riots and student unrest. The Polish uprising of 1863 set Alexander II against the idea of a representative government,[1] and was responsible for a conservative government movement. Alexander II nevertheless continued the politics of modernisation. In 1863, a new university statute restored the autonomy of the universities, the following year edicts reformed the courts of law, and established provincial and district councils of self-government – 'zemstvos', and the 1865 regulation of the press relaxed oppressive censorship in great measure.

The complex socio–political processes accompanied the breaking down of the old social structures as well as established traditions and values. These painful transformations found their reflection in Russian social thought, literature, art, and music and set in motion the search for ways to solve complex problems. Politically conscious youth, especially students, sought ways to 'pay the debt to people'[2] and to solve the problems not by reforming but crushing the existing system and rejecting the values on which the system rested. This was the credo of the so-called 'nihilists' (from Latin *nihil*, nothing) depicted by Ivan Sergeevich Turgenev (1818–83) in his famous novel *Otsy i deti* [*Fathers and Sons*], written in 1862. Turgenev expressed his despair over Russia's 'backwardness' and his scepticism towards the capability and usefulness of those young men who believed in nothing while disclaiming conventional spiritual values, moral principles, and forms of social life.

The political situation produced controversial reaction from the government. Oppression of students and revolutionists increased, and a secret police force, the so-called Third Division, persecuted popular socialist writers, and a variety of revolutionary socialists, radicals, and anarchists. At the same time, the Government made steps to get support from society in the struggle against 'kramola', resistance to lawful authority. The growing reactionary sentiments in society were expressed by the influential *Moskovskie vedomosti* [*Moscow Gazette*] of M.N. Katkov (1818–87). In 1880, despite growing revolutionary movement, Alexander II, who by that time had survived seven attempts made on his life, relaxed the police suppression, restored a measure of press freedom, and even agreed to a proposed elected body to advise the Council of State. His assassination in 1881 by an member of the terrorist group 'Narodnaia Volia' ['People's Will'] ended an era.[3]

The politics of modernisation during the turbulent 1860s seemed to work quite efficiently in the sphere of education. The educational reforms of 1863 reflected the state's interest in supporting modernisation of the entire system of education and, in particular, institutions of higher learning. The Imperial universities began to acquire more independence as well as financial assistance from the government to introduce new programmes based on practical training which, in the context of the economic need to stimulate the development of science and technology, began growing in importance. There was a slow but steady increase in state-funded, university-based research and teaching institutes and laboratories – well developed in the German lands but virtually absent from Russian universities prior to the 1860s.

A couple of decades later, laboratories would become an essential part of university education, particularly medical training, but in the early 1860s it was not yet possible to introduce the new teaching research laboratories even at significant medical faculties such as Dorpat and Moscow,[4] to say nothing about the small medical faculties at Kazan, Kharkov, and Kiev universities. Medical faculties, including the celebrated Dorpat school, were simply too weak to compete with the Medico-Surgical Academy, which had come under the War Ministry and affirmed its position as the country's flagship medical educational institution.

After the Crimean War, due to a series of successful institutional reforms, the Medico-Surgical Academy began to acquire even greater prominence as the modernised centre for medical teaching and research. The Academy possessed uniquely favourable conditions for the development of medical science that could not be emulated at any Russian university. Subordination of the Academy to the War Ministry appeared to be quite advantageous. The military, traditionally the most powerful faction in the Russian government, pushed for the substantial funding required to modernise the existing and create the new research and teaching facilities at its educational institutions. At the same time, the government, even after the university reform of 1863, was still too slow to finance the introduction of practical training and construction of laboratories at the medical faculties of the universities. The physiological laboratories at the physico–mathematical faculties of St Petersburg and Odessa Universities never achieved the level of funding of the laboratories at the Medico-Surgical Academy. As an élite institution, the Academy had special status and traditions. Like other such institutions, among them the private schools, military academies, and specialised institutes attached to the ministries other than Ministry of Education, it was saved from the constantly shifting views of the ministers and bureaucrats responsible for educational policy. These institutions traditionally enjoyed privileges that allowed much better conditions and more freedom for the faculty and the students,[5] and did not experience rigid control and the petty guardianship of the ministry officials, typical for the style of governance at the Russian universities.

The reformation of the Medico-Surgical Academy started in 1857, when Alexander II instructed the military medical authorities to launch a programme aimed at modernising the country's military medical education, 'in order to put medical science in Russia on the same ground of perfection as in France and Germany'.[6] To implement this formidable task, the War Ministry had a suitable candidate among professors of the Medico-Surgical Academy in Petr Aleksandrovich Dubovitskii. A wealthy landlord from Riazan province, Dubovitskii had once chosen the career of military surgeon, not typical for the social-estate class to which he belonged, but he had to abandon surgery for teaching because of an injury to his hand.[7] He taught in Kazan and, from 1840, at the Medico-Surgical Academy. A corresponding member of the Paris Academy of Medicine (1846), and author of numerous works on field surgery, Dubovitskii was an influential member of the St Petersburg military medical establishment well positioned at the Academy. Having accepted the post of President of the Academy in 1857, Dubovitskii laid down conditions that proved crucial for his successful reformist administration. He requested that the post of Trustee [popechitel'] be abolished so that the Academy was directly subordinated to the War

Image 7.1

Petr A. Dubovitskii.
Courtesy of the Karl Sudhoff Institute for the History of Medicine and Natural Science, Leipzig University.

Ministry. Dubovitskii knew only too well the dependence of Russian educational institutions on Trustees, appointed to execute an intermediary role between the institution and the Ministry.[8] Another important request was that a council of professors, the Conference, was to exercise full control over the academic, administrative, and financial matters at the Academy.

The core of the Conference constituted the so-called 'triumvirate',[9] Dubovitskii and two his close associates, Academic Secretary N.N. Zinin and newly appointed Vice-President I.T. Glebov. It was a coherent team of leaders, who, in view of their accomplishments and social connections, combined administrative experience, pedagogical talent, wide scientific outlook, and an aspiration for modernisation of the Academy. Nikolai

Image 7.2

Ivan T. Glebov.
Courtesy of the Karl Sudhoff Institute for the History of Medicine and Natural Science, Leipzig University.

Nikolaevich Zinin (1812–80) in Sechenov's words 'represented hard science in the triumvirate'. A renowned organic chemist, who had successfully promoted a Liebigian model of preparing chemists in Kazan in the 1840s, Zinin was a strong advocate of the idea that basic sciences should constitute an important component in the preclinical training of a physician. The leading professor of anatomy and physiology from Moscow, Ivan Timofeevich Glebov (1806–84), was an outspoken supporter of the idea that physiology was the foremost among the medical sciences, and that the introduction of laboratory courses in experimental physiology would strengthen its position and prestige among medical disciplines. Soon after his appointment, Glebov visited some European clinics, anatomical and physiological institutes, to report on achievements in medical training and research. In *Kratkii obzor deistvii Imperatorskoi Meditsinsko-Khirurgicheskoi Akademii* [*A Brief Survey of the Activities on the Improvement of the Imperial Medical Surgical Academy*] (1860), Glebov shaped a programme of reforms and worked out a detailed plan for their implementation. The programme was focused on five major goals: introduction of basic science courses to develop a 'scientific outlook'; setting up laboratories for research and teaching; improvements in teaching the core clinical disciplines making them more practical; establishing a system of training qualified teaching and research personnel; and incorporating a scientific approach in teaching specific aspects of military medical practice.[10]

The reformers believed that the existing structures of the Medico-Surgical Academy could be adapted to the new goals and demands. 'No other educational institution in Russia and not many in Europe could boast of such favourable conditions and means for training physicians' as the Academy:

> Usually the specialised institutions are scattered at quite a distance from one another, for example, in Paris, the syphilis hospital Hôtel du Nord from the typhus hospital Hôtel de St Louis, the Pitié from the Hôpital Necker, and the Hôpital des Enfants Trouvés from Charenton. We have all hospitals and faculty clinics concentrated in one place in the territory of the Academy; our teaching hospitals have lecture halls and libraries.[11]

The existing facilities however required considerable expansion and reconstruction to house new teaching clinics, departments, institutes, and laboratories. The vast territory occupied by the Academy also needed improvement, since it had been neglected from the time of its foundation at the end of the eighteenth century. Situated on the banks of the Neva and Bol'shaia Nevka rivers, the location was swampy in some places and had only two groups of stone buildings. Through the tireless efforts of Dubovitskii, all old wooden buildings were torn down and the territory became a well-planned, neat park with shady alleys, numerous flowerbeds, and ponds. Dubovitskii's 'passion for construction' was embodied by the finely designed spacious buildings of the Mikhailovskii (Vil'e) hospital, the Natural History and Sciences Institute, the Botanical Institute with a greenhouse, the Anatomo-Physiological Institute, and the Veterinary Institute. The 'building campaign' required a great deal of investment, and Dubovitskii succeeded in persuading the War Ministry that, with new laboratories, clinics and institutes, the Medico-Surgical Academy would rank among the best in Europe. Dubovitskii and N.O. Sukhozanet (1794–1871), the former War Minister in 1856–61, generously donated to the wealth of the Academy.

Dubovitskii also committed his salary for all his years of service at the Academy to create a special fund for needy students.[12]

One of the major priorities for the reformers was scientisation of the curriculum:

> [N]atural sciences should be equal in their importance to clinical subjects, the scope of teaching sciences should be wide and full, and the way of teaching should be theoretical as well as practical.... All phenomena and processes that take place in nature are similar to those in the human body. And the bodily processes in miniature follow the same mechanical (physical) and chemical laws as in nature.... Here is the link between medicine as a science of human beings and the natural sciences. Therefore natural sciences are fundamental and not auxiliary in teaching medicine.[13]

To ensure practical and rational training in sciences, the Conference facilitated the arrangement of three new departments: natural sciences (physics, meteorology, climatology, and physical geography); botany; and practical anatomy. The cost of new departments and new programmes entailed a substantial capital investment, and the War Ministry allotted 5,185 roubles requested by the Conference. Additional funds were provided for two new clinical chairs, one in nervous and mental diseases and the other in ophthalmic diseases.[14]

What ultimately made instruction practical was the introduction of the new laboratory courses in microscopy, basic chemical and physical techniques, and experimental physiology; these added new skills to those gained in the teaching clinic, dissecting room, and botanical garden. The laboratory training of a future physician allowed integration of the methods of basic sciences combined with traditional anatomo–clinical methods into practical medicine. Among the laboratory courses, experimental physiology, which by that time had achieved spectacular results, was to acquire a central place in the Academy's curriculum:

> Introduction of a physiology course with lecture demonstrations and laboratory training in experimental methods and techniques has a fundamental importance for the study of medicine. Arrangement of the physiological laboratory for teaching purposes and for investigations that lead to understanding and discovering laws of organic life is one of the most fruitful ways to develop science at our Academy, and at the same time to acquaint the students to the principles of scientific medicine.[15]

The reformers understood that a laboratory course in physiology would reinforce didactic instruction, however, systematic laboratory training for all students was not at that time their major concern. The rhetoric of the

Conference stressed research components for the faculty and graduates preparing for an academic career. For both medical reformers and medical scientists, the laboratory was a symbol of modernity and scientific method, a sign indicating the beginning of a gradual shift in medical instruction and research from the teaching clinic to the laboratory. The reformers recognised the great potential of the laboratory in the development of science; on the other hand, they also realised that there would be no immediate benefit for practical medicine from exiting and challenging advances in experimental physiology. The physiological and clinical laboratories set up at the Medico-Surgical Academy offered the appropriate physical setting for research and instruction, to prepare skilled professionals, and to encourage the development of medical science. The new physiological laboratory was a separate structure from the old Anatomo-Physiology Institute, founded by Nikolai Pirogov in the 1840s and renovated in 1859. The objectives of the old Institute, as outlined in its statute, were:

> [T]o provide adequate conditions for practical studies in anatomy and independent research in comparative anatomy, physiological anatomy, and experimental physiology, and to prepare lecturers in anatomy and physiology not only for the Academy but for other medical educational institutions of Russia.[16]

The new Anatomo-Physiological Institute, which would house a spacious and excellently equipped physiological laboratory, was completed after Dubovitskii's death, in 1871.

A pressing issue for the reformers was the selection of professors capable of conducting original research. Through the advocacy of Glebov, young scientists such as Sechenov, Botkin, Bekkers, and Junge were called to the Medico-Surgical Academy. Medical graduates from Moscow, they continued their studies in German laboratories and clinics, pursuing various research paths – Sechenov in experimental physiology, Botkin in clinical medicine, Bekkers in surgery, and Junge in ophthalmology. The reformers understood, however, that the Academy needed a structure which ensured postgraduate training for the future faculty. In 1858, such a structure, the so-called Institute of Physicians, was created. The Medico-Surgical Academy's graduates were selected to spend three years at the Second Land Forces Hospital and to pursue thesis research at one of the Academy's departments. After their vivas, they were sent abroad for two years' postgraduate studies. The Institute's mission was:

> [T]o disseminate exciting achievements in medical science and to spread scientific knowledge and methods among undergraduates of the Academy, as well as in medical communities throughout Russia; to introduce principles

Table 7.1

Financial Accounts of the Medico-Surgical Academy, 1859 and 1864.

Roubles	1859	1864
Maintenance costs of the Academy	**151,874**	**208,526**
Maintenance costs of the research institutes	**7,768**	**28,837**
Libraries (purchasing books, subscription to the foreign scientific journals, etc.)	**3,000**	**6,000**
Practical courses in physiology and anatomy	**350**	**700**
Foreign travel expenses for research	**1,800**	**7,200**
Publication expences for the faculty	**400**	**1,401**
Funding for graduates studying abroad	**600**	**2,333**

of scientific medicine into clinical and practical medicine; to establish an adequate research environment for independent development of Russian medical science.[17]

The Institute's product was medical scientists active in research and capable instructors. For twenty years, the Institute was, in the words of Alexander Borodin:

> [A] breeding ground for Russian scholars and achieved educational strength both in medicine and natural science. Among 129 Institute graduates, 58 became professors at our Academy and other universities, while the rest became distinguished practitioners or medical administrators.[18]

The most renowned graduates of the Institute of Physicians included professors of the Medico-Surgical Academy pathologist V.V. Pashutin, physiologist I.P. Pavlov, and neurologist V. M. Bekhterev (1857–1927).

The Medico-Surgical Academy, which embodied the entire system of military medical education, served science as well as state interests. Dmitrii Miliutin, military reformer and an enlightened member of the immediate circle of Alexander II, envisioned that modernisation and increased funding for military academies including the Medico-Surgical Academy, would position these institutions well within national political and economic ambitions and expectations. Table 7.1 illustrates the stable growth in research funding at the Academy throughout a half-decade from 1859 to 1864.[19]

Years later, V.V. Pashutin wrote about the Medico-Surgical Academy of the 1860s:

Image 7.3

Dmitrii A. Miliutin.
Courtesy of the Karl Sudhoff Institute for the History of Medicine and Natural Science, Leipzig University.

> War Minister D.A. Miliutin, who executed the immediate will of the Tsar, supported Dubovitskii in every possible way. In ten years, Dubovitskii reformed the Medico-Surgical Academy in all aspects and put it on the level of the best European educational institutions.[20]

The powerful influence of Dubovitskii was pervasive, not only because of his untiring administrative activities, managerial abilities and social connections, but also because his governing style was in accordance with the spirit of the 1860s with its liberal upsurge and aspiration for scientific advances. The 'golden decade' of 1860–70 marked quite a spectacular transformation of the Medico-Surgical Academy into a modern institution with physiological, chemical and clinical laboratories, and specialised teaching clinics, a centre for scientific medicine and advanced experimental inquiry. Its superior conditions and material resources proved pivotal for the research and teaching careers of Nikolai Zinin, Alexander Borodin, Sergei Botkin, Ivan Sechenov, and Elie Cyon, who radically shaped new teaching fields, experimental physiology, laboratory-based clinical medicine, and chemistry applied to medicine, and eminently represented the emerging national scientific medical élite of an entirely new sort with teaching, research, and publications as a holistic professorial standard.

Notes

1. According to the Vienna Congress Agreements of 1815, the Russian Tsar Alexander I received the kingship of the new 'independent' Poland, which in fact became an heritable Romanov territory. Some Polish territories went to Austria and to Prussia, see A.J. Slavin, *The Way of the West* (Lexington: Xerox College Publishing, 1975), 2 vols, Vol. II, 232.
2. A. Polunov, *Russia in the Nineteenth Century: Autocracy, Reform, and Social change, 1814–1914* (Armonk: M.E. Sharpe, 2005), 142
3. Slavin, *op. cit* (note 1)., Vol. II, 413–4.
4. On the history of the medical faculty of Moscow University, see A.M. Stochek, M.A. Paltsev, S.N. Zatravkin, *Meditsinskii fakultet Moskovskogo universiteta v reformakh prosveshcheniia pervoi treti 19 veka* (Moscow: Meditsina, 1998), 7–11.
5. R.G. Eimontova, *Russkie universitety na grani dvukh epokh: ot Rossii krepostnoi k Rossii kapitalisticheskoi* (Moscow: Nauka, 1985), 31–3, 40–5; see also S. Kassow, *Students, Professors, and the State in Tsarist Russia* (Berkeley: University of California Press, 1989), 18–20.
6. I.T. Glebov, *Kratkii obzor deistvii Imperatorskoi meditsinsko-khirurgicheskoi akademii za 1857, 1858 i 1859 gody v vidakh uluchsheniia etogo zavedeniia* (St Petersburg: Voennoe ministerstvo, 1860), 4–5.
7. I.L. Shevchenko (ed.), *Professora Voenno-meditsinskoi (meditsinsko-khirurgicheskoi) akademii (1798–1998)* (St Petersburg: Nauka, 1998), 12. For Dubovitskii's biography, ee also V.N. Beliakov, *P.A. Dubovitskii* (Leningrad: Voenno-meditsinskaia akademiia im. S.M. Kirova, 1976).
8. Eimontova, *op. cit.* (note 5), 77–8, see also N.I. Pirogov, *Questions of Life: Diary of an Old Physician* (Canton: Science History Publications, 1991), 278.
9. I.M. Sechenov, *Autobiographical Notes,* D.B. Lindsley (ed.), K. Hanes (trans.) (Washington: American Institute of Biological Sciences, 1965), 100.
10. Glebov, *op. cit.* (note 6), 6.
11. *Ibid.*, 16. On location, arrangement, and staff of the Paris hospitals, see A. Muehry, *Observations on the Comparative State of Medicine in France, England, and Germany during a Journey into these Countries in the Year of*

1835, E.G. Davis (trans.) (Philadelphia: Waldie, 1838), 17–23; and E. Ackerknecht, *Medicine at the Paris Hospital* (Baltimore: Johns Hopkins University Press, 1967), 15–25: 20–1.

12. The Mikhailovskii hospital was built on the funds left by I.V. Vil'e, President of the Academy in 1808–38. The rest of his considerable fortune Vil'e willed to Nicholas I, see N.I. Ivanovskii (ed.), *Istoriia Imperatorskoi voenno-meditsinskoi (byvshei Medico-khirurgicheskoi) akademii za 100 let. 1798-1898* (St Petersburg: Tipografiia Ministerstva vnutrennikh del, 1898), 524–45. A 'special income' of the Medico-Surgical Academy constituted interest from private endowments, see Voennoe ministerstvo, *Smeta na 1866 god po Imperatorskoi S.-Peterburgskoi meditsinsko-khirurgicheskoi akademii* (St Petersburg: Voennoe Ministerstvo, 1867), 30.
13. Glebov, *op. cit.* (note 6), 6–7.
14. *Rossiiskii gosudarstvennyi voenno-istoricheskii arkhiv*, Moscow, fond 1, opis' 1, 1860, no 24894.
15. 'Proceedings of the Conference of the Academy, 26 March 1860', in K.S. Koshtoyants, *et al.* (eds), *Nauchnoe nasledstvo: I.M. Sechenov: neopublikovannye raboty, perepiska i dokumenty* (Moscow: Akademiia Nauk, 1956), 25.
16. *Rossiiskii gosudarstvennyi voenno-istoricheskii arkhiv,* Moscow, fond 1, opis' 1, 1859, no 24527.
17. *Ibid.,* fond 879, opis' 1-4, 1861, no 1062; Glebov, *op. cit.* (note 6), 33; Ivanovskii, *op. cit.* (note 11), 540.
18. A.P. Borodin, 'Speech Delivered on February 9, 1880 at Zinin's funeral', cited in N.A. Figurovskii and I.I. Solov'ev, *Borodin: A Chemist's Biography* (Berlin: Springer Verlag, 1988), 150.
19. *Rossiiskii gosudarstvennyi voenno-istoricheskii arkhiv,* Moscow, fond 316, opis' 60, 1864, no 271; Voennoe ministerstvo, *op. cit.* (note 12), 24–5.
20. V.V. Pashutin, *Kratkii ocherk Imperatorskoi voenno-meditsinskoi akademii za 100 let eia syshchestvovaniia* (St Petersburg: Izd-vo Ricker, 1898), 7, 19.

8

The 'Medico-Chemical Academy': Zinin's Laboratory

Even early in his career, the young Zinin was a fundamental influence on the development of the nascent chemistry community in Russia, not least through the first Russian research school of organic chemistry. The first decade of his tenure at Kazan University, to around 1847, he devoted to research on aromatic compounds and to implementing the Liebigian model of training chemists, the achievements for which he became famous. From 1847, his career was connected to the Medico-Surgical Academy where he, together with Dubovitskii and Glebov, launched the reforms that drastically changed medical teaching and research. Zinin successfully applied his experience and influence to the re-organsation of the natural science departments and the arrangement of chemical and physical laboratories. There is no better way to understand the development of natural sciences at the Medico-Surgical Academy than to examine Zinin's investigative and teaching activities.

Zinin was born into the family of an officer in diplomatic service in Shushsa, the Caucasus, in 1812. Both parents died when he was still young, and he was sent to his uncle in Saratov, in the Volga region. He entered the local gymnasium where, according to the Gymnasium statute of 1804, he studied logic, psychology, ethics, aesthetics, law, political economy, ancient and modern languages, physics, and mathematics. He showed exceptional excellence in mathematics and Latin, and wanted to study further at the St Petersburg School of Engineering and Communication, a prestigious institution founded in 1810 by de Betancourt as competition to the Paris *École des Ponts et Chaussées.* But the sudden death of his uncle left Zinin without fortune, and he had to enter the mathematical department in Kazan, the closest university to Saratov.

In 1833, Zinin graduated from university. His thesis, on the perturbation of the elliptic movement of the planets, supervised by Nikolai I. Lobachevskii (1792–1856), was awarded a gold medal. The same, year Zinin took the position of assistant at the chair of physics and, in 1835, passed the masters exam in mathematics. Although Zinin was keen on astronomy and physics, the faculty council decided he should specialise in chemistry, in view of the lack of chemistry instructors at the University.[1] In

Image 8.1

Nikolai N. Zinin.
Courtesy of the Karl Sudhoff Institute for the History of Medicine and Natural Science, Leipzig University.

1836, Zinin passed the masters exam in chemistry. His thesis, on the phenomena of chemical affinity and Berzelius's theory of constant proportions versus the chemical statics of Berthollet, boldly criticised both theories for the lack of harmony and explanatory power as compared to physical theories.[2] In 1837, Zinin was appointed adjunct professor to the chemistry chair at which point the Trustee of the Kazan Educational District, Count M.N. Musin-Pushkin, who valued Zinin, interceded with the Ministry of Education, and Zinin was sent to Germany to prepare for an academic career.

In Berlin, Zinin became a close associate of a group of Russian medical students, and with them began to attend the lectures of Johannes Müller and visit the university clinics. He attended nearly all the science courses Berlin University offered – mathematics with Peter Gustav Dirichlet (1805–59), physics with Georg Simon Ohm (1789–1854), and crystallography and minerology with Gustav Rose, who had visited Kazan with the Alexander Humboldt expedition to Siberia and the Urals in 1829. He took the courses of inorganic and analytical chemistry with Heinrich Rose, and a private course, the so-called 'privatissime', in organic chemistry with Eilhard Mitscherlich. In his report to Kazan University, Zinin pointed out that laboratory studies in Berlin were organised in a rather primitive way, and that chemistry was taught at an elementary level, primarily for pharmaceutical students.[3]

His cherished wish was to get the permission of the Trustee to move elsewhere and become acquainted with other universities. His plan was first to visit Liebig's laboratory in Giessen, and then Göttingen to hear the lectures of the 'prince of mathematicians', Carl Friedrich Gauss (1777–1855) and to see his famous observatory, natural for a former student of Lobachevskii, who by administrative decision had to turn his interests to chemistry. But once in Liebig's laboratory, Zinin decided to stay. Liebig had a fine laboratory and a growing group of trainees, the Praktikanten, from Germany, France, and Britain, and Zinin would stay a year. In 1835, when the University of Giessen finally began to finance and expand its laboratory, the number of students increased. Some of Liebig's Praktikanten were advanced students or collegial guests, so the Giessen institute began to operate as a factory for training chemists. By 1838, when Zinin first came, Liebig had begun to use all those hands in a concerted and organised fashion to further his own research agenda. His laboratory was renovated and expanded and, by 1841, had fifty students working within its walls. At first, Liebig had attracted students by promoting his institute as a training school for pharmacists, but with time he tended more and more to tout practical instruction in chemisty as a necessary component of education in a wide variety of fields.[4] Liebig's innovations were adopted elsewhere in Germany, such as in the Marburg laboratory of Robert Bunsen. The new style of group laboratory teaching was also eventually adopted outside the chemical community, both in other fields, such as physiology, and in other countries, France and Russia in particular. Zinin would introduce the new style of laboratory teaching in Russia first in Kazan, and then in the St Petersburg Medico-Surgical Academy – where he powerfully promoted this model not only for his primary discipline of chemistry but also for physiology and clinical medicine.

Zinin's 1838 report to the Trustee tells much about the rich scientific life in Liebig's laboratory. Zinin liked Liebig for his willingness to share his experience and give valuable advice, and for his kind disposition to those working at his laboratory, who were frequently welcomed to Liebig's apartment, situated conveniently in the same building. Most of all, Zinin was fascinated by the training of chemists. He felt that the laboratory was unique in its own way – students from different countries were mastering the improved method of elementary analysis by burning organic substances using the Kaliapparat. Liebig's ingenious invention allowed the researcher to determine quickly how much carbon a given organic substance contained. Berzelius was reported to have spent a month analysing a single organic compound, whereas Liebig was proud to report that his laboratory could perform about twenty analyses a month. Moreover, the Kaliapparat method was superbly suited to Liebig's pedagogical strategy, ie. the instruction of groups of students in pharmaceutical and chemical practica, a strategy which developed into a characteristic German style of group research.[5]

There was another implication – analysis became the universal, omnipresent, omnipotent tool for research and teaching. It became the major procedure, taught at Liebig's laboratory immediately after the first skills were acquired in the purification of a substance. Analysis affected even the appearance of those who worked in Liebig's laboratory – there were no gas burners at the time, and chemists working in the laboratory used to wear felt or paper caps and gowns to protect themselves from the coal dust that was dispersed into the air by the bellows. The university wits called Liebig's students 'sinil'shchiki' [sinil'naia kislota, poisonous prussic or hydro-cyanic acid, HCN].[6] Liebig's method, although a delicate and labour intensive operation, appeared so simple that soon it became widespread in many laboratories both in Germany and France.[7]

By the time Zinin came to Giessen laboratory, Liebig had synthesised a number of organic compounds and successfully studied their properties. Now Liebig was working on benzaldehyde, an easily acquired organic substance obtained by heating the oil of bitter almonds. As early as 1832, Liebig, together with Wöhler, obtained a substance which he called benzoin, by the reaction of alkali on benzaldehyde, and found by analysis that benzoin is isomeric to benzaldehyde. Zinin, on the advice of Liebig, was working on the synthesis of benzoin, and first showed that its formation depends on the presence of hydrocyanic acid in the oil of bitter almonds, and that the catalyst for the reaction is potassium cyanide.[8]

Meanwhile, Zinin requested a one-year extention to his sojourn and, in the autumn of 1839, returned to Berlin to study technology, metallurgy, and mining according to the instructions of the Trustee. Apart from the courses in assigned disciplines, Zinin continued to attend Müller's lectures, and this time he made friends with Jakob Henle (1809–85)[9] and Theodor Schwann (1810–82), two of Müller's students and collaborators, who, from 1835, had been actively involved in microscopical studies. Fascinated by Schwann's research, Zinin asked him to give him additional lessons. They studied in Schwann's private lodgings, since there was no additional space in Müller's institute, overcrowded as it was with museum and laboratory materials. To justify his 'deviation from technology', Zinin reported to the Trustee in Kazan about the particular importance and novelty of microscopic studies.[10] Indeed, Schwann had just recently published his widely acclaimed *Mikroskopische Untersuchungen*, where he advanced the idea that the cell was the fundamental unit of plants and animals.[11] Schwann at that time was also actively engaged in the study of the chemistry of digestion, and had published his work on pepsin in *Müllers Archiv* and *Poggendorffs Annalen*. He showed that the membranes of the stomach produced the active digestive mixture with acid, and that the active substance present in gastric juice, named by him 'pepsin', was activated by hydrochloric acid.[12] About this time Liebig, too, began his investigations on the chemistry of nutrition, respiration, metabolism, and movement that would appear in his famous *Die organische Chemie in ihrer Anwendung auf Physiologie und Pathologie* in 1842.

Zinin's communications with Schwann and Liebig created his interest in the emerging discipline of physiological chemistry. Though he would never be directly engaged in problems concerned with the chemistry of living organisms, Zinin may have believed, as his mentor prophetically did, that chemistry and physiology would be no more separate than chemistry and physics were in his day.[13] Zinin would assert this idea at the Medico-Surgical Academy, repeatedly stressing the importance of the simultaneous development of chemistry and physiology. His broad interests in comparative anatomy, physiology, and histology proved essential in shaping his approach to science education for medical researchers and practitioners. Certainly his concern over deficiencies in medical education, especially in teaching natural sciences, became more acute while he was in communication with Liebig, Müller, and Schwann.

In April 1839, Zinin returned to Giessen and from there in the autumn, with Liebig's recommendation, went to Paris, where he attended the lectures of such luminaries of chemistry as Jean Baptiste André Dumas (1800–84) and Joseph Louis Gay-Lussac (1778–1850). From Paris, he proudly reported to the Trustee, that he had presented his research on the synthesis of benzoin carried out in Liebig's laboratory to the French Academy of Sciences.[14] Zinin spent the winter in Paris working in the laboratory of Jules Pelouze (1807–67), a close associate of Liebig. Pelouze consciously followed Liebig's lead – his 1838 professorship at the École Polytechnique enabled him to open a private teaching and research laboratory, adjacent to his official residence at the Mint. Many fine chemists would train or pursue original research there aside from Zinin, including Claude Bernard (1813–78), and Charles Frédéric Gerhardt (1816–56). In 1839, the capacity of Pelouze's laboratory based on the Giessen model was increased from six to twelve workers.[15] It was essential for Zinin to see the new model transplanted into a different tradition under the different conditions of the French educational system. Ten years later, Zinin, beginning his professorship at the Medico-Surgical Academy, would set up his own private laboratory for research and training for his students.

Returning to Russia in 1841, Zinin defended his doctoral thesis on benzoyl compounds and the new bodies of the benzoyl group at St Petersburg University.[16] His name was by no means unknown to the thesis committee, for he had already published his paper on the topic – important by any measure. The committee included all the notable St Petersburg chemists – Professor of the University and the Artillery School, Germain Henri Hess (1802–50), who by that time had published his research in organic chemistry, and a series of important thermochemical investigations, and Carl Julius Frietzsche (1808–71) of the Academy of Sciences, who had worked with Eilhard Mitscherlich (1796–1863) in Berlin in the 1830s, and who had produced aniline by distilling indigo, the expensive natural dye, with caustic potash. Also present was Alexander A. Voskresenskii (1809–80), Professor of the Main Pedagogical Institute, the first Russian student at Giessen, who had published on some organic compounds (theobromine, quinine) which he first obtained in Liebig's laboratory.

Soon after, Zinin attempted to get a position in chemistry at Kharkov University. Why he favoured Kharkov over Kazan, where he was expected to continue his academic career, is unclear. The reaction from the Trustee, Musin-Pushkin, was immediate – he apparently wanted the promising young scientist back to Kazan University. In his note to the Ministry of Education, Musin-Pushkin emphatically concluded that it was to Kazan University that Zinin owed his education, and it was there that Zinin would gain the fame his talents deserved.[17] The shrewd Trustee was right: it was in Kazan that Zinin performed his famous syntheses and lay grounds for the first research school of the new type, the pride of Russian organic chemistry.

In 1842, Zinin returned to Kazan to become an ordinary professor. The University possessed chemical and physical laboratories, housed in the handsome two-storey building designed and built by Lobachevskii in 1836. The chemical laboratory occupied six rooms on the ground floor and the basement. The interior was luxurious – carved cupboards, redwood furniture. Two fireplaces served for ventilation. There was a good collection of reagents and apparatus acquired during the first four years of the existence of the laboratory at a cost of 10,000 roubles. Zinin's senior colleague, Professor of Chemistry, Carl Ernst Claus (1796–1864), was an excellent experimenter of the old school, along with Carl Wilhelm Scheele (1742–86), and Johann Tobias Lovitz (1757–1804). [18]

Claus, who would go on to isolate platinum in 1844 exerted considerable influence on Zinin. Claus and Zinin spent most of their time in the laboratory, as their research required, and Zinin organised practicum for interested students following Liebig's method, mixing instruction and research in a way that encouraged students to start their own projects. A friendly working atmosphere, the availability for those who were interested and the order of laboratory life resembled that of the laboratory at Giessen. It was an exciting novelty, and Alexander Butlerov, the future notable organic chemist, recalled how powerfully he was impressed by the continuous and fascinating experiments performed by Claus and Zinin at the laboratory.[19]

In Kazan, Zinin became increasingly interested in organic compounds. His work on benzaldehyde and benzoin could not continue because the import of the poisonous oil of bitter almonds was prohibited by the Russian customs regulations. So he started research on naphthalene, a fundamental radical, which he had earlier examined while working in Liebig's laboratory, and which was available from the laboratory collection of reagents. By reducing nitronaphthalene with ammonium sulphide, he made alpha-naphthylamine, which he called naphthalidam.[20] The next synthesis was with nitrobenzene. The reduction of nitrobenzene with ammonium sulphide yielded a substance, which Zinin called benzidam.[21] Benzidam, as Zinin realised, was identical to aniline, obtained by Fritzsche's 1841 method of treating indigo with caustic potash. Zinin's reaction for the conversion of the aromatic nitro compounds into amines (aniline, naphthylamine, etc.) in 1842, gained him esteem among European organic chemists. August Wilhelm Hofmann (1818–92), whom Zinin had met in Liebig's laboratory, and for whom aniline had become the primary substance of his research career, soon began to apply Zinin's reaction of the conversion of nitrobenzole into aniline in his laboratory at the Royal College in London. Years later, in 1880, Hofmann commemorating Zinin at the *Deutsche Chemische Gesellschaft* said that if Zinin had done nothing else but this reaction, his name would still be written in golden letters in the annals of chemistry.[22] In Russia, Zinin's method for obtaining cheap aniline laid the foundation for the dye industry.

But in 1842, the work of the chemical laboratory and the rest of the University was interrupted by a disastrous fire. It destroyed half of Kazan, taking with it some of the fine University buildings, designed by Lobachevskii, including the anatomical theatre and the observatory, his pride and joy.[23] Fortunately, the instruments and the library were saved. Lobachevskii had the University buildings restored, and Zinin resumed his work at the laboratory. He was promoted to ordinary professor in 1845, only to have his research again interrupted, this time by drastic changes in the atmosphere at the University. Lobachevskii resigned, and the Trustee Musin-Pushkin left for a new position in St Petersburg. Despite his obvious attachment to Kazan University, Zinin decided to move to St Petersburg, and in 1847 applied for chemistry and physics chair at the St Petersburg Medico-Surgical Academy. The influential Musin-Pushkin interfered to make Zinin's transfer from Kazan to St Petersburg easy and to facilitate his appointment at the Medico-Surgical Academy.

Having assumed the position in 1848, Zinin implemented some important changes to the curriculum, offering courses of five lectures a week in inorganic and analytical chemistry. A course in organic chemistry with its application to physiology and pathology was shifted to the second year. Zinin took all the courses in chemistry. His lectures were accompanied by spectacular demonstrations, and he used tables and graphs to visualise the phenomena he discussed. He usually reported his investigations, and the research of his foreign colleagues in organic chemistry, explaining to students the mysteries of chemical transformations, new concepts and theories. He also discussed the place and relation of chemistry to other sciences, and stressed a particular usefulness of mathematical knowledge for a natural scientist. Here one can trace the university tradition: natural science courses were offered at the Faculty of Physical and Mathematical Sciences. At Kazan University natural sciences, especially botany and zoology, were treated with particular respect. Zinin believed, as did his mentor, Lobachevskii in Kazan, that mathematics should be used as a methodological tool in all the natural sciences.[24] Zinin's lectures, with a broad natural sciences context and stressing the importance of chemistry as one of the bases for modern medical education, was quite a novelty for the institution traditionally narrowed to the conservative reign of anatomy and surgery. Nothing could honour the Academy in the matter of chemistry instruction more than courses taught by one of the most distinguished scientists in the field.

The average size of the audience in Zinin's early years of instruction is not known, but figures from 1860 indicate a usual attendance of about three hundred students.[25] This is how Alexander Borodin, Zinin's student, pictured his teacher:

> Due to his vast intelligence and phenomenal memory, he was a living walking reference encyclopaedia on every branch of knowledge. They [the students] turned to him for information about new discoveries in chemistry, physics, technology, pharmacy, physiology, comparative anatomy, mineralogy, etc.; they sought his advice on various scientific questions, research topics, reference sources; they asked for practical help on how to use some new piece of equipment, and even on how to make an injection on some crayfish or turtle, etc.[26]

However, Zinin tended to offer much of his instruction in the laboratory rather than the lecture hall. From the very beginning of his career at the Medico-Surgical Academy he resolved to establish a modern facility for a school of chemistry. What he found there was frustrating in the extreme. The chemical laboratory had inadequate space, only some good apparatus, a collection of reagents, and a modest budget of thirty roubles per year. Zinin understood that the time had not yet come to appeal to the Academy's authorities for funding a research and teaching laboratory modelled on the Giessen facility. He also understood the severe limitations of his plan for proper funding – firstly, chemistry was not a medical discipline *per se*, and secondly, the Academy was a medical school with the a basic curriculum unchanged for centuries, so it was highly unprobable that he would be able to easily introduce chemical practicum to it. Finally, laboratory training for medical students, even in German universities in the late 1840s to early 1850s, was still in the early stages of development.

Under these conditions, Zinin had only one way out – he arranged a private laboratory in his apartment on Shpalernaia Street. Borodin reported that Zinin performed his famous syntheses of 1853–5 in his private laboratory. It was refined and elegant work, though conditions of his home laboratory were not suitable for experiments of such class. Borodin, a frequent visitor to Zinin's private lodgings pictured Zinin's working place as an 'organised chaos' – a lot of tables of different sizes with all kinds of apparatus, bottles, and test-tubes labelled with the description of the course of the experiment.[27] Here, in his private laboratory, Zinin taught a group of selected students the technique of organic analysis, discussed with them the new theory of August Laurent (1808–53) and Charles Gerhardt (1816–56), and treated his guests with dinner. The most distinguished among Zinin's student was N.N. Beketov (1827–1911), future professor of Kharkov University and notable physical chemist.[28] Later, Zinin acquired two other capable students, Borodin and Khlebnikov, both future professors at the Academy.

Zinin's research productivity slowed down significantly during the first years in St Petersburg. Teaching and running both his private and the Medico-Surgical Academy's laboratories was time-consuming, and with his new appointment as Secretary of the Academy, in 1852, his responsibilities were even more increased. As a result, his first research of his St Petersburg period appeared only in 1852. The investigation concerned synthesis of amines – Zinin discovered azoxybenzene (azobenzide) and benzidine by reducing azobenzene dissolved in alcoholic ammonia by hydrogen sulphide.[29] In 1855, independently from Marcellin Berthelot (1827–1907), Zinin synthesised mustard oil (allyl thiocyanate).[30] In these works, Zinin referred to the organic bases as analogues to ammonia, indicating his acceptance of the theoretical postulates of Laurent and Gerhardt, which were strongly opposed by the most influential chemists on both sides of the Rhine – Justus von Liebig and Jean Baptiste Dumas.

During the first five years at the Academy, Zinin acquired the reputation of being a leader, who, as the Academy wits put it, transformed the Medico-Surgical Academy into the Medico-Chemical Academy. That the Vice-President of the Academy, Glebov, a thoroughly medical man became interested in the unitary system of types advanced by Gerhardt and Laurent in the 1850s, is particularly telling. Borodin later related, in his speech at the jubilee of I.T. Glebov in 1880, that Glebov asked him, at that time Zinin's student and assistant, to explain Gerhardt's new theory of types. After several evenings of studies, Glebov argued against the main point of Gerhardt's theory, which stated that every reaction is double decomposition. Borodin, an admirer of Gerhardt, defended the theory with youthful ardour before Glebov. In his speech at the jubilee, Borodine concluded with his usual wit:

> Twenty years had passed since Professor Glebov happened to be my student, and now we know, that not every reaction is double decomposition, there are also reactions of addition and subtraction of elements. That is how the logical thinking and intuition of the mature student surpassed the professional approach of the young teacher.[31]

With diverse interests and skills, Zinin was engaged in the work of various councils and societies, such as the Manufacture Council at the Ministry of Finance, and the Society of Russian Physicians, to name just two. On one occasion, in 1854, during the Crimean War, Zinin, together with the lieutenant V. Petrushevskii, instructor of chemistry at the Pavlovskii Cadet Corpus, undertook an investigation sponsored by the War Ministry into the powerful explosive, nitroglycerine discovered by Ascanio Sobrero (1812–88), in 1846. Zinin and Petrushevskii were working in the old forge near Zinin's country house at Petergof. It so happened that Zinin's neighbour at Petergof was the Russian industrialist Emmanuel Nobel (1801–72), inventor of submarine mines and owner of the St Petersburg mechanical works. In summer 1854, Nobel's son, Alfred (1833–96), was visiting his father in Petergof. Alfred had just arrived from Paris, where he had studied in the laboratory of Pelouze, and where he may have heard of Zinin and another student of Pelouze, Sobrero. Alfred became interested in Zinin's experiments with nitroglycerine, and they spent much time discussing the matter. Zinin's experiments decisively proved that pure nitroglycerine could not be used safely in grenades. The War Ministry took into consideration Zinin's results and production of nitroglycerine as an explosive at the St Petersburg munitions was discontinued. But Alfred Nobel, the future founder of the Nobel Prize, would go on to found the industrial production of nitroglycerine in Sweden and elsewhere.[32]

Zinin's extra-Academy activities, however, did not push back his major plan to create modern facilities for teaching and research in experimental sciences, chemistry in particular. His plan began to materialise slowly after the Crimean War, in 1858. Together with Dubovitskii and Glebov, Zinin recommended a general re-organisation of the Academy to the military medical authorities, and a reconstruction of the major buildings which housed the Academy's clinics. Most of all, however, Zinin was concerned to provide additional support for improving teaching natural sciences. He insisted on the creation of separate chairs of chemistry, physics, comparative anatomy, and botany, and proposed the renovation and completetion of the natural science collections and libraries.[33] Zinin was, of course, painfully aware of the deficiencies of science instruction for future medical practitioners, and of the place traditionally ascribed to natural science courses in the medical curriculum:

> Medicine, like science, represents only an application of natural science to the problem of the preservation and restoration of health. Natural science, therefore, must play a most important role in medical education, by no means must it be a supplementary and an auxiliary aid. The physician needs to master the general system of science, the method of thinking, and devices and methods of investigation rather than many fragmentary facts of applied natural sciences. Therefore, teaching natural sciences in the Academy must be... necessarily full, not restricted to the cramped limits of applied knowledge. And for sound learning and a true evaluation of what has been done in science by others, for a clear knowledge of which path science grows by and [how] scientific data are cultivated and increased, it is necessary for each student to work directly and independently at something, according to his speciality, in some branch of natural science.[34]

For Zinin, as for the advocates of scientific medicine, and for the physiologist–physicists, '[o]nly physics and chemistry hold the key to the explanation of all those complex and infinitely different physiological and pathological processes, which occur in the organism.'[35] Crucial to Zinin's programme of improving science education was the simultaneous development of the experimental disciplines. He clearly saw great advantages in incorporating laboratory studies into the medical curriculum, and he exerted decisive support in the creation of two new laboratories, the physiological of Sechenov and the clinical research of Botkin.

Shaping a programme for chemical instruction at the Academy, Zinin had to adjust it to the needs of a medical institution. He understood that to ensure its relevance to medical practice, laboratory training in chemical diagnostic procedures and in quantitative chemistry should be available for the students. The resources and facilities needed for practical training in chemistry were far beyond what he could afford at the Academy's old laboratory. This is how Borodin portrayed the old laboratory of 1862:

> [C]onditions in the Department of Chemistry were at that time in a very sad state. Thirty roubles a year were appropriated to chemistry.... We had to make all necessary parts such as rubber connections, etc. ourselves, because sometimes it was impossible to find even a test tube in St Petersburg.... The Academy's laboratory had two dirty, gloomy rooms with arches, stone floors, several tables, and empty cupboards. Because of the lack of hoods, the distillations, evaporations, etc. had to be performed outside, even in winter. Organised practical training did not exist. But even in such conditions, Zinin always found a drive for work. Five or six fellows were always working in the laboratory, partly on their own and sometimes under Zinin's supervision.[36]

Zinin was an advocate of pure chemistry, and his ambitious plan to create a separate institute for research, as well as teaching, was directed toward the advancement of the science of chemistry. He was well aware that Liebig's model had caught on in Germany – it had been adopted for instance by Wöhler in Göttingen, and Bunsen in Marburg and later in Heidelberg. Obviously, Zinin favoured the model of chemical practicum, which exploited a set of apparatus and techniques, including Liebig's famous Kaliapparat technique, for teaching organic analysis. Zinin's position as the key figure of the 'triumvirate' at the Medico-Surgical Academy gave him a powerful voice. He suggested to the military medical authorities that, for a separate building for the Natural History and Sciences Institute and the equipment of the chemical and physical laboratories, he needed the capital investment of 45,000 roubles. To these capital expenses he added an annual outlay of 2,000 roubles for apparatus and reagents as well as an annual materials budget for glass- and porcelain-ware, metal parts, and fuel.

In 1863, just such a new institute was opened – Zinin built a proper research facility, equipped with gas burners, an innovation that began to be adopted throughout Europe in the 1850s, and all necessary appliances, including furnaces, wooden furniture and cabinets, and instruments stonework and metalwork.

Doubtless, some members of the Conference of the Academy were envious of the successful resolution of Zinin's plan to set up a separate spacious and well-equipped research facility for chemistry. Sceptical about the propriety of a laboratory for pure chemical research in a school of medicine, they were critical of all these unnecessary novelties and expenses. Zinin was shielded from these attacks by the official power of his position as a Secretary of the Academy and by the trust and support of Dubovitskii and the War Minister. Zinin's experience echoes that of Adolphe Wurtz (1817–84), one of the most distinguished organic chemists. As early as 1853, Wurtz built his private laboratory at the Faculté de Médecine where he held the professorial position. Having followed the Liebigian model in its private phase, Wurtz sought the second step – the state support of his laboratory. Wurtz's procedure was indeed quite contrary to regulations; some medical professors of the Faculty expressed doubts on the necessity of having any such laboratory at the medical faculty, but the dean and higher authorities of the medical faculty closed their eyes to it and did not interfere.[37] In the late 1850s, Wurtz's laboratory had become one of the most attractive places for foreign independent researchers, particularly from Russia. The laboratory was brought under official aegis only by the end of the 1860s, and then only after numerous appeals and petitions.

Zinin, however, had an excellent precedent to follow in St Petersburg. The Artillery Academy had already completed construction of the chemical laboratory building, lavishly trimmed and equipped, with an excellent supply of reagents. The total cost of the laboratory amounted to 100,000 roubles, with another 18,000 roubles spent on the construction of the gas plant, which supplied gas to both Academies, Artillery and Medico-Surgical. The laboratory at the Artillery Academy was built for and due to the endeavour of Leonid N. Shishkov (1830–1909). Shishkov's research on fulminate of mercury[38] and his collaborative work with Bunsen on the burning of gunpowders brought the Russian chemist the esteem of French chemists, Antoine Ballard (1802–76) and Jean Baptiste Dumas, both authorities on the chemistry of explosives. Even more importantly, Shishkov's research of 1855 drew the attention of the high military authorities, which resulted in the immediate and lavish funding for the construction and equipment of the new chemistry facilities at the Artillery Academy. Butlerov, who had visited nearly all leading French and German chemical laboratories, reported, in 1861, the best chemical laboratory he had ever seen was the laboratory at the St Petersburg Artillery Academy.[39]

In 1864, Zinin left the Medico-Surgical Academy for the St Petersburg Academy of Sciences, but his ties with the Medico-Surgical Academy remained close and he continued to serve for another ten years as Director of Chemical Works, the position, created specially for him, and preserved his membership at the Conference participating in all its important decisions.[40] Zinin made sure that Borodin, his best student and collaborator, was his successor and continued the task to which Zinin had devoted seventeen of his most productive years of research and teaching.

Notes

1. On chemistry at Kazan University, see S.N. Vinogradov, 'Chemistry at Kazan University in the Nineteenth Century: A Case of Intellectual Lineage,' *Isis,* 56 (1965), 168–73.
2. The text of the dissertation has not survived, although the conclusions are available in the 'Lichnoe delo Zinina' in the *Rossiiskii gosudarstvennyi voenno-istoricheskii arkhiv,* Moscow.
3. N.N. Zinin, 'Report to the Trustee of Kazan Educational District, 1837', cited in V.R. Polishchuk, *Chuvstvo veshchestva* (Moscow: Znanie, 1981), 32.
4. Allan Rocke offers a fine picture of Liebig's laboratory and demonstrates what events and trends contributed to Liebig's rise and his international reputation, see A.J. Rocke, 'Celebrity Culture in Parisian Chemistry', *Bulletin for the History of Chemistry,* xxvi, 2 (2001), 81–91: 82.
5. A.J. Rocke, *Nationalizing Science: Adolphe Wurtz and the Battle for French Chemistry* (Cambridge: MIT Press, 2001), 65, 85.
6. Polishchuk, *op. cit.* (note 3), 44. On the famous lithograph by W. Trautschold and H. von Ringen, reproduced in W.H. Brock, *Justus von Liebig: The Chemical Gatekeeper* (Cambridge: Cambridge University Press, 1997), 55, one can see students wearing top hats. The suggestion is that such costume made the practice of chemistry a dignified and respectable profession, see T.H. Levere, *Transforming Matter* (Baltimore: Johns Hopkins University Press, 2001), 130.
7. On the new challenges for organic analysis in the 1820s and the introduction of the Kaliapparat, see Rocke, *op. cit.* (note 5), 30–41; on the art of analysis in international perspective see Rocke, *op. cit.* (note 5), 64–71; and A.J. Rocke, 'Organic Analysis in Comparative Perspective: Liebig, Dumas, and Berzelius, 1811-1837', in F.L. Holmes and T.H. Levere (eds), *Instruments and Experimentation in the History of Chemistry* (Cambridge: MIT Press, 2000), 273–310; on the role of Liebig's students in the immediate aftermath of the invention of Kaliapparat, see F.L. Holmes, 'The Complementarity of Teaching and Research in Liebig's Laboratory', *Osiris,* ii, 5 (1989) 121–64: 139–42.
8. N. Zinin, *Annalen der Chemie und Pharmazie,* xxxi, (1839), 329; *idem, Annalen der Chemie und Pharmazie,* xxxiv, (1840), 186.
9. Henle's excellence in miscroscopic anatomy would play a crucial role in his future ingenious physiologico–clinical observations. He continued as anatomist at Zürich University and published his *Allgemeine Anatomie,* in which he successfully related histology to the cell theory. Later he was called to Heidelberg, where he began to edit the *Zeitschrift für rationelle Medizin,* and to work on his textbook *Handbuch der rationellen Pathologie.* In this book, he attempted a systematic approach to physiopathology, based not on

descriptive but scientific and empirical treatment of problems, emphasising the principle that pathology is the physiology of the diseased individual. Like his close friend from his time in Zürich, Carl Ludwig, Henle promoted the idea of the close relationship between experimental physiology and clinical medicine. See K.E. Rothschuh, *History of Physiology*, G.B. Risse (trans. and ed.), (Huntington: R.E. Krieger Publishing Co., 1971), 244–5.

10. Zinin's Report to the Trustee, 1838, cited in Polishchuk, *op. cit.* (note 3), 55.
11. T. Schwann, *Mikroskopische Untersuchungen über die Übereinstimmung in der Struktur und dem Wachstume der Thiere und Pflanzen* (Leipzig: W. Engelmann, 1839).
12. T. Schwann, 'Über das Wesen des Verdauungsprozesses', *Annalen der Physik und Chemie*, xxxviii (1836), 358–64.
13. Cited in P.F. Cranefield (ed.), S. Lichtner-Ayèd (trans.), *Two Great Scientists of the Nineteenth Century: Correspondence of E. du Bois-Reymond and C. Ludwig* (Baltimore: Johns Hopkins University Press, 1982), 124, n.3.
14. Zinin's Report to the Trustee, 1839, cited Polishchuk, *op. cit.* (note 3), 57.
15. On the Pelouze laboratory, see Rocke, *op. cit.* (note 5), 150; see also *idem*, *op. cit.* (note 4), 83.
16. N.N. Zinin, 'O soedineniiakh benzoila i ob otkrytykh novykh telakh, otnosiashchikhsia k benzoilovomu riadu', (St Petersburg: Tipografiia Akademii Nauk, 1841).
17. Cited in Polishchuk, *op. cit.* (note 3), 60–1.
18. On Claus's laboratory in Kazan University, see Polishchuk, *op. cit.* (note 3), 62–6; 77–80. On Claus's biography, see N.N. Ushakov, *K.K. Klaus* (Moscow: Nauka, 1972), see also J.R Poggendorff (ed.), *Biographisch-literarische Handwörterbuch zur Geschichte der exacten Wissenschaften* (Leipzig: Barth, 1863), Vol. I, 425; Vol. III, 278. In 1840, Claus began his famous research on the native Russian platinum, which had been mined in the Urals since 1819. The platinum fields in the Urals were so rich that platinum–iridium alloys had been used for coinage for the eleven years since 1828. The Mining Department had long ago encouraged investigation of the Russian native platinum, sending large portions to Humphry Davy, Alexander Humboldt, and Jöns Berzelius. Claus was happy to obtain from the St Petersburg Mint, a small sample of platinum residue for his investigations and, by 1842, he found in it 'a new body'. It took him another two years to work up, in total, 15lb of platinum residue to isolate the new metal, which he named ruthenium (from Latin, Ruthenia – Russia). Russian native platinum is insoluble in aqua regia (a mixture of nitric and hydrochloric acids, which dissolves gold and platinum).
19. The founder of the Kazan school of chemistry, Butlerov, acquired his interest in chemistry from his teachers, Claus and Zinin. In 1857, Butlerov became professor, and, in 1861, Rector of Kazan University.

20. N. Zinin, 'Organische Salzbasen aus Nitronaphthalos und Nitrobenzid mittelst Schwefelwasserstoff entstehend,' *Annalen der Chemie und Pharmacie,* xxxxiv (1842), 283–7.
21. *Bulletin de la classe physico-mathématique de l'Académie Impériale des Sciences de St.-Pétersbourg,* x (1842), 272–85 (no. 18), *Zeitschrift für praktische Chemie,* xxvii (1842), 140–9.
22. A. Hofmann, 'N. Zinin', Nekrolog, *Berichte der deutschen chemischen Gesellschaft,* (1880), xiii, 449–50: 449; see also V.V. Markovnikov, 'Moskovskaia rech' o Butlerove. 1886', in A.M. Butlerov, *Po materialam sovremennikov* (Moscow: Nauka, 1978), 48–66: 64.
23. A. Vucinich, 'N.I. Lobachevskii', *Isis,* 53 (1962), 465–81: 480.
24. *Ibid.*, 477.
25. N.I. Ivanovskii (ed.), *Istoriia Imperatorskoi voenno-meditsinskoi (byvshei Medico-khirurgicheskoi) akademii za 100 let. 1798–1898* (St Petersburg: Tipografiia Ministerstva Vnutrennikh Del, 1898), 97–8.
26. Cited in B.N. Menshutkin, *Zinin, ego zhizn' i nauchnaia deiatel'nost'* (Berlin, 1821), 86–9; and in N.A. Figurovskii and I.I. Solov'ev, *Aleksandr Porfir'evich Borodin: A Chemist's Biography,* C. Steinberg and G.B. Kauffman (trans.), (Berlin: Springer Verlag, 1988), 19.
27. A.P.Borodin and A.M. Butlerov, 'N.N. Zinin. Vospominaniia o nem i biograficheskii ocherk', *Zhurnal fiziko-khimicheskogo obshchestva,* xii (5), (1880), 215–64: 220.
28. G.V. Bykov, 'N.N. Beketov', in C.C. Gillispie (ed.), *Dictionary of Scientific Biography* (New York: Charles Scribner, 1970–present), Vol. 1, 579.
29. N. Zinin, 'Über die Einwirkung des ätherischen Senföhls auf die organischen Basen', *Annalen der Chemie und Pharmazie,* 84 (1852), 346–9.
30. *Bulletin de la Classe Physico-mathématique de l'Académie Impériale des Sciences de St.-Pétersbourg,* xiii (1855), 288; *Annalen der Chemie und Pharmazie,* xcv (1855), 128.
31. A.P. Borodin, 'Rech' na iubilee I.T. Glebova, 1880', in S.A. Dianin (ed.), *Pis'ma A.P. Borodina, 1878–82* (Moscow: Gos. muz. izd-vo, 1849), Vol. III, 128–9.
32. On Zinin and Alfred Nobel, see R. Sohlman and H. Schück, *The Life of Alfred Nobel* (Stockholm: Nobelstiftelsen, 1929), 95. Even serious disasters at his plants, which took many lives, did not stop Alfred Nobel in his search for the safe use of nitroglycerine as an explosive. In 1867, he took out a patent for its use when absorbed in kieselguhr, known as dynamite. On the discovery of nitroglycerine and on the first dynamite, see O. Guttmann, 'Die Industrie der Explosivstoffe', in P.A. Bolley *et al.* (eds), *Handbuch der chemischen Technologie* (Brunswick: Vieweg, 1895), 6 vols, Vol. VI, 28–34.
33. *Rossiiskii gosudarstennyi voenno-istoricheskii arkhiv,* Moscow, fond 316, 1863, opis' 46, delo 373.

34. Cited in Ivanovskii, *op. cit.* (note 25), 529.
35. Cited in *ibid.,* 531.
36. Menshutkin, *op. cit.* (note 26), 58–9.
37. Rocke, *op. cit.* (note 5), 184.
38. L. Schischkoff, 'Über die Zusammensetzung des Kuallquecksilbers, so wie einige Zersetzungsprodukte desselben', *Bulletin de la classe physico-mathématique de l'Académie Impériale des Sciences de St.-Pétersbourg,* xix, 17 (1855), 97–112. For acknowledgment of Shishkov's research by the General Staff of the Russian army see, *Zhurnal Ministerstva narodnogo prosveshcheniia,* vii, 6 (1857), 106–7. On Shishkov's participation in the foundation of 'la Société Chimique' in 1857 in Paris, which was soon to become France's national chemical society, see Rocke, *op. cit.* (note 5), 216–8.
39. Cited in N.A. Figurovskii, *Istoriia estestvoznaniia v Rossii* (Moscow: Akademiia Nauk SSSR, 1962), Vol. II, 78
40. Ivanovskii, *op. cit.* (note 25), 90.

9

Synthesis and Symphonies: Borodin's Laboratory

One of the most interesting Russian scientists of the nineteenth century was Alexander Borodin (1833–87). Today, he is principally remembered as a brilliant composer. His epic opera *Prince Igor* adorns the repertoires of the best opera companies the world over. What is less known is that he was a capable chemist and played a central role in the chemistry community of St Petersburg, acquiring deserved respect as one of the best professors at the Medico-Surgical Academy, where he introduced large-scale laboratory instruction in chemistry. Borodin grew up in an age of reform that expressed its passion for national independence through science and music. Unlike the musicians of the group to which he belonged, he never abandoned science for his music. For Borodin, research and teaching innovative practices in the chemistry laboratory and composing new musical forms were two different keys of the same reforming and creative theme.

Borodin was an illegitimate son of Prince Luka Gedeianov and Avdotia Antonova, the young daughter of a soldier. The child was registered as the lawful son of one of the Prince's serfs, a customary procedure in such cases. Avdotia was passionately fond of Alexander, but she never officially recognised him as her son, and always referred to herself as his aunt. Before he died, Prince Gedeianov comfortably provided for Avdotia and granted freedom to their son. Avdotia gave Alexander an excellent home education, including both French and German. From an early age, Alexander developed a lively interest in chemistry – his mother's house smelled of sulphur and was full of various containers of liquids upon which Alexander performed his experiments. From the age of fourteen, music began to take as important a place in his life, as did his studies in science, and he tried his hand at composition. Together with his boyhood friend Mikhail R. Shchiglev (1834–1903), he took piano lessons from a second-rate German professor, and also learned the cello and flute. The boys played four-hand arrangements of the symphonies of Haydn and Beethoven, and enthusiastically attended the concerts given in Petergof and Pavlovsk.[1]

At seventeen, Alexander entered the Medico-Surgical Academy, a choice supported by Avdotia under the influence of her legitimate husband, who was an army doctor. At the Medico-Surgical Academy, in addition to his

Image 9.1

Alexander P. Borodin.
Courtesy of the Karl Sudhoff Institute for the History of Medicine and Natural Science, Leipzig University.

medical courses, Borodin began to specialise in chemistry. In this, Zinin was his most capable mentor. Zinin was also his guide in a variety of fields, such as botany, anatomy, and crystallography, which fascinated the young Borodin. Orphaned in early childhood, Zinin had particular feelings for Borodin and regarded him as his adopted son. Borodin was devoted to experimental work in the laboratory and soon became Zinin's teaching and research assistant. As far as Borodin's music was concerned, Zinin was aware that his assistant was passionately fond of composing music but he seemed not to approve of Borodin's musical activities:

> It would be better if you give less thought to writing songs. I have placed all my hopes in you to be my successor one day. You waste too much time thinking about music. A man cannot serve two masters.[2]

It was true: Alexander did spend all his free time either at concerts or playing chamber music with his acid-burned fingers.

On graduating from the Academy,[3] Borodin served for some time at the Second Land Forces Military Hospital, but he felt that he was not at all suited to medical practice. He later recalled with horror his short experience as an army doctor, when once he had fainted three times during the procedure of treating a soldier hardly alive after flogging. The only episode connected with his service at the hospital which he remembered with delight was one evening in 1856, when he, a second house-physician, first met Modest Mussorgskii (1839–81), the seventeen-year-old ensign of the élite Preobrazhenskii Guards, on duty. They soon met again at one of the parties where Mussorgskii played the piano with exquisite grace and expression.[4] At that time they pursued their respective careers – one medical, one military – but for both music would become something more than mere passion.

Zinin, however, thought Borodin had an 'unquestionable talent both in teaching and research', and his true vocation should be chemistry. In the autumn of 1859, following the decision of the Conference of the Medico-Surgical Academy, Borodin left for Europe to prepare for an academic career in chemistry. Most of that time Borodin spent working at the private laboratory of Emil Erlenmeyer in Heidelberg but he also visited the laboratories of the leading French chemists, Adolf Wurtz and Marcellin Berthelot, and attended the lectures of Claude Bernard and Louis Pasteur in Paris. At the Paris École de Médecine, he took lessons in glassblowing – a skill fundamental to a chemist, especially in Russia. Yet in all this chemistry Borodin did not forget his music and in Heidelberg, inspired by the young pianist Ekaterina Protopopova, his future wife, he composed several pieces.

Music did not interfere with his intensive studies, however. In Erlenmeyer's laboratory, Borodin pursued his research on benzidine derivatives and published seven research papers primarily in the *Zeitshrift für Chemie und Pharmazie.* We know that during the 1840s and 1850s, Zinin was intensively involved in research on aromatic compounds, and it is not surprising that Borodin followed the research path of his mentor. Borodin found a method for obtaining from an organic acid a halide containing one carbon atom less. He also investigated the structure of derivatives of benzidine and valeraldehyde, and prepared fluorobenzene.[5] Borodin regarded his Heidelberg studies as decisive in his development as a chemist and Zinin as his principal mentor who inspired and supported him as a scientist.[6]

Upon returning to St Petersburg in 1862, Borodin had bright prospects: he was appointed adjunct to the chair of chemistry. The Medico-Surgical Academy provided an apartment near the laboratory in the new building of the Natural History and Sciences Institute, where Borodin settled with his wife, Ekaterina. Zinin let it be known that he would soon retire from his professorship, and that Borodin would succeed him. The new institute was close to its completion and Borodin took upon himself a great deal of work in the arrangement and equipment of the new laboratory. He had purchased much of the laboratory equipment himself while abroad, especially the precision instruments 'which one cannot order but which must be selected and examined in person'.[7] Now, besides his teaching responsibilities, he was busy with all kinds of organisational and financial matters, 'writing, calculating', submitting all kinds of requests and reports.

The new laboratory did not avert Borodin from his other cherished interest, although he had no notion that he might become a serious musician. 'I am a Sunday composer who strives to remain obscure,' said Borodin to Milii Balakirev (1836–1910) whom he met in 1862 at a party given by Sergei Botkin, his colleague at the Medico-Surgical Academy. Balakirev at that time headed a circle of young composers, Modest Mussorgskii, Nikolai Rimskii-Korsakov (1844–1908), and César Cui (1835–1918), who believed they had a vocation to assert the originality of their national music, and to continue the tradition laid down by Mikhail Glinka (1804–57), Russia's great composer. Together with Borodin they became known as the 'Mighty Five' in the words of Vladimir Stassov (1824–1906), a writer on art and music and their most ardent champion, or 'Les Cinq Russes', as they were called in France.[8] From that time on Borodin began to attend Balakirev's musical gatherings, where he became acquainted with the pieces, composed by Mussorgskii and Rimskii-Korsakov. Their music impressed Borodin powerfully: 'I was struck by the brilliance, sensibility, and the energy of their playing, and by the great beauty of the work itself.'[9]

Young, but already an experienced musician, Balakirev recognised Borodin's genuine talent, this despite Borodin's early compositions belonging to the genre of chamber music which Balakirev so 'condemned'. The influence of Balakirev and his associates was not in any sudden, imaginary revolution but in their powerful assertion to Borodin that his 'true vocation is music'. Ekaterina Borodina wrote to her mother:

> His recent acquaintance with Balakirev has quickly born fruit; the effect is fabulous. I could not help but notice the change… this Westerner, this 'ardent Mendelssohnist', was playing me nearly the whole of the first Allegro of his E Flat Major Symphony.[10]

Borodin would dedicate his first symphony to Balakirev.

For Borodin, a love for chemistry and a love for music were interwoven into typical nationalist passions of the 1860s. As a musician he felt that Russia should create her own symphonic music. As a scientist he believed that Russia should develop her own science, for which she needed research institutions like those in Europe. For Borodin, his fellow chemists, who set up new laboratories, pursued independent research, and developed their discipline, and his fellow musicians, who strove to create new music and to found a new national musical school, were similarly fascinating. Comfortable in both environments, he felt it impossible to make a choice of preference one for the other.

In the autumn of 1863, the Natural History and Sciences Institute was opened with a spacious chemical laboratory. Zinin had constructed an expensive enterprise. Although the laboratory was funded generously, the additional expenses required for the organisation of practical studies for large groups of students were considerable. In the summer, Borodin had written to Butlerov in Kazan:

> The laboratory will be good, but what can we do if we do not increase the staff? Day after day Zinin pleads with all his might, but it seems to be to no avail. We have only three attendants.... The number of students will certainly reach fifty (this was the condition without which they would not agree to give us money for the laboratory). Although Zinin and I insist that one professor cannot both lecture and conduct a practical course, nothing comes out of it.... About 2,000 roubles a year is allowed for maintenance of the laboratory, excluding materials costs, such as acids, soda, potash, alcohol, etc. This allowance is not enough for the practical classes of a large number of students, providing that we are not allowed to deduct anything for the materials and broken dishes. The students are not able to pay for the laboratory.... One cannot impose on the students the obligation of any fee, for even without this, they pay fifty roubles a year to the Academy. All this greatly complicates future work in the laboratory.[11]

Despite frustrating financial limitations, with the new institute Zinin could finally implement his long-awaited plan to incorporate into the curriculum laboratory training in chemistry for a large number of students and to create research possibilities for advanced students and interested doctors. The construction of the Institute, as Zinin had planned, was predicated on a major shift in the orientation towards science-based medical instruction at the Academy. The Academy now possessed a standard chemical laboratory, with a separate budget, adequately equipped for the work of 104 students during their mandatory scheduled two hours. The laboratory was available all day for those interested students who could now assist their mentor and participate in his investigations. Besides, physicians could now get additional training to pursue the part of their research, which required chemical methods and techniques.

In 1864, Zinin resigned his professorship and Borodin filled the vacancy. Dubovitskii and Glebov seemed to have shared Zinin's opinion that Borodin was an ideal candidate for a position – he had spent so many years at the laboratory, first as a student and then as an assistant. He was a promising, able researcher and an experienced teacher. Besides which, he appeared a skilful and flexible laboratory leader.[12] A.P. Dobroslavin, Borodin's student and later close friend, in a report written for Vladimir Stassov, vividly portrays Borodin in the lecture hall and in the laboratory. His lecture courses in organic chemistry, in chemistry applied to physiology and pathology and in the history of chemical theories were no less popular than lectures by Zinin, who had been a skilful orator. The atmosphere in the laboratory was friendly and warm. Borodin frequently invited his co-workers to dinner in his private lodgings adjacent to the laboratory. He was liked for his humour and kindness. Only rarely would he become irritated if someone carelessly spoiled the instruments or used the 'clean' laboratory instead of a special 'black' room for large-scale chemical procedures. Anyone could come to him to discuss his ideas, research project, and the like.[13]

Rimskii-Korsakov, a frequent visitor in the mid-1860s, noted Borodin's equal devotion to the laboratory and to his music, synthesising new organic compounds and creating new musical forms:

> We discussed music a great deal; he played his projected works and showed me the sketches of the symphony. Borodin was an exceedingly cordial and cultured man, pleasant and oddly witty to talk with. On visiting him I often found him working in the laboratory, which adjoined his apartment. When he sat over his retorts filled with some colourless gas and distilled it by means of a tube from one vessel into another, I used to tell him that he was 'transfusing emptiness into vacancy'. Having finished his work, he would go with me to his apartment, where we began musical operations and conversations, in the midst of which he used to jump up, run back to the laboratory to see whether something had not burned out or boiled over; meanwhile he filled the corridor with incredible sequences of successions of ninths or sevenths. Then he would come back, and we proceeded with the music or interrupted conversation.[14]

This period was the most fruitful for Borodin's laboratory research. In 1864, he completed his work on the compounds of aldehydes, which led him to the discovery of aldehyde resins, aldol and some other compounds. In his 'Über die Einwirkung des Natrium auf Valeraldehyd' ['Investigations of the Action of Sodium on Valeraldehyde'], he presented a general method for the condensation of aldehydes in the presence of metallic sodium and potassium and caustic alkalis.[15] He continued his studies of the products of condensation of valeraldehyde, his most important research during the following years. He also carried out experiments of a physico-chemical nature, on the attraction of water by deliquescent substances, and on the evaporation of aqueous solutions. As Borodin wrote to the physical chemist P.P. Alekseev, '[t]his work has turned out to be frightfully difficult beyond all expectations, especially concerning temperature conditions. I do not yet know what will result, or what will come of it.'[16] Simultaneously, Borodin was revising the score of his First Symphony for public performance. It seems that Borodin treated his laboratory research and his work on musical composition as something quite similar: the same complexity, requirements of accuracy and precision, and finally the same kind of uncertainty regarding the end result – the course of reaction and substances yielded in the laboratory, alongside the first performance of the symphony and its acceptance by the public:

> It's a general rule that you never know where you are with a symphony until it is performed.... Besides the intentional harmonic peculiarities, I came across most extraordinary sequences of minor seconds, diminished fifths, clashes of major and minor, and the devil only knows what... I never realised correcting parts could be such a hellish business.[17]

This work provoked sighs of exasperation from the composer. A frequently quoted excerpt from one of Borodin's letters reveals that intensive work on composition was followed by periods when:

> Music sleeps; the altar of Apollo goes out; ... the muses weep, tears flow over the brim of the urns into a stream, the stream babbles and with sadness tells of my cooling to the art today.[18]

His chemistry schedule went on without interruption – lectures, practical courses, and his research on valeraldehydes. He regularly participated in official and unofficial activities of the Russian chemical community. In 1868, he took part in the foundation meeting of the Russian Chemical Society – it 'was a very nice and happy occasion for me', he wrote to Alekseev.[19] Zinin became the first President of the Society. Borodin recalled that, in the late 1850s, Zinin's laboratory at the Academy 'was like a miniature chemical club, and at its improvised sessions bubbled up the life of Russian chemistry.'[20] In the early 1860s, St Petersburg chemists used to gather in the private laboratory of A.N. Engelgardt (1828–93) and N.N. Sokolov (1826–77), who funded the publication of the first chemical periodical, *Zhurnal khimicheskogo i fizicheskogo obshchestva* in Russia and later donated their laboratory to St Petersburg University. For Borodin, the meetings of the Russian Chemical Society and its periodical became convenient outlets for communicating the results of the investigations performed in his laboratory. His close colleague of that time was Alexander Butlerov with whom he remained life-long friends; he was in touch with Dmitrii Mendeleev, his old friend from the Heidelberg circle; with Alexander M. Zaitsev in Kazan and Vladimir V. Markovnikov in Moscow, both notable disciples of Butlerov, and with Peter Alekseev in Kiev.

The success of the first public performance of his First (E Flat Major) Symphony at a Free Musical School concert at the beginning of 1869 was decisive for Borodin the composer. He almost immediately set to work on the Second (B Minor) Symphony, and soon thereafter began to entertain the idea of composing the opera *Prince Igor* on a theme from Russian history. Stassov, a connoisseur of ancient Russian chronicles, supplied Borodin with the necessary materials, including epic Russian songs and some authentic Polovtsian songs. In the abundance of her folk songs Russia had no rival, although this wealth lay untouched.[21] 'The Mighty Five', founders of the new Russian musical school and advocates of national trend in music, were regarded as revolutionaries in decisive opposition to the so-called Moscow group, headed by Nikolai Rubinstein (1835–81) and Peter Tchaikovsky (1840–93). In St Petersburg, Anton Rubinstein (1829–94), the founder of the Imperial Russian Musical Society, and Alexander Serov (1820–71), the Russian 'Wagnerian' composer and influential musical critic, as well as some of the musicians of both of the Conservatories in Moscow and St Petersburg, were hostile to the Balakirev group.[22] These established music circles did not acclaim Borodin's first symphony. In a letter to his wife, Borodin left us a striking verbal portrait of the audience and concerts of the two rival camps, the Imperial Russian Musical Society, patronised by Grand Duchess Elena Pavlovna, and the Free Musical School, organised by Balakirev and Stassov:

> [A]ll the fashionable people were there… the entire entourage of Elena Pavlovna.... Everything about the concert put one in mind of a salon: Italian coloratura singing... epaulettes, sabres, shockingly low-necked dresses, and so on and so forth.... Immediately afterwards at the very same hall at the rehearsal of a Free Musical School concert, everything was suddenly transformed as if by magic. The upper-class *beau monde* had vanished, and the air was filled with the tremendous sounds of Berlioz's *Lélio*.[23]

Despite being so powerfully attracted to the activities of the new musical school, Borodin nevertheless admitted that chemistry still held him very tightly. At the beginning of the new academic year of 1869, he was busy with the arrangement and equipment of additional laboratory premises.[24] He spent much time at the laboratory, trying to bring his research on the condensation of aldehydes to a conclusion, since he had become aware that his work was close to that of August Kekulé. Borodin was furious when he heard that Kekulé accused him of borrowing his idea:

> Kekulé is claiming that my work on valerian aldehyde (which I am doing now) has been borrowed from him (that is to say, not so far as the factual side is concerned, but the very idea itself). He has printed this accusation in the *Berichte* of the Berlin Chemical Society. This compelled me there and then to make a statement of the true facts of the case, and to prove that I have been working on this problem since 1865, whereas Kekulé entered the field only as late as August last year. So much for German honesty![25]

This episode drove him to step up his research as he mentioned to his wife, 'I am caught up in the laboratory work'.[26] Soon he communicated the results of his investigation on the condensation products of valeraldehyde at the meeting of the Russian Chemical Society.[27] While studying the condensation reaction of acetaldehydes, Borodin obtained a chemical compound known as aldol. From an article by Adolphe Wurtz in the *Comptes rendus* of 1872, he soon learned that the French chemists had obtained an identical compound, and therefore decided not to develop further research in this direction,[28] and only to publish a short article about his work on aldol in the *Berichte*.[29] M. Iu. Goldstein, Borodin's student and assistant, believed that Borodin simply could not pursue research of such class because of scarcity of means and lack of trained assistants, 'whereas Wurtz has enormous means and works with twenty hands, since he does not hesitate to load his assistants with dirty work.'[30]

Despite obvious disappointment, Borodin seems at that time to have enjoyed his ability to balance chemistry and music. These were days when he worked all day in the laboratory and was in 'a completely happy period of laboratory activity', and still found time to 'supply the musicians with the first number of *Prince Igor* where Yaroslavna's dream appears so charmingly.'[31] At one point, however, he lost interest in opera; he felt that he was more 'symphonist and lyricist by nature', whereas for opera, which appeared to him as 'enormous trouble and waste of time', he had not 'enough experience, ability, or time'.[32] Not surprisingly, Borodin's musical friends expressed annoyance at his meagre music output, in particular with reference to *Prince Igor*. This created a certain amount of ill-feeling:

> Our musicians never stop abusing me. They say I never do anything, and won't drop my idiotic activities, that is to say, my work at the laboratory, and

so forth. They really think that I cannot and should not have any activities except music.[33]

In 1874, Zinin retired from the position of Director of Chemical Works, and responsibility for practical courses passed to Borodin. Now, practical courses in analytical chemistry were given for three-to-four hundred students as compared to classes for just a hundred students a decade earlier. The organisation of large-scale laboratory training required a good deal of managerial skills. Borodin secured an additional space, necessary equipment and supplies, and adequate funding, and at the same time was responsible for all administrative and financial matters. As in any teaching laboratory, the assistants were supposed to give introductory lectures followed by demonstrative 'practical' instruction. They were also responsible for the supervision and guidance of the students in order to avoid wasting time, labour, materials, and vessels, and to eliminate the possibility of explosion or poisoning, from careless or ignorant use of dangerous chemicals used in even basic experimental work. The perennial problem of lack of laboratory assistants became especially acute with the large classes for practical studies. Borodin had to hire a second assistant and to supplement the insufficient salaries of the two attendants in the laboratory from his private account.

Students had only two compulsory hours for laboratory work. They needed a flexible schedule since it was impossible to perform some chemical operations within a two-hour period. Time should be considered for each individual case, for example, it was not possible to collect some precipitates when they were not completely washed or to calcine something before it was dry. The work of the laboratory was organised in such a way that it was available for the students and the graduates and physicians any time in the day and evening. Alexander Dianin, Borodin's assistant and collaborator recalled: 'People closely associated with Borodin at that time knew what amount of time, energy, trouble, and often, personal expense the introduction of large practical classes cost him.'[34] Despite obvious limitations, the work of the laboratory was well co-ordinated and organised, and Borodin managed to efficiently supervise several research works performed in his laboratory. In recognition of his service the Conference, in 1877, awarded Borodin with the title of Academician of the Medico-Surgical Academy.

Fellow chemists also appreciated Borodin's activity as a laboratory leader. In a letter to his wife, Borodin mentioned the cordial reception, respect and esteem from Zaitsev, Butlerov, and Mendeleev at the Conference of Naturalists and Physicians in 1873 in Kazan:

> In our chemical section there were many interesting reports, and among them, I say without boasting, mine were remarkable. The quality and number of them (seven works) greatly impressed all members of the section and advanced our laboratory in the opinion of the chemists and even non-chemists.[35]

These were the works on the structure of complex organic compounds performed by Borodin and his students: Dr Shalfeev (1845–1910), the future professor of physiological chemistry at Warsaw University, investigated derivatives of pelargonic acid; Dianin studied the effect of ferric salts on naphthols; and P.G. Golubev (1838–1914) worked on the reduction of nitrobenzoin by tin in hydrochloric acid. Borodin continued his major project on the composition of the products of the condensation of aldehydes.[36]

The mid-1870s were productive in both research and composition. After an interval of four years Borodin's thoughts again turned to his opera. In 1875, he wrote the sketches to his famous *Polovtsian Dances* from *Prince Igor*. The next year, *The Chorus of Praise* from the opera had great success at a concert of the Free Musical School. Borodin's letters reveal his peculiarly ambivalent attitude towards his musical achievements:

> It [the success] had a special significance as far as the fate of my opera is concerned. But I ought to say that I am still a composer in search of oblivion; and I always feel shy to tell that I am composing. For somebody it is so simple to admit that music is his vocation, an end in life; but for me it is a recreation, an idle pastime, which provides diversion from my real work, my work as a professor and scientist. It is for this reason that although I want my opera to be completed, I nevertheless have qualms of conscience that I am getting too involved in my opera and that may encroach on my other activities… but in any case now I have to finish my opera whether I want it or not.[37]

In a letter to his wife in 1876, Borodin admitted that he was in a unique situation for a professor of the Academy – he completed his Second (B Minor) Symphony, and now his two symphonies were to be performed at the Russian Musical Society concerts.[38] He published his research on nitrosoamarine and his clinically oriented investigation on conversion of nitrogen in the organism.[39] He continued to supervise research of his students, regularly reporting at the meeting of the Russian Chemical Society. His own research career in organic chemistry, however, ended with the last work on nitrosoamarine.

In subsequent years, most of his time, freed from official duties at the Medico-Surgical Academy, Borodin devoted to the women's courses for training midwives, which had been organised at the Academy in 1872. In the letters of this period, he frequently mentions that his involvement with these courses were time consuming, leaving no space either for research or music. Experimental work and composition required continuous and concentrated work in the laboratory or at the piano. He simply had not the time for either. Rimskii-Korsakov's portrayal of Borodin's life in the late 1870s conveys deep regret and disappointment. By then a professor at the St Petersburg Conservatory, Rimskii-Korsakov had abandoned his promising navy career for music. However, he may have approved of Borodin's engagement in research and teaching. What seemed to Rimskii-Korsakov unacceptable was that his friend carelessly wasted his talent on such apparently trivial things as the women's courses:

> During these years his affairs and surroundings had changed as follows: Borodin, who had always given but little time to music and who often said (when reproached for it) that he loved chemistry and music equally well, began to devote still less time to music than before. Yet it was not science that enticed him. He had become one of the prominent workers in establishing medical courses for women and had begun to participate in various societies for the aid and support of students, especially women.... Rarely did I find him in his laboratory, still more rarely at musical composition or at piano.... Moreover, knowing well his kind and easy-going nature, medical students and students of all sorts besieged him with every manner of solicitation and request, all of which he tried to fulfil with characteristic self-denial.... His inconvenient apartment never allowed him any privacy. His home life, due to a continual illness of his wife, was a one unending disorder.[40]

Indeed, Borodin abandoned his research in organic chemistry and could spare even less time for his music, but his academic duties still remained the major sphere of his activities:

> I am completely absorbed in the pool of academic life: examinations, re-examinations, lectures, the laboratory, the Conference, committees, reports, messages, all this immediately envelops me.... But I, like an old war-horse hearing a trumpet, prick up my ears and rush with fresh strength into academic life.[41]

In the Report of 1886, which happened to be his last report to the Conference of the Medico-Surgical Academy, Borodin summed up his thirty-year service to the Academy. His major concern was the conditions of teaching chemistry at the Academy. He pointed out that at university medical faculties, practical courses in physiological chemistry were given by the medical faculty, while the general courses in organic, inorganic, and analytical chemistry were given by the physico–mathematical faculty. At the Kazan medical faculty, chemistry courses were taught by four professors, and each professor had at least one assistant. The chemistry department of the Medico-Surgical Academy department had only two professors, N.N. Sokolov and himself, and only one assistant for both of them.[42] In general, the rhetoric of the Report, with customary complaints about the lack of budget and personnel, was essentially the same as in Zinin's time but with one important difference – the chemical practicum had acquired a different shape, from a small laboratory of the late 1850s for a few students, to mandatory laboratory practical classes for four-hundred students in the mid-1870s. In this transition Borodin's contribution had been remarkable.

In musical matters, the early 1880s brought Borodin a success in Europe. His First Symphony was performed at the Allgemeine deutsche Musikverein concerts in Baden-Baden in 1880 and in Leipzig and Dresden in 1882–3, and his Second Symphony in Antwerp, Liège, and Paris in 1884. Borodin believed that he owed much of his success in Germany to Franz Liszt (1811–86). In Belgium and in France, his music had the enthusiastic and influential support of Liszt's friend, the Countess Louise de Mercy-Argenteaux, and the French composer Camille Saint-Saëns (1835–1921), who introduced Borodin to the Parisian Société des Auteurs.[43] In the 1840s, Liszt, himself a celebrated pianist and composer, made several triumphant tours in Russia. Some thirty years later, he enthusiastically supported the new Russian musical school in Germany. Borodin was closely acquainted with Liszt, and left fascinating recollections which vividly portray Liszt as a handsome, artistic, and brilliant conversationalist. The recollections are full of comments on the Maestro, his students from different countries, known as 'Lisztianer' and 'Lisztianerinnen', on Liszt's daughter Cosima von Bülow and Richard Wagner, and on the notable musical events associated with Liszt's activities.[44] To Liszt, Borodin dedicated his symphonic poem of 1880, *In the Steppes of Central Asia,* in which he explored a Russian vision of the Orient, a great attraction for Russian musicians at that time. Liszt's interest and active support of the Balakirev group, Rimskii-Korsakov and Borodin in particular, was of great importance since at home their music usually received, at best, a cool reception. In a letter to his wife, Borodin mentioned that a performance of his First Symphony in Baden-Baden initiated and arranged by Liszt, brought him esteem among the musicians of the Moscow Conservatory.[45] Borodin was well aware that recognition at home, both in science and in art, depended on winning a reputation in Europe. Whereas Russian science, chemistry, mathematics, and experimental physiology had already won acclaim in Europe, 'prejudices against Russian music are still very strong', wrote Borodin to Louise de Mercy-Argenteaux in 1885, 'and it is difficult to overcome these opinions. We Russians, the so-called Northern bears, etc., too long have been regarded as consumers to be now accepted as creators.'[46]

His major musical creation, the opera *Prince Igor* was still moving slowly. During an academic year as usual Borodin could not even dream about composing:

> I have little time for anything else. My troubles with the women's courses take much time, but it is a success. I am not able even to look through the correct proofs of the First Symphony. Nevertheless, I am able to compose Igor a little in the mornings.[47]

In one of the letters of that time, he mentions that he could compose his opera only 'when I am not teaching and do not work in the laboratory because I am sick.'[48] Even at home, usually full of guests, relatives, or students, he did not have quiet hours for music. It only accentuated a dissonance, which during his final years grew into a crescendo. Only once, a year before his death, does one his letters to his wife break from care for her poor health and pessimistic moods to make a tired and disappointed note:

> But time still rushes by, absolutely flies, and here's me thirty years a professor, and all the time slaving away.... And what... could I do about it, even if I wanted to break off my official appointment to go and live as I pleased! After all, one has to eat. My pension won't cover everything, and you certainly can't make a living out of music.[49]

Borodin died suddenly from a heart attack at the Academy's ball in February of 1887. It was a shock to all his colleagues and to his musical friends. Alexander Dianin succeeded him in the chemistry chair at the Medico-Surgical Academy, and preserved the most detailed recollections on his mentor and father-in-law. Rimskii-Korsakov and Stassov finished, orchestrated, and set in order all that had been left behind by Borodin. The first performance of *Prince Igor* took place at the St Petersburg Mariinskii Opera Theatre in 1890, three years after his death. It was an enormous success.[50]

It is ironic that Zinin's assertion that man cannot serve two masters appeared not to apply to his most talented student. Borodin did serve two masters, chemistry and music and to the utmost of his abilities. True, he did not surpass his mentor in original research in chemistry; his talents were brought to bear on the development of laboratory chemistry within the medical education system, and his superb orchestral works and his opera, extraordinary in its rhythmic patterns and colourful scenes, continue to fascinate us with their charm and sensibility.

Notes

1. S.A. Dianin, *A.P. Borodin,* R. Lord (trans.), (Oxford: Oxford University Press, 1963), 10–11. The following biographical details are taken from the principal biography by S.A. Dianin, the son of Alexander Dianin, Borodin's student, assistant and successor to the Medico-Surgical Academy's chair. Alexander Dianin was married to the Borodins' adopted daughter, and his son Sergei grew up at the Borodins' home.
2. Cited in *ibid.*, 16.
3. A.P. Borodin, 'Ob analogii mysh'iakovoi i fosfornoi kislot v khimicheskom i toksilogicheskom deistvii' (unpublished MD dissertation, St Petersburg, 1859).
4. A.P. Borodin, 'Vospominaniia o Musorgskom. 1881', in S.A. Dianin (ed.), *Pis'ma A.P. Borodina, 1883–87* (Moscow: Gosudarstvennoe muzykal'noe izdatel'stvo, 1949), Vol. IV, 297–9: 297. This volume also contains Borodin's correspondence with other chemists over additional years.
5. A.P. Borodin, 'Einwirkung des Jodäthyls auf Benzidin', *Zeitschrift für Chemie und Pharmazie,* iii (1860), 533–6; 'Über einige Derivate des Benzidins', *Zeitschrift für Chemie und Pharmazie,* iii (1860), 641–3; 'Faits pour servir à l'histoire des fluorures et preparation du fluorure de benzoyle', *Comptes rendus des séances de l'Académie des Sciences (Paris),* 55 (1862), 553–6. For the complete list of Borodin's chemical works, see N.A. Figurovskii and I.I. Solov'ev, *Aleksandr Porfir'evich Borodin: A Chemist's Biography,* C. Steinberg and G.B. Kauffman (trans.), (Berlin: Springer Verlag, 1988), 121–6.
6. A.P. Borodin, 'Rech', proiznesennaia 9 fevralia 1880 na pokhoronakh N.N. Zinina' in S.A. Dianin (ed.), *Pis'ma A.P. Borodina, 1878–82* (Moscow: Gosudarstvennoe muzykal'noe izdatel'stvo, 1949), Vol. III, 85–8.
7. On purchasing instruments for the Academy's laboratory, see Borodin to Mendeleev, letter dated 5 March 1861, Paris, in Dianin, *op. cit.* (note 4), Vol. IV, 247. See also Borodin to Protopopova, letter dated September 1862, Heidelberg, in S.A. Dianin (ed.), *Pis'ma A.P. Borodina, 1857–71* (Moscow: Gos. izd-vo, muzykal'nyi sector, 1927), Vol. I, 56. Borodin's wife, Ekaterina Sergeevna Protopopova, the daughter of a physician at the Golitsin hospital in Moscow, suffered from lung disease from her youth. The damp St Petersburg climate did not suit her, and she used to spend most of the time in Moscow at her mother's home. To those frequent separations, we owe a fascinating correspondence between Borodin and his wife.
8. V.V. Stasov, *A.P. Borodin: ego zhizn', perepiska, i muzykal'nye stat'i. 1834–87* (St Petersburg: Izd. Suvorina, 1889), 27. Stasov, the son of the famous St Petersburg architect V.P. Stassov was an ardent proponent of realism and the national trend in art and music. He was acquainted with virtually all leading European composers and musicians. He also wrote on archaeology, history,

and philology. In 1900, Stasov became an Honorary Member of St Petersburg Academy of Sciences.

9. A.P. Borodin, 'Vospominania o Musorgskom', in Dianin, *op. cit.* (note 4), Vol. IV, 298.
10. Cited in Dianin, *op. cit.* (note 1), 41–2. Jakob Ludwig Felix Mendelssohn-Bartholdy (1809–47), the German composer, conductor, organist, and pianist, founded Leipzig Conservatory in 1843. In Russia, the followers of the trend imposed by Leipzig Conservatory were in opposition to Balakirev's group, as were both Russian Conservatories, in St Petersburg and in Moscow (founded in 1862 and in 1866 respectively). Different in their social status and approach to composition the musicians of the Balakirev group were united by a passion for the authentic and the national in art, and for realistic truth, which the 'mighty handful' wanted to use against the so-called German party, which, in the words of Balakirev, was trying to hold the entire development of music down to a few outworn procedures, originally fostered by the Leipzig conservatory, a place on which Liszt poured scorn.
11. Borodin to Butlerov, letter dated 15 July 1863, St Petersburg, in Dianin, *op. cit.* (note 4), Vol. IV, 262.
12. P.A. Dubovitskii, 'Khodataistvo o naznachenii Borodina', *Rossiiskii gosudarstvennyi voenno-istoricheskii arkhiv*, Moscow, delo Borodina.
13. Stassov, *op. cit.* (note 8), 149–50; see also, Dianin, *op. cit.* (note 1), 40.
14. N.A. Rimskii-Korsakov, *My Musical Life,* C. van Vechten (ed.), J.A. Joffe (trans.), (New York: Alfred A. Knopf, 1947), 57–8.
15. A. Borodin, 'Über die Einwirkung des Natrium auf Valeraldehyd', *Zeitschrift für praktische Chemie,* 93 (1864), 413–25.
16. Borodin to Alekseev, letter dated 23 December 1868, St Petersburg, in Dianin, *op. cit.* (note 4) Vol. IV, 273.
17. Borodin to Balakirev, letter dated December 1868, St Petersburg in Dianin, *Pis'ma A.P. Borodina, 1857–71, op. cit.* (note 7), Vol. I, 141.
18. Borodin to E.S. Borodina, letter dated September 1868, St Petersburg, in Dianin, *ibid.*, 62.
19. Borodin to Alekseev, letter dated 23 December 1868, St Petersburg, in Dianin, *op. cit.* (note 4), Vol. IV, 272.
20. A.P. Borodin and A.M. Butlerov, 'N.N. Zinin: vospominaniia o nem i biograficheskii ocherk', *Zhurnal russkogo fiziko-khimicheskogo obshchestva,* xii, 5 (1880), 217.
21. V.I. Serov, *The Mighty Five: The Cradle of Russian National Music* (New York: Allen, Towne & Heath, 1848), 7–12.
22. Alexander N. Serov (1820–71) was a friend of V.V. Stasov and a close associate of Balakirev. From 1864, however, Serov was in opposition to the Balakirev group. Serov was an ardent admirer of Richard Wagner, whom he had met in 1858 in Paris. In 1862–3, Wagner accepted a number of concert

appearances, travelling as an orchestra conductor to Moscow, St Petersburg, Prague, and Vienna. As a musical critic, Serov is known for his classical essays on the Russian composers Mikhail Glinka (1804–57) and Alexander Dargomyzhskii (1813–69), on Russian and Ukranian folk music, and on Richard Wagner (1813–83) and Ludwig van Beethoven (1770–1827). See, A.N. Serov, *Stat'i o muzyke* (Moscow: Muzyka, 1984), 6 vols; see also N. Slonimskii and L. Kuhr (eds), *Baker's Biographical Dictionary of Musicians* (New York: Schirmer Books, 2001), 6 vols, Vol. 6, 3818–28

23. Borodin to E.S. Borodina, letter dated 3 November 1869, St Petersburg, in Dianin, *Pis'ma A.P. Borodina, 1857–71, op. cit.* (note 7), Vol. I, 161–2; see also Dinin, *op. cit.* (note 1), 66. Hektor Louis Berlioz (1803–69), the French composer, conductor, innovator of the new musical and harmonic forms in orchestration and instrumentation. *Lélio, ou le retour à la vie* is the sequel of the famous *Symphonie fantastique.* The public score of the *Symphonie* is dedicated to the Russian tsar Nicholas I. That is explained by the fact that Berlioz was well received in Russia in 1847. See Slonimsky and Kuhn, *ibid.*, Vol. 1, 314–8. Of great interest are Borodin's articles: 'On the Concerts of the Russian Musical Society and the Free Musical School', see A.P. Borodin, 'Muzykalnye zametki. 1869', published in Dianin, *op. cit.* (note 4), Vol. IV, 273–93.
24. Borodin to E.S. Borodina, letter dated 8 September 1869, St Petersburg, in Dianin, *Pis'ma A.P. Borodina, 1857–71, op. cit.* (note 7), Vol. I, 146–7; see also Borodin to E. S. Borodina, letter dated 19 October 1869, St Petersburg, in *idem,* 153–4.
25. Borodin to E.S. Borodina, letter dated 9 March 1870, St Petersburg in *ibid.,* 202, see also Dianin, *op. cit.* (note 1), 67.
26. Borodin to E.S. Borodina, letter dated 8 November 1870, St Petersburg in Dianin, *Pis'ma A.P. Borodina, 1857–71, op. cit.* (note 7), Vol. I, 262.
27. A.P. Borodin, 'Produkty uplotneniia al'degida', in *Zhurnal russkogo fiziki-khimicheskogo obshchestva,* ii, 4 (1870), 90–2. Abstracted by V. Richter in *Berichte der deutschen chemischen Gesellschaft zu Berlin,* iii (1870), 423–5.
28. For a discussion on investigations of the condensation of aldehydes by Borodin and by his powerful rivals in Germany and France, Kekulé and Wurtz, see I.D. Rae, 'The Research in Organic Chemistry of Alexandr Borodin (1833–1887)', *Ambix,* 36 (1989), 121–37.
29. A. Borodin, 'Über einen neuen Abkömmling des Valerals', *Berichte der deutschen chemischen Gesellschaft zu Berlin,* vi (1873), 982–5; abstracted by H.E. Armstrong in *Journal of the Chemical Society,* xxvii (1874), 145.
30. M.I. Goldstein, 'A.P. Borodin', in K.K. Arsen'ev and F.F. Petrushevskii (eds), *Entsiklopedicheskii slovar* (St Petersburg: F.A. Brokgauz & I.A. Efron, 1893), Vol. IV, 440.

31. Borodin to E.S. Borodina, letter dated 3 October 1869, St Petersburg, in Dianin, *Pis'ma A.P. Borodina, 1857–71, op. cit.* (note 7), Vol. I, 150–1.
32. Borodin to E.S. Borodina, letter dated 4 March 1870, St Petersburg in Dianin, *ibid.*, 200–1.
33. Borodin to E.S. Borodina, letter dated 9 March 1870, St Petersburg in *ibid.*, 203; see also, Dianin, *op. cit.* (note 1), 68.
34. A.P. Dianin, 'A.P. Borodin: Biograficheskii ocherk i vospominaniia', in *Zhurnal russkogo fiziko-khimicheskogo obshchestva,* xx, 4 (1888), 367–79: 373.
35. Borodin to E.S. Borodina, letter dated August 1873, St Petersburg, in S.A. Dianin (ed.), *Pis'ma A.P. Borodina, 1872–77* (Moscow: Gosudarstvennoe muzykal'noe izdatel'stvo, 1947), Vol. II, 38–9.
36. Figurovskii and Solov'ev, *op. cit.* (note 5), 82–3.
37. Borodin to Karmalina, letter dated 1 June 1876, St Petersburg, in Dianin, *op. cit.* (note 35), Vol. II, 109.
38. Dianin, *op. cit.* (note 1), 96. Borodin's Second Symphony, dedicated to Ekaterina S. Borodina was first performed in St Petersburg in February of 1877.
39. A. Borodin, 'Über Nitrosoamarin' in *Berichte der deutschen chemischen Gesellschaft zu Berlin,* viii (1875), 933–6, abstracted by G.T. Atkinson in *Journal of the Chemical Society,* xxix (1876), 269–70; 'O novom sposobe opredeleniia azota v moche', in *Protokoly obshchestva russkikh vrachei v St. Peterburge,* i (1875–6), 278; 'O novom sposobe kolichestvennogo opredeleniia mocheviny', in *Zhurnal russkogo-chimicheskogo obshchestva,* viii, 5 (1876), 145.
40. Rimskii-Korsakov, *op. cit.* (note 14), 194–5. In summer, Borodin usually did not compose because of the inconveniences of his life in the countryside, namely the lack of a piano, which he sometimes failed to bring with him, and most of all, the condition of health of his wife; see *idem,* 242–3, 247.
41. Borodin to E.S. Borodina, letter dated September 1880, St Petersburg, in Dianin, *op. cit.* (note 6), Vol. III, 107.
42. A.P. Borodin, 'Otchet v Konferentsiiu Akademii. 1886', in Dianin, *op. cit.* (note 4), Vol. IV, 302–4.
43. Borodin to de Mercy-Argenteaux, letter dated 15 September 1884, Moscow, in *ibid.,* 82–4. In 1878, Borodin visited Jena on the occasion of the arrangement of a doctoral examination for his assistants, Alexander Dianin and Mikhail Goldstein (1853–1905), who, after graduating from the Medico-Surgical Academy, decided to obtain a PhD in chemistry abroad. From Jena, Borodin moved to Weimar to visit Liszt. In the early 1880s, Borodin visited Liszt in Weimar again and also met him at the Marburg Musical Festival. Through Liszt Borodin became acquainted with some of the influential German musicians and publishers, such as Carl Riedel (1827–88), professor of the Leipzig Conservatory and Chair of the

Allgemeine deutsche Musikverein, and with Countess Louise de Mercy-Argenteaux (1837–90), a close friend of Liszt, influential musical activist and ardent supporter of the new Russian musical school. She published a book about César Cui in France. Borodin visited the Countess in her estate in Belgium several times in connection with concerts, which she arranged, where Borodin's symphonies were performed. On de Mercy-Argenteaux, see C. Bronne, *La comtesse de Mercy-Argenteaux et la musique russe* (Paris: Librarie des Champs-Élysées, 1935).

44. A.P. Borodin, 'Moi vospominaniia o Liste. 1878', in Dianin, *op. cit.* (note 6), Vol. III, 13–39; *idem*, 'Prodolzhenie Listiady. 1881', in *idem,* 167–79; and *idem*, 'List u sebia doma v Veimare: iz lichnykh vospominanii A.P. Borodina. 1883', in Dianin, *op. cit.* (note 4), Vol. IV, 13–7. See also A. Habets, *Borodin and Liszt,* R. Newmarch (trans.) (New York: Ams Press, 1977).
45. Borodin to E.S. Borodina, letter dated 18 December 1880, Moscow, in Dianin, *op. cit.* (note 6), Vol. III, 135.
46. Borodin to de Mercy-Argenteaux, letter dated 18 January 1885, St Petersburg, in Dianin, *op. cit.* (note 4), Vol. IV, 110–1.
47. Borodin to E.S. Borodina, letter dated 5 November 1883, St Petersburg, in Dianin, *ibid.*, 53–4.
48. Cited in Stasov, *op. cit.* (note 8), 6–7.
49. Borodin to E.S. Borodina, letter dated 1 February 1886, St Petersburg, in Dianin, *op. cit.* (note 4), Vol. IV, 180.
50. Rimskii-Korsakov, *op. cit.* (note 14), 283–4.

10

'Scientific Medicine': Botkin's Teaching Clinic and Laboratory

The mid-nineteenth century saw spectacular developments in experimental physiology and cellular pathology. These radically new fields refuted excepted dogmas and began to powerfully influence clinical medicine. The emerging trend in clinical medicine found a voice through the journals *Zeitschrift für die wissenschaftliche Medizin* [*Journal of Scientific Medicine*] and *Zeitschrift für rationelle Medizin* [*Journal of Rational Medicine*]. Widely accepted as scientific medicine, the new movement emphasised the experiment, along with its methodology and apparatus, facilitating the precise determination and verification of changes in a diseased organism to formulate a clear concept of disease. Clinician–experimenters shifted their activities from the hospital ward to the laboratory. The introduction of laboratory methods, tests and examinations as new diagnostic tools required innovative changes in medical education. At Berlin University in the late 1850s, three courses out of eighty focused on laboratory training: physiology, pathological histology, and medical chemistry.[1] At the St Petersburg Medico-Surgical Academy, laboratory courses in physiology, histology and cellular pathology, and medical chemistry began to be offered in the early 1860s. These alone promised to provide practical and scientific instruction as well as basic skills and experience in laboratory research.

The important changes in clinical medicine were closely associated with the introduction of the laboratory into the clinic. In this respect, Sergei P. Botkin, the clinician–experimenter and a powerful advocate of the laboratory in the St Petersburg medical establishment, exerted notable and lasting influence on the development of clinical laboratory research and teaching in Russia.

After studies with Virchow and Traube in Germany, Ludwig in Vienna, and Bernard in Paris, Botkin returned to St Petersburg to an appointment as adjunct to a clinical chair at the Medico-Surgical Academy. The reformist administration of the Medico-Surgical Academy saw in Botkin, a German-educated internist with a background in physiology, an ideal candidate for the introduction of the ideas and practices of scientific medicine into clinical teaching. This programme was designed to stimulate clinico-physiological and clinico-pathological research, and to train the students in quantitative diagnostic, chemical, and microscopic methods, and clinical use of instruments.

In the autumn of 1860, at his first lecture at the Medico-Surgical Academy, Botkin pointed to the importance of natural sciences education for a medical practitioner:

> Practical medicine should be placed among natural sciences, and the methods employed in examination, observation and treatment of the patient must be the methods employed by a natural scientist, who establishes his conclusions on carefully observed and scientifically proven facts.[2]

For Botkin, only exact, measurable and objective experience obtained in the laboratory was valid in clinical practice. He believed that clinical medicine should be orientated towards the organisational forms and investigative practices of laboratory sciences, physiology, medical chemistry, and cellular pathology. Therefore, the hospital should acquire a specialised laboratory for training the students in scientific methods relevant to medical practice, as well as for carrying out clinically oriented research.[3]

The same year, Botkin, supported by Dubovitskii and Zinin, obtained necessary premises for the laboratory in the thirty-four-bed faculty clinic within the former marine hospital. He received funding of 1,471 roubles for laboratory arrangements and 1,500 roubles for instruments. An additional 1,000 roubles for equipment for Botkin's study-room were granted by special request from Tsar Alexander II.[4] The new clinical laboratory was well-equipped, its inventory included: microscopes; different kinds of scales (physiological, pharmaceutical, chemical); water, sand and air baths with thermometers; spirit lamps; drying apparatus; equipment for urinalysis; air pump with brass plates and bell-glass; du Bois-Reymond galvanometer and and induction apparatus; polarization apparatus; Hoppe-Seyler apparatus for estimating haemoglobin content of the blood; dynamometer, a device for measuring forces exerted by muscles; insufflators, a device for blowing gases or substances into a bodily cavity; aerometer [Gr. *araios* thin, rare + *-meter*], or hydrometer for measuring density of liquids; chemical glassware, materials and reagents; porcelain crucibles and cups for evaporation; and batteries for strong currents.[5]

From the list of instruments, it can be inferred that a standard set of methods to examine bodily fluids and tissues could be employed, such as the decomposition of haemoglobin to obtain haematin to demonstrate its crystalline form; a water-bath could be used for experiments on digestion, or for evaporating at a constant temperature for different purposes. Du Bois-Reymond's galvanometer, which by that time had become standard equipment for a physiological laboratory, acquired its place in the clinical laboratory of Botkin too. If not used for clinical purposes *per se,* the galvanometer was a piece of technology that represented the ideals of physiology – precision, quantification, and objectivity – which became a model for scientific medicine. The microscope, another important piece of technology at the disposal of the clinician, became emblematic of the modern developments in medicine that offered explanations of bodily functions and malfunctions on a cellular level.

Botkin was well aware that what he was doing was novel. There were no such laboratories in any of the clinics he had visited in Paris or Vienna. In the Berlin Charité, the clinical professor Ludwig Traube used the fine Hoppe-Seyler laboratory at the Virchow Pathological Institute for his research. Essentially, Hoppe-Seyler's laboratory was a research centre for physiological chemistry. Located within the favourable environment of the Charité hospital, the laboratory was also orientated to the study of clinical problems, and in this way it was an early prototype of a specialised clinical laboratory. From the mid-1860s, following the pursuit of the two leading Charité clinicians, Traube and Theodor Friedrich von Frerichs (1819–85), many German internists made it a point of honour to establish a laboratory in their clinics. In France, Claude Bernard's advocacy of the laboratory reached a crescendo between 1865 to 1878, when the goal of reform was to improve and create laboratory facilities and support research.[6] Speaking about the clinical laboratory, Bernard pointed out:

> The laboratory of a physiological physician must be most complicated of all laboratories, because the experimental analyses to be made there are the most complex of all, requiring the help of all other sciences.[7]

To set up and run a clinical laboratory was not an easy task. Botkin's laboratory did not have a separate budget, and Botkin had to use the funds allotted to the clinic to arrange laboratory training for the students.[8] Another difficulty was the absence of assistants skilful in clinical laboratory work. Practical training at the Medico-Surgical Academy's clinics was deficient – the students sometimes did not even know how to use a thermometer on a patient, not to mention the application of physical and chemical methods in diagnosis. The first year of his appointment, Botkin spent most of his time in the laboratory as he had to do everything himself – the physicians who assisted him were not prepared to teach experimental methods.[9] The work at the laboratory was organised in such a way that students could gradually advance from basic microscopic and chemical methods, laboratory tests of blood and urine to assigned research projects. Botkin believed that studies in the teaching clinic and laboratory allowed students to unite bedside experience and basic experimental investigation of the disease. Physicians with laboratory training would be able to use these methods and results in their practice – in diagnostics at least, if not in treatment.

Botkin's commitment to 'scientific medicine', alongside his general belief in the improvement of life and social progress, made his lectures well regarded and popular among the students. On the other hand, his lectures differed markedly from those of the clinical professors of the older generation, being business-like discussions without unnecessary eloquence, sometimes with original improvisations, and accompanied by carefully prepared demonstrations on patients. He presented disease not as an abstract entity but as an individual case, and in doing so he taught how to treat the patient not the disease.[10] In his teaching clinic, Botkin opened a dispensary available for the civilian population of St Petersburg. The dispensary provided access to a wider variety of new cases, and Botkin spent much time there, coming to see patents five days a week. The first walk-in clinic in St Petersburg, the dispensary became very popular, and the number of outpatients increased from 958 in 1861 to 4,574 in 1881.[11]

After the resignation of the elderly Professor P.D. Shipulinskiii (1808–72) in 1861, Botkin was expected to succeed the chair. Although Botkin was strongly supported by Dubovitskii, Zinin, and Glebov, the most powerful members of the Conference, there was strong opposition to Botkin's appointment as ordinary professor from the influential 'German faction', headed by the clinical professors B.E. Ekk and V.V. Besser. For them, Botkin's 'novelties' were useless and even harmful. They treated the young adjunct sceptically, as a person who spent too much time in the laboratory and who relied on methods not yet widely accepted in practical medicine. Clinical practice was still firmly based on an empirical approach in diagnostics and therapeutics, and the introduction of experimental studies of the disease meant a shift in medical practice from the bedside to the laboratory. The clinical professors of the 'old school' were reluctant to cede their authority to the young enthusiastic clinician–researcher. At one of the meetings of the Conference, Nikolai Iakubovich, Professor of Histology, pointed out that Botkin had done more in one year than the majority of his fellow clinicians had done during their whole careers at the Medico-Surgical Academy. This statement, as Botkin mentioned in a letter to his brother, brought him many enemies and resulted in bitter attacks from some members of the Conference. Not surprisingly, Botkin's appointment to ordinary professor in 1861 did not go smoothly. Botkin was ready to resign in the face of possible black-balling. Nevertheless, the petition of students and young physicians in support of Botkin influenced the voting in Botkin's favour.[12]

Broadly educated and cultivated, Botkin was a gregarious man of his time and fascinating to his contemporaries. From 1860, when Botkin settled with his family in St Petersburg, he would receive his close associates from the Medico-Surgical Academy every Saturday. Sechenov, Wencheslav L.

Gruber, the old Professor of Anatomy, and Evgenii Wencheslavovich Pelikan (1828–84), Director of Medical Department in War Ministry were the most frequent guests.[13] Starting as a small circle of intimate colleagues who held similar views, Botkin's Saturday gatherings grew into crowded, animated and boisterous events. As the guests were of different professional backgrounds, medical problems were seldom discussed. Through his brothers, Vasilii, a musical and literary critic, and Mihkail, professor at the St Petersburg Academy of Fine Arts, Botkin was acquainted with most of the local artistic and musical community. Indeed, Borodin first met Balakirev, the leader of the new musical school, at one Botkin's gatherings. Botkin himself was a proficient violinist and great lover of music. With his appointment as personal physician to the Tsar's family in 1872, and his second marriage to E.A. Mordvinova,[14] a close associate of the Grand Duchess Elena Pavlovna, the circle of Botkin's guests gradually changed. Besides his fellow professors and artistic élite, many members of educated high society and officials of the medical and military establishment frequented the parties. The Saturdays had their own history, their flourishing periods and decline; they reflected the intellectual and cultural life of Russia at the time.[15]

Among the numerous responsibilities Botkin held at the Medico-Surgical Academy, teaching at the hospital remained his first priority throughout his academic career. His schedule at the teaching clinic was busy – rounds, a laboratory course in microscopy and chemical analysis for the students and physicians, lectures, and work in the dispensary. His method as a clinical teacher was to examine carefully a patient before students in a variety of ways. Sometimes it took two or more lectures to analyse and diagnose a case. He used to cover every detail that indicated deviations from the normal functioning, grouping these deviations, and then arrive at a diagnosis based on the known microscopic and chemical examinations. Among his colleagues he was known as an extraordinarily astute diagnostician.

A clinical teacher of the new generation, Botkin ascribed a great importance to methods used in pathological anatomy, experimental physiology, and physiological chemistry. These methods, in his opinion, could serve as the basic tool for recognising, understanding, and diagnosing disease.

In 1867, Botkin published the first volume of his *Kurs kliniki vnutrennikh boleznei* [*Clinical Course in Internal Diseases*], which was, in essence, a detailed analysis of just one case of arteriosclerosis. In the preface to the volume, he explained the design and the main idea of his course:

> A living organism, affected by a large variety of the external conditions, presents itself in diverse physiological and pathological manifestations. It is

impossible to cover all these manifestations in a limited time during a clinical course. Therefore, a clinical teacher should teach a method which a future physician may apply in his practice in addition to his theoretical knowledge. One may reprove me that I devoted the whole volume just to one case instead of a number of analogous cases. But my aim here is not to accumulate 'causative' material, but to relate my methods of investigation and way of reasoning, worked out during my clinical and pathological studies of disease.[16]

In 1868, Botkin published the second volume, which covered just one case of typhus fever.[17] He continued to work on the *Clinical Course*, his major and most important publication, and, in 1875, the third and final volume appeared in two parts.[18] At first glance it seems that a unifying title for these three separate volumes does not correspond to its content, the analysis of three different clinical cases. One might expect a systematic account of a standard range of topics in the course of internal diseases. However, Botkin's idea of presenting three cases instead of all nosological forms is reasonable, if one takes into account that German internal medicine, of which Botkin was an ardent proponent at that time, lacked interest in the description of complexes of symptoms, or identification of the 'specifics' of nosological form to build up the clinical picture empirically. Contrary to accumulating 'causation', the Berlin school sought to introduce new methods of investigation and examination – whether graphic, microscopic, or chemical – into clincal medicine. Botkin's *Clinical Course* is a revealing example of the clinical research literature of the new school.[19]

In 1871, Botkin's laboratory research was disrupted by fire, which damaged the hospital building, its clinic and laboratory. In 1874, Botkin's facility was moved permanently to the Mikhailovskii hospital, which housed a clinic for forty beds, a new laboratory, a lecture hall, premises for experimental animals, and a vivarium.[20] We may catch a glimpse of Botkin's laboratory for animal experiments in pharmacology through the eyes of Ivan P. Pavlov, future famous physiologist, who managed Botkin's laboratory from 1878 to 1890:

Despite something unfavourable about that laboratory, mainly, of course, scarcity of means, the time I spent there was very useful for my scientific career. Firstly, total independence and the a possibility of doing only laboratory work (I had no duties in the clinic). I worked there without discrimination between what was mine and what was not. For months and years, I spent my laboratory labours taking part in the works of others. However, there was always a personal advantage: I had more and more practice in physiological thinking in a wider sense and in laboratory

> technique. Moreover, I always had interesting and instructive (though unfortunately extremely rare) discussions with Sergei Petrovich Botkin. Here, I did my thesis on the nerves of the heart; and here, after I had returned from abroad, I began my work on digestion. Both projects I worked out independently.[21]

It was in Botkin's laboratory that Pavlov developed his distinctive abilities and enthusiasm for running a laboratory, and obtained a unique clinico–physiological experience to become a first-class physiologist. Despite obvious difficulties and disappointments, which Pavlov went through during his ten-year work at Botkin's laboratory,[22] Pavlov always appreciated Botkin's talent as a 'magic doctor' and strong belief in the exceptional value of the laboratory for clinical practice, research and teaching

Botkin's interest in the role of the nervous system in the pathogenesis of disease determined the focus of experimental research carried out in his laboratory. These were primarily works related to the study of the nervous system and its effect on pathophysiological conditions in the organism. Botkin, however, could not spend as much time in the laboratory as he used to do during the first decade of his professorship. According to Pavlov, Botkin rarely appeared in the laboratory because of enormous responsibilities in his clinical and private practice. He did, however, assign dissertation topics to the physician–interns working at his laboratory. The scope of experimental investigations performed in Botkin's laboratory during 1869–83, as shown in Table 10.1 overleaf, is revealing.[23]

Pavlov interpreted the main feature of Botkin's concept of disease as 'nervism', a 'physiological trend aiming at extending the influence of the nervous system to the majority of the organism's functions.' In the conclusion to his doctoral thesis carried out in Botkin's laboratory, Pavlov wrote:

> I was surrounded by the clinical ideas of Professor Botkin and I acknowledge with a hearty gratitude the fruitful impact of Botkin's concept of nervism on my work and in the whole on my physiological views; this concept was deep and broad, and frequently anticipated the experimental facts, and which, in my opinion, was a great service to physiology.[24]

The major points of Botkin's 'neurogenic' theory of disease were presented in his *Clinical Course*. The first volume of the *Clinical Course* contains a description of a transient form of heart dilatation, which Botkin called *distensio*. Botkin surmised that it was caused not by anatomical changes in the valves of the heart or in the myocardium, but by temporary dilatation of the heart chambers overloaded by the blood. He supposed further that it was a transient intermediate form between dilatation due to

Table 10.1

Investigations Performed in Botkin's Laboratory during 1869–83

Title	Investigator	Year
Nerve endings in skin epithelium	**A.B. Podkopaev**	**1869**
Imitation of the diaphragm paralysis	**V. Alushevskii**	**1870**
Localisation of the brain	**I.P. Lebedev**	**1871–3**
Neuralgia and neurosis	**P.I. Uspenskii**	**1871**
The effect of the splenic nerves on spleen constriction	**A.S. Bochechkarov**	**1874**
Innervation of the isolated heart with artificial circulation in the cold-blooded animal	**V. I. Drozdov**	**1875**
Sensory and motor nerves in the cervical ways of the brain	**K.V. Voroshilov**	**1877**
Nervous mechanisms governing equilibrium of the blood pressure	**I.P. Pavlov**	**1877**
Heat excitation of the nerves	**I.P. Pavlov**	**1879**
The effect of the excitation of the heart on its function and nourishment	**N.P. Simanovskii**	**1881**
Reflex excitation of the spinal cord and its relation to clinical manifestation in typhus	**V.T. Pokrovskii**	**1882**
The intensifying nerve of the heart	**I.P. Pavlov**	**1883**

morphological changes, and a reflexogenic form of dilatation, which occurred during percussion in the heart area.[25]

The focal point of the second volume is pathogenesis of fever. The prevailing theory of fever, most fully described by the German physician Carl von Liebermeister (1833–1901), stated that contagious elements which penetrated the blood contained 'pyrogenic substances'. These substances excited the so-called heat centre, which accelerated the oxidising processes. The products of incomplete burning that were formed during accelerated dissociation inhibited the 'centre of cooling', thus production of heat was no longer balanced by cooling; resulting in excessive warming up of the organism, the fever. Botkin called this theory 'chemical', and assumed that it did not explain all clinical cases of fever. In particular, he referred to cases when considerable elevation of temperature occurred during neurological breakdown: after apoplectic stroke; after brain haemorrhage due to embolism or thrombosis; in cases of gall-stone, nephritic or enteric colic. All these cases, as Botkin assumed, could not be explained by the presence of a contagious agent. Aetiology of such fevers was different from 'pyrogenic' fevers, but the mechanism underlying psychogenic, haemorrhagic, and 'pyrogenic' fevers may be the same – of reflex character. Therefore, 'nervous centres, which govern the processes of bodily cooling are of utmost importance in the explanation of any abnormal elevation of temperature.'[26] This was, in essence, Botkin's 'neurogenic' theory of fever.

The third volume of the *Clinical Course*, published in 1875, consists of two parts, the first entitled 'On the Contraction of the Spleen' and the second 'On the Reflex Phenomena in the Skin Vessels and on Reflex Perspiration.' Botkin continued the theme of establishing the role of the nervous centres in the pathological processes. He described clinical cases when contraction and dilation of the spleen had occurred not due to a pathological process but as a result of nervous excitation and inhibition. He determined marked contractions of the spleen during percussion or in the course of pregnancy. Experimental studies by I.R. Tarkhanov, professor of physiology at the Medico-Surgical Academy confirmed Botkin's assumptions. In a series of experiments with the excitation of the central ends of the vagus and sciatic nerves and of the medulla oblongata, Tarkhanov showed that the spleen contracted more vigorously during the excitation of the medulla, providing that its connection with the abdominal ganglia was preserved. Botkin concluded that 'in the brain there may exist nervous centres controlling the spleen muscles and its vasomotor actions.'[27]

In the second part of the volume, Botkin examined a case of dysfunction of perspiration. The excepted view held that perspiration was connected with a vasomotor centre since perspiration was accompanied by intensive dilation of the blood vessels of the sudoriferous glands. In the case under examination, Botkin could not find any signs of the dilation of these vessels. He assumed that perspiration may be controlled by a separate nervous centre.[28]

The recurring technique within Botkin's *Clinical Course* is the presentation of carefully selected cases from his clinical practice with pertinent experimental data. These cases suggested that various 'nervous centres' are involved in different normal and pathological processes – for example, centres controlling circulation or haemopoiesis, or specific centres for an organ such as the one controlling the vessels and muscles of the spleen (as first described by Tarkhanov and Botkin) or the centre controlling vessels and muscle of the heart (the aortic depressor nerve, as first described by Ludwig and Cyon). It should be noted that, before Pavlov's works on higher nervous activity, the 'nervous centre' was understood localistically; the distributional role of the cortex was not yet known. However, the theory of the nervous centre was a step further in understanding the pathogenesis of disease.[29]

The vulnerable point in Botkin's 'neurogenic' theory of disease, as he himself acknowledged, was that in the majority of described cases, manifestations of possible pathological changes were of extremely short duration. Neurogenic fevers and perspirations are momentary; neurogenic changes in the spleen are even more transient, they occur during percussion and palpation and cease just after it. The neurogenic theory, however, did not apply to severe durable pathological conditions such as typhus or aneurysm of the aorta. Botkin understood that short neurogenic forms, strictly speaking, could not be considered pathological forms, but forms that are on the border of physiological and pathological conditions. He assumed that short neurogenic changes and durable clinical forms might be two different phases of the same pathological process. He pointed out, however, that the connection between these two phases was debatable unless these transient forms were clearly defined.[30]

Botkin's theory of pathogenesis was criticised mostly by pathologists, in particular by V.V. Pashutin. In the early 1860s, Pashutin began his research career at Sechenov's laboratory at the Medico-Surgical Academy. The most promising among Sechenov's students, Pashutin assisted his mentor in numerous experiments aimed at clarifying the reflex action in the brain and spinal cord. However, Pashutin's research interests shifted to experimental pathology after studying with Felix Hoppe-Seyler in Strasbourg, In the 1870s, Pashutin set up a laboratory of experimental pathology first in Kazan University, and then at the Medico-Surgical Academy. An influential pathologist, Pashutin opposed a physiological approach to disease. He adhered to a morphological, localised approach, and believed that a cellular mechanism was the key to understanding the course of all complex phenomena in healthy and morbid states. Like Virchow and Hoppe-Seyler, Pashutin attached little importance to the reflex mechanisms in pathogenesis, and refuted Botkin's neurogenic explanations of fever or venous haemostasis.[31]

Yet, regardless of the theoretical value of his concept of 'nervism', Botkin produced the clinico–experimental studies that equalled those of Ludwig Traube and Friedrich Theodor Frerichs (1819–85), the best clinicians of the German school. The *Clinical Course* was translated and published in France and Germany, and was well received in both Berlin and Paris.[32] But what marks Botkin's professional career even more prominently is the powerful influence that his clinical school had on the development of laboratory medicine in Russia. One hundred and three physicians studied at Botkin's clinic and laboratory, eighty-five obtained their doctoral degrees there, thirty-seven physician–interns from his clinic and laboratory became professors, including twenty-five in clinical medicine. Most of the clinical chairs at the Medico-Surgical Academy were headed by Botkin's disciples, such as V.A. Manasein (1841–1901), I.T. Chudnovskii (1843–96), M.V. Uspenskii (1833–1900), and at the university clinics in other cities, including K.N. Vinogradov (1847– 1906) in Kazan, V.T. Pokrovskii (1838–77) in Kiev, V.G. Lashkevich (1835–88) in Kharkov, L.V. Popov (1845–1908) in Warsaw. Seventeen of the physician–interns who worked at Botkin's laboratory received their doctoral degrees in disciplines other than internal medicine: Pavlov in physiology; T.I. Bogomolov (1843–97) in medical chemistry; V. Verkhovskii (1843–1916) in otolaryngology; V.N. Reitz (1838–1904) in paediatrics; and A.G. Polotebnov (1838–1908) in dermatology and venerology.[33] All of them became notable scientists and teachers.

The St Petersburg clinical school had a no less eminent rival in Moscow, the clinical school of Professor Grigorii A. Zakhar'in (1829–97). Zakhar'in had studied in Berlin with both Frerichs and Virchow, but adhered to Frerichs's school. Before his professorship at the Charité, Frerichs had studied physiological chemistry with Liebig, and was keen in chemico-physiological research as well as in pathological anatomy. In clinical medicine, Frerichs valued and emphasised independent bedside observations and was opposed to a predominance of pathological anatomy. Both schools, of Frerichs, and of Traube and Virchow were equally scientific in their approach and methods. At the same time, the divergence between these schools was so great that numerous and eminent disciples of Frerichs and Traube formed separate camps, usually in opposition to one another.[34] This was the case with Botkin, who was Traube's disciple and his rival, Zakhar'in, a student of Frerichs.

Zakhar'in taught at the medical faculty of Moscow University and had a reputation as a brilliant clinical teacher and physician who remained faithful to a classical Hippocratic tradition and to the so-called anamnesis method based on the questioning of the patient. A keen observer at the bedside and careful examiner of the patient, Zakhar'in ascribed minor importance to the results obtained in the laboratory and to the experimental studies of disease. Zakhar'in taught his students to start with careful questioning of the patient, and only as a last resort did he use laboratory findings. He believed that if a physician made a preliminary diagnosis based on theory or any hypothesis, he would be biased during the questioning and examination of the patient.[35] Botkin's method, on the contrary, was centered around objective information – analysis of the blood, urine, and sputum. Botkin believed that a patient could not relate particulars about his disease and a physician should be ready to ask leading questions while questioning the patient.

Botkin once said that practical medicine is divided into art and science.[36] The difference between the two approaches lies in this division: Zakhar'in saw in medicine art and Botkin, science. Zakhar'in represented traditional, more clinical approach to disease and cure, Botkin represented a novel attempt to solve clinical problems in the laboratory. In a preface to the French edition of Zahkar'in's *Klinicheskie lektsii po zheludochno-kishechnym zabolevaniiam* [later published in English under the prosaic title *Clinical Lectures*], the Paris clinician Henri Huchard summed up the difference between the Moscow and St Petersburg schools:

> Due to the influential activities of the two renowned clinicians, Botkin in St Petersburg, and Zakhar'in in Moscow… two quite different schools were formed… Botkin's school was always more theoretical than practical.... The school of Zakhar'in, on the contrary, based itself on examination, exact knowledge of anamnesis and aetiology, questioning of the patient, which he has elevated to an art, and on therapy, so skilful that in the hands of Zakhar'in it became close to an exact science.[37]

N.F. Golubov, Zakhar'in's disciple, provided a critical outline of the rival school. Golubov claimed that Botkin largely subordinated clinical studies to pathological anatomy and experimentation.[38] Indeed, Virchow's idea that pathology should be based on a combination of clinical observation and experiment had made a great impact on Botkin. He shared Virchow's threefold approach to research on disease and cure:

> The first is the clinic: examination of the sick with all the means of physics and chemistry according to the major concepts of physiology and anatomy. The second is experiment: induction of the disease in the animal and study of the effect of the drug on the disease. The third, finally, is microscopy: the study of the (dead) body and its isolated parts with the scalpel, microscope and reagents.[39]

There was, however, a marked difference in Botkin's approach to experiment. Throughout his career Botkin remained a bedside physician, and for him the experiment was only a means for studying clinical problems. Clinical experiment, he believed, 'is aimed at answering the question of how to help the suffering patient, while a physiological experiment is aimed at answering the question, why?' For Botkin, clinical experiment was guided by:

> [C]areful observation of the patient during the course of disease, comparison of the initial diagnosis with that of at the discharge, taking into account the results of the applied therapeutic means. This bedside experience may suggest a topic for experimental research. In the majority of cases, the main object of the experiment is an animal, and the clinician can use the results within certain limits.[40]

Golubov pointed to the strong influence of Virchow on Botkin, although he admitted that Botkin's clinical thinking was broader then Virchow's. Whilst ascribing an important place to cellular pathology in the study of the disease, Botkin also believed that an anatomical, localistic approach was too narrow and one-sided to dominate clinical medicine:

> A physician should proceed from pathological changes as the basis but not the essence of disease. Bedside experience and the data and results obtained by physiology and allied sciences are of particular importance for the understanding of disease.[41]

He viewed disease as a process spreading all over the organism; he argued that the cause of catarrhal jaundice, for instance, was not a mucous plug in the common bile duct – Virchow's mechanistic explanation – and interpreted the disease as infectious – infectious jaundice in Russia is still called 'Botkin's disease'. Furthermore, Botkin recognised the importance of the environment in the development of disease: 'Our conception of disease is closely connected with its cause, which is always conditioned by environment, acting on a diseased organism either directly or indirectly.'[42]

Golubov maintained that Botkin's *Clinical Course* bore a marked influence of the Traube's school, which overemphasised pathological anatomy in clinical studies, and whose contribution to the understanding of the disease was limited to anatomical local diagnosis.[43] The Soviet clinician F.R. Borodulin argued that Botkin's concept of disease was, in essence, physiological or functional, since Botkin held that pathological processes may be of reflexogenic character. In contrast, cellular pathology maintained that any diseased condition is caused by a certain local morphological change. A physiological or functional approach ascribes a major role to a reflex action in the development of a pathological process. Reflex, a mechanism governed by the nervous system, can be regarded as a functional system, in which various elements of the nervous system as well as various parenchymatous organs are involved. It is in this sense that a physiological approach, in contrast to cellular, is termed functional. Finally, since a reflex is a general function of the nervous system that performs communication and control over the entire organism, a physiological or functional approach can be considered holistic. The cellular approach, in contrast, is centred around the cell and the organ, and takes into account only local morphological changes.[44]

By the late 1870s, the dominance of pathological anatomy in German clinical medicine has been undermined by the so-called 'functional approach' that emphasised traditional clinical methods and ascribed less importance to experimental data. The overwhelming authority of the mighty Virchow–Traube school became less pronounced. This change was powerfully seen in 1880, when the aging Frerichs and a group of younger clinicians at the Charité began to publish the new *Zeitschrift für klinische Medizin*, intended to assert independence of clinical medicine from pathological anatomy and to oppose the tendency for the theoretical branches of science to dominate in medical teaching and research.[45]

There was one more important point for which the adherents to the Moscow school criticised Botkin: the so-called 'therapeutic nihilism'.[46] Indeed, Botkin was known to attach less importance to 'symptomatic' therapy. In the treatment of the sick, he attributed greater importance to adequate diet and proper care for the patient. At his lectures, Botkin pointed out that therapeutics was in most cases empirical and arbitrary, subjected to various and frequent changes. Despite the efforts of physicians of all generations to put therapeutics on a rational footing, it still lacked strictly scientific foundations, as compared to diagnostics. Botkin's deep disappointment with the results of application of remedies available for treatment, and their meagre curative effect is best expressed in his letter to Belogolovyi:

> I have a sad conviction that our therapeutic means are powerless. Half of patients whom I receive in the dispensary spend their money on medicines at the apothecary, which may relieve them for some twenty-four hours, but would not cure their ailments.[47]

It was an obvious limitation of therapeutics that forced Botkin to attribute much importance to the 'pharmaceutical' experiment on animals in the laboratory and to seek new ways to combat disease. It was in his laboratory and clinic that the action of various medicines was tested, such as lychnis (*herba adonis vernalis*), lily of the valley (*tinctura corvallariae*; still sometimes refered to as 'Botkin heart drops' by physicians of the older generation), strophanthus (*semen strophanthi*), and caffeine (*coffeinum*). He believed that it was important to find specific or 'cutting off' means for each disease such as quinine for malaria, and mercury for syphilis.[48]

Botkin felt that the advance made by experimental physiology, pathological anatomy, and physiological chemistry did not match with the pace of developments in clinical medicine – the innovations in practical medicine were still rare, and therapeutic means remained too limited. 'Laboratory' clinicians, such as Botkin and Traube largely exploited methods of the allied sciences in the clinical studies and made notable contributions to diagnostics. In general, however, the gains of clinical medicine from the successes in the laboratory sciences were few. Pathological anatomy could provide evidence of the diseased organ or tissue and confirm diagnoses by post-mortem examination. Chemical analysis could help in diagnosing only certain conditions and could also be used for more accurate preparation of remedies. The possibilities for applying results obtained through experiments on animals as well as therapeutic evaluation of the efficacy of the remedies available for treatment were still restricted. The obvious limitations of therapeutic means, as well as an inability to comprehend the complexity and individuality of a diseased organism in the majority of cases, determined contemporary medical practice.[49]

Physiological, or 'scientific', medicine stressed experimental data obtained in the laboratory and, consequently, the laboratory itself became a symbol of a new relationship between medicine and science. Despite few observable gains, scientific medicine came to be accepted as a powerful means in conquering disease.[50] For Botkin, 'scientific' medicine with its ideals of quantification, precision, and objectivity became a model. His belief in relevance of the exact sciences to medicine bears a distinct mark of its time – a reductionist–physicalist approach to a living organism and its functions powerfully enunciated by experimental physiology. One of the recurrent themes of his *Clinical Course* of 1867 is the dichotomy of medicine, its division into art and science. In the preface, he wrote:

> The ideal science of medicine would use mathematical laws to describe and determine all bodily processes. But mathematical laws now cannot be used in dealing with a living organism in either a healthy or morbid state. That is why a physician relies on his bed-side experience and intuition, which comprises his gift or art, to cure, but not on his knowledge of physical and chemical regularities underlying normal and pathological processes in the organism. Hence, medicine, being one of the natural sciences, is not yet an independent science, it is only on the way to acquire its independence.... Physical and chemical mechanisms in a living organism are so complex that it is still not possible to bring various manifestations of healthy and diseased organisms under mathematical laws despite all attempts of the human intellect. It is quite easy for those keen on algebra to solve an equation with an unknown value; it is another matter to solve a problem in practical medicine. You may be acquainted with physiology, pathology, and therapeutics, but still you need skills, intuition, and the gift for applying your knowledge to an individual patient. And these comprise an art, hence we talk about practical medicine as non-exact science.... Here lies a difference between a natural scientist and a physician. The first will not make an arbitrary conclusion, he can wait and verify his assumption by an experiment; a physician, however, cannot wait, he must diagnose a case even if he has meagre data, when logical analysis is impossible, and only the course of the disease or post mortem will show whether the hypothesis was true or not.[51]

Contemporary medicine was still closer to an art or craft, rather than to science, and it was difficult to reconcile them in practical medicine. It was for this very reason some of Botkin's fellow physicians, at the dawn of their career, had chosen science – Sechenov, experimental physiology, and Borodin, chemistry. Committed to mathematics, in which he was an accomplished student in his youth, Botkin too felt that his true vocation was science. The necessity to submit logic to intuition, to allow empiricism in diagnostics and treatment, and the severe limitations of existing therapeutic means made medical practise for Botkin a burden. He wrote to Belogolovyi, 'from all my activities, teaching is the only one that interests and revives me; medical practise is irksome to me.'[52] In his private correspondence, Botkin admitted that his enormous responsibilities as a practicing physician frequently did not leave time to elaborate and analyse each case and to compare them with other cases, and to distinguish particular features of disease in a certain patient. The only imperative here was to help the patient and to relieve him from suffering. He could not do much in severe cases, and this brought him feeling of helplessness that at times turned his private practice into a heavy obligation. Despite certain disappointments in the potential of practical medicine, Botkin remained a faithful bedside physician widely known for his admirable abilities for cure and care, and for kind and helpful attitude to his patients.

Ironically, Botkin could not escape an accusation of political nihilism in connection with the outbreak of plague in the Astrakhan region of the Volga in 1879. Soon after the first cases of plague were reported in Astrakhan, Botkin, at one of his lectures, demonstrated a patient with the preliminary diagnosis of plague. Fortunately for the patient the recovery occurred on the twenty-eighth day, but for Botkin this ill-fated case brought in anxiety and disappointment. A special commission examined the case, and diagnosed it as an idiopathic bubo – a swelling of the lymph nodes – presumably caused by syphilis, but could not explain such signs as skin eruption. Belogolovyi at the end of the 1890s, after Botkin's death, wrote that Botkin was well aware that the patient previously had syphilis. Belogolovyi suggested that it might have been a case of *tularemia bubonica*, not known at that time. Botkin's enemies were exultant – the St Petersburg luminary could not tell plague from syphilis. But the worst consequence was that reactionary newspapers incriminated Botkin as giving grounds for spreading rumours about the beginning of plague epidemics in St Petersburg. These rumours caused panic at the stock exchange and the setting up of a sanitary cordon at the Austro–Hungarian–Russian border. The influencial conservative newspaper of M.N. Katkov emphasised that Botkin's attempt to secure the non-existing outbreak of plague in St Petersburg resulted in losses, which could be compared to the damage caused by nihilist movement.[53] Interestingly enough, the conservative press associated Botkin's strictly medical activities directly with nihilism. Botkin was known among his friends to be indifferent to any politics.[54] By that time, the echo of the once-disturbing nihilist movement had already vanished, but similar movements such as anarchism were under way. The rhetoric of the conservative press was clear – to attract public attention to old–new evils. In this case, Botkin appeared a convenient target. He belonged to a generation of natural scientists of the 1860s closely associated, by the conservative establishment, with the destructive influence of nihilism on Russian society.

This accusation was a heavy blow for Botkin. His reaction to all these attacks was particular painful as his continuous attempts to improve epidemiological conditions in Russia were well known. In 1864, when an outbreak of recurrent typhus struck St Petersburg among the poor, Botkin tried to draw attention of the medical authorities to malnutrition and the unacceptable living conditions of the poor that were largely responsible for high mortality from the illness.[55] Within the next two years, Botkin initiated the foundation of the epidemiological society through the medical faculties of the universities. One of the aims of the society was to provide medical and financial help to victims of epidemics through physicians. In the *Arkhiv sudebnoi meditsiny i obshchestvennoi gigieny* [*Archive for Forensic Medicine and Public Hygiene*], edited by Botkin's friend Dr Lovtsov, Botkin published the proposed statutes of the society.[56] This idea did not materialise, but at least useful information on epidemiology and hygiene and sanitary measures was available for both physicians and the public through the *Epidemiologicheskii listok* [*Epidemiological Leaflet*] published by Botkin and Lotsov for two years. Botkin's initiative to create epidemiological society had found neither support nor understanding from both medical faculties and governmental officials at the State Medical Department. In was only in 1886, nearly two decades later, that the Government had to turn its attention to epidemiological control and measures. The International Sanitary Conference in Venice pointed out that Russia, with her recurrent epidemics of cholera and typhus, should take urgent measures to sanitise her ports and border regions, otherwise commercial relations with her would be reconsidered. After such a serious reprimand, the Government was pushed to take serious steps to improve sanitary conditions in the country. This time Botkin headed the newly founded Commission for the Improvement of Sanitary Conditions and Decrease of Mortality in the State Medical Council.[57]

Botkin enjoyed a successful academic career at the Medico-Surgical Academy. He published about seventy-five works in various fields, therapy, infectious diseases, experimental pathology, and pharmacology. A recognised founder of military field therapy in Russia, he adapted treatment of internal diseases to military conditions. In his youth, he worked in the military hospital in the Crimea and, during the Russo–Turkish campaign of 1877, spent seven months at the Balkan front. As a personal physician to Alexander II, Botkin moved with the Tsar's headquarters from place to place and, by his own initiative, inspected the military hospitals and lazarettos. The difference in medical services during the two campaigns, the Crimean and the Russo–Turkish, was remarkable. More hospitals were arranged and they were adequately equipped and supplied. The evacuation of the wounded was performed much more efficiently. The services of the Sisters of Mercy and generous private donations were exerted on a larger scale. Volunteers of the Red Cross, which did not exist during the Crimean War, also brought relief to the wounded. Botkin also acknowledged that medical and sanitary personnel were better trained and more skilful. At the same time, he was overwhelmed and frustrated when faced with poor administration, indifference and the superficial attitude of the medical authorities to the problems of the hospital, the ineffectiveness and helplessness of the medical personnel in severe cases, and the suffering and painful distress of the wounded. Botkin's letters to his wife, written in the form of a diary, related his experiences during the Balkan campaign.[58] Botkin's military services as well as his public and philanthropic activities were highly appreciated by the military command and the Tsar.[59]

Botkin's later career concentrated on building and administering the St Petersburg hospital system, with special emphasis on the control of infectious diseases. Here, his students and disciples dominated, through appointments as hospital directors arranged through Botkin's influence. All these hospitals were well equipped with laboratories for clinical testing and research.[60] His major contribution to Russian medicine was the introduction of the laboratory to the clinic and continuous efforts to develop the scientific approach in practical medicine. His commitment to the construction of 'scientific medicine' at the Medico-Surgical Academy was the part of a general sentiment of the Russian scientific and medical élite who believed that society would benefit from scientific knowledge and practical education of specialists in various fields, medicine in particular. Botkin encouraged science-based studies of the clinical problems in his laboratory, although he knew only too well that even comprehensive analysis of the processes in diseased organisms would not draw him nearer to finding the cure. He fostered experimental research on drugs, although he believed that therapeutics, in contrast to pathology and diagnostics, could not be improved very much through the application of scientific methods. But he did believe that physiology and experimental medicine had limitless potential to advance the understanding of disease and its cure. Therefore

Botkin promoted laboratory training in physiology and basic science in the education of future physicians. In the inspiring words of Pavlov:

> Botkin was the best embodiment of the rightful and fruitful union of medicine and physiology – those two kinds of human activities that raise the building of science of the human organism.... He sent his students to study at the laboratory, and this high esteem of the experiment by a clinician is, in my opinion, of no less importance to the fame of Sergei Petrovich [Botkin] than his clinical activity, which is known all over Russia.[61]

Notes

1. H.H. Euler, *Die Entwicklung der medizinischen Spezialfächer an der Universitäten des deutsches Sprachgebietes* (Stuttgart: Fischer, 1970), 89. On the introduction of scientific medicine into the clinic in Heidelberg University, see A. Tuchman, 'From the Lecture to the Laboratory: the Institutionalization of Scientific Medicine at the University of Heidelberg', in W. Coleman and F.L. Holmes (eds), *The Investigative Enterprise: Experimental Physiology in Nineteenth-Century Medicine* (Berkeley: University of California Press, 1988), 65–99: 73.
2. S.P. Botkin, 'Iz pervoi klinicheskoi lektsii', *Meditsinskii vestnik*, 41 (1862), 391–3: 392.
3. *Ibid.*, 393.
4. *Rossiiskii gosudarstvennyi voenno-istoricheskii arkhiv,* Moscow, fond 316, list 4–5; 8, 1860.
5. *Ibid.*, fond 316, opis' 31, delo 573, list 3, 19 October 1860.
6. W. Coleman, 'The Cognitive Basis of the Discipline: Claude Bernard on Physiology', *Isis* 76 (1985), 49–70: 62–3.
7. C. Bernard, *Introduction à l'étude de la médecine expérimentale* (Paris: J.B Baillière, 1865), English translation, C. Bernard, *Introduction to the Study of Experimental Medicine,* H.C. Green (trans.) (New York: Dover Publications, 1957), 149. On Bernard and scientific medicine, see M. Hagner, 'Scientific Medicine', in D. Cahan (ed.), *From Natural Philosophy to the Sciences: Writing the History of Nineteenth-Century Science* (Chicago: University of Chicago Press, 2003), 49–87: 54–5.
8. S.P. Botkin 'Report to President of the Academy, 1860', in *Rossiiskii gosudarstvennyi voenno-istoricheskii arkhiv,* Moscow, fond 316.
9. V.N. Sirotinin, 'Biograficheskii ocherk', in S.P. Botkin, *Kurs kliniki vnutrennikh boleznei* (St Petersburg: Izd-vo Panteleeva, 1912), 1–29: 20; 25–6.
10. N.A. Belogolovyi, *S.P. Botkin: ego zhizn' i meditsinskaia deiatel'nost'* (St Petersburg: Tipografiia I.N. Erlikha, 1892), 30–2. The following

biographical deatails are taken from the principal biography of Botkin, first published a decade after Botkin's death by his close friend and fellow physician, Nikolai A. Belogolovyi.

11. *Rossiiskii gosudarstvennyi voenno-istoricheskii arkhiv,* Moscow, fond 316, opis' 32, delo 3, 182–4.
12. S.P. Botkin to V.P. Botkin, letter dated 10 December 1861, St Petersburg, in Belogolovyi, *op. cit.* (note 10), 26–7.
13. I.M. Sechenov, *Autobiographical Notes,* D.B. Lindsley (ed.), K. Hanes (trans.) (Washington: American Institute of Biological Sciences, 1965), 112.
14. Botkin had five sons and a daughter from his first marriage, and six daughters from his second marriage. The fate of Botkin's son Evgenii (1865–1918) was tragic. A physician to Tsar Nicholas II, he was murdered with the royal family in Ekaterinburg in 1918. See T. Botkine, *A la mémoire de mon père le docteur Eugène Botkine, médecin de la famille impériale russe* (Paris: B. Grasset, 1985). On Doctor Evgenii Botkin, see also N.A. Sokolov, *Ubiistvo tsarskoi sem'i* (Moscow: Sovetskii Pisatel', 1991, reprint of the Berlin edition of 1925), 283–94.
15. Belogolovyi, *op. cit.* (note 10), 36–8, 45–6.
16. S.P. Botkin, *Kurs kliniki vnutrennikh boleznei* (St Petersburg: Izd-vo voennogo ministerstva, 1867), Vol. I, 17. *Kurs* was published in Germany as S. Botkin, *Medicinische Klinik in demonstrativen Vorträgen: zur Diagnostik, Entwicklungsgeschichte und Therapie der Herzkrankheiten* (Berlin: A. Hirschwald, 1867).
17. S.P. Botkin, *Kurs kliniki vnutrennukh boleznei* (St Petersburg: Izd-vo Voennogo ministerstva, 1868), Vol. II; published in Germany as S. Botkin, *Medicinische Klinik in demonstrativen Vorträgen: über das Fieber im allgemeinen Flecktyphus* (Berlin: A. Hirschwald, 1869).
18. S.P. Botkin, *Kurs kliniki vnutrennukh boleznei* (St Petersburg: Izd-vo voennogo ministerstva, 1875), Vol. III; published in Germany as S. Botkin, *Die Kontraktilität der Milz und die Beziehung der Infektionsprocesse zur Milz, Leber, den Nieren und dem Herzen* (Berlin: A. Hirschwald, 1874); and *idem.*, 'Über die Reflexerscheinungen im Gebiete der Hautgefässe und über der reflektorischen Schweisse', *Berliner klinische Wochenschrift,* vii (1875), 1–32; continued in viii (1875), 1–28.
19. On the Berlin clinical school, see K. Faber, *Nosography: The Evolution of Clinical Medicine in Modern Times* (New York: AMS Press, 1930), 59–94.
20. L.I. Kutsenko, *Istoricheskii ocherk kafedry akademicheskoi terapevticheskoi kliniki Voenno-meditsinskoi akademii, 1810–98* (St Petersburg: Izd-vo K. Ricker, 1898), 232–3.
21. I.P. Pavlov, 'Tovarishcheskaia pamiatka vrachei vypuska 1879 g., izdannaia ko dniu 25-letiia so dnia okonchaniia kursa' in I.P. Pavlov, *Izbrannye trudy* (Moscow: Meditsina, 1999), 24–6: 25.

22. D. Todes, *Pavlov's Physiology Factory: Experiment, Interpretation, Laboratory Expertise* (Baltimore: Johns Hopkins University Press, 2002), 61, 301.
23. Kutsenko, *op. cit.* (note 20), 229–31, see also F.R. Borodulin, *S.P. Botkin i nevrogennaia teoriia meditsiny* (Moscow: Medgiz, 1949), 97. In 1869, Botkin began to edit *Arkhiv kliniki vnutrennikh boleznei [Archive for the Clinic of Internal Diseases by Professor Botkin],* the first specialised journal of its kind in Russia. Thirteen volumes were published during twenty yeatrs of its existence.
24. I.P. Pavlov, 'Sovremennoe ob'edinenie v eksperimente glavneishikh storon meditsiny na primere pishchevareniia,' in *Trudy ob-va russkikh vrachei v S.-Peterburge za 1899-1900 gg.,* 67 (1900), 197–242, in I.P. Pavlov, *Polnoe sobranie sochinenii* (Moscow–Leningrad: Akademia Nauk, 1946), 6 vols, Vol. I, 364–5.
25. Botkin, *Kurs kliniki, op. cit.* (note 16), Vol. I, 17.
26. Botkin, *Kurs kliniki, op. cit.* (note 17), Vol. II, 156–7.
27. Botkin, *Kurs kliniki, op. cit.* (note 18), Vol. III, 61.
28. *Ibid.,* 117.
29. Borodulin, *op. cit.* (note 23), 78.
30. Botkin, *Kurs kliniki, op. cit.* (note 17), Vol. II, 77.
31. V.V. Pashutin, *Kurs obshchei eksperimental'noi patologii* (St Petersburg: Ricker, 1885), 2 vols, Vol. I, 19.
32. Belogolovyi, *op. cit.* (note 10), 17.
33. Borodulin, *op. cit.* (note 23), 96; 127.
34. Faber, *op. cit.* (note 19), 76.
35. G.A. Zakhar'in, *Klinicheskie lektsii i izbrannye stat'i* (Moscow, 1894), vii.
36. Belogolovyi, *op. cit.* (note 10), 70.
37. H. Huchard (intr.), 'Exposé de l'enseignement clinique par Zakharin', in G. Zakhar'in, *Leçons cliniques sur les maladies abdominales et sur l'emploi interne des eaux minerales* (Paris: Octave Doin, 1893), iii; English-language edition: A. Rovinsky (trans.), *Clinical lectures: Delivered Before the Students of the Imperial Moscow University* (Boston: Damrell & Upham, 1899).
38. N.F. Golubov, 'O napravleniiakh v russkoi klinicheskoi meditsine', in Zakhar'in, *op. cit.* (note 35), 27.
39. K. Wenig, 'Introduction', in K. Wenig (ed.), *Rudolf Virchow und Emil du Bois-Reymond: Briefe 1864-1894* (Marburg: Basilisken Presse, 1995), 1–39: 31.
40. Botkin, *Kurs kliniki, op. cit.* (note 16), Vol. I, 17.
41. *Ibid.,* 4.
42. S.P. Botkin, *Obshchie osnovy klinicheskoi meditsiny,* Rech' proiznesennaia na torzhesvennom akte v Voenno-meditsinskoi akademii, 7 dekabria 1886 (St Petersburg: Izd-vo K. Ricker, 1887), 8.
43. Golubov, *op. cit.* (note 38), 49.

44. Borodulin, *op. cit.* (note 23), 77.
45. On functional diagnosis as a trend in clinical medicine, see Faber, *op. cit.* (note 19), 112–71. In Russia, despite the immense authority of pathological anatomy, the functional approach was essentially reflected in the works of Botkin and other notable clinicians, Nikolai Sklifosovskii (1836–1904) and Alexander Ostroumov (1844–1908). Russian clinician E.E. Eikhvald wrote in 1871:

    > It is necessary to take not an anatomical but a physiological viewpoint. The anatomical viewpoint as the basis of modern pathology gives medicine a somewhat gloomy character. From the anatomical viewpoint all disorders of blood circulation are incurable.... The physiological viewpoint is more consoling and no less important.

46. Golubov, *op. cit.* (note 38), 7.
47. Belogolovyi, *op. cit.* (note 10), 42.
48. Borodulin, *op. cit.* (note 23), 81.
49. On the development of nineteenth-century therapeutics, see E.H. Ackerknecht, *Therapie von den primitiven bis zum 20. Jahrhundert* (Stuttgart: Enke, 1970), 95–121; see also P. Diepgen, *Geschichte der Medizin: Die historische Entwicklung der Heilkunde und des ärztlichen Lebens* (Berlin: W. de Gruyter, 1949–55), 2 vols, Vol. I, 163–6; for a useful discussion on therapeutic reasoning in the mid-nineteenth century see G. Weisz, *The Medical Mandarins: The French Academy of Medicine in the Nineteenth and Twentieth Centuries* (New York: Oxford University Press, 1995), 159–67.
50. On the relationship between medicine and science and its influence on nineteenth-century medical teaching, see J. Bleker, 'Die Idee der Einheit von Theorie und Praxis in der Medizin und ihr Einfluss auf den klinischen Unterricht im 19. Jahrhundert,' *Arzt und Krankenhaus,* 55 (1982), 232–6; see also T. Lenoir, 'Laboratories, Medicine and Public Life in Germany 1830–1849', in A. Cunningham and P. Williams (eds), *The Laboratory Revolution in Medicine* (Cambridge: Cambridge University Press, 1992), 14–71: 36–7; and A. Tuchman, *op. cit.* (note 1), 85–6.
51. Botkin, *Kurs kliniki, op. cit.* (note 16), Vol. I, 3–4.
52. N.A. Belogolovyi, 'Vosponimaniia o S.P. Botkine', in *idem, Vospominaniia i drugie stat'i* (St Petersburg: Tipografiia K. Aleksandrova, 1897), 251–375: 333.
53. Belogolovyi, *op. cit.* (note 10), 55.
54. *Ibid.,* 58.
55. S.P. Botkin, 'K etiologii vozvratnoi goriachki', *Meditsinskii vestnik,* i (1865), 4–5.
56. S.P. Botkin, 'Proekt ob uchrezhdenii epidemiologicheskogo obshchestva', *Arkhiv sudebnoi meditsiny i obshchestvennoi gigieny,* iv (1865), 24–6.

57. Borodulin, *op. cit.* (note 23), 94.
58. Botkin to his wife, letter dated 17 June 1877, in S.P. Botkin, *Pis'ma iz Bolgarii, 1877 god* (St Petersburg: Izd-vo Panteleeva, 1893), 30; on Botkin at the Balkan front see, Belogolovyi, *op. cit.* (note 10), 51– 4; Belogolovyi, *op. cit.* (note 52), 344–7.
59. 'O sluzhbe ordinarnigo professora akademii Botkina: Apr. 4, 1862 – Dec. 17, 1890', in *Rossiiskii gosudarstvennyi voenno-istoricheskii arkhiv,* Moscow, fond 316, op. 60, d. 373, listy 1–200.
60. Aleksandrovskaia barachnaia bol'nitsa [Aleksandrovskii Hospital for Infectious Diseases], headed by Botkin's student N.I. Sokolov, was one of the best in St Petersburg. It was built according to Botkin's plan – each department was housed in a separate pavilion to secure strict isolation of the infectious patients (hence the name 'barachnaia' or consisting of barracks). The wards in the pavilions were spacious and light. There were two laboratories – one, specially equipped for testing water, the other for clinical research – and a post-mortem theatre. Another of Botkin's disciples, A.A. Nechaev, arranged two specialised laboratories and a post-mortem theatre at the Obukhovskii hospital, which he headed. In 1880, the Society of Russian Physicians of St Petersburg donated funds for the construction of the hospital for the poor. This hospital acquired the newest sanitary equipment such as a steam disinfection chamber for patients' clothes. According to his colleagues and the city's medical authorities, Botkin put all his heart into this hospital, it was his pride and joy, and he usually invited his foreign colleagues to visit the place. Professor Leiden at one of the meetings of the Berlin Medical Society was reported to say:

    > Botkin executed great influence on the filling of medical appointments in St Petersburg hospitals by his disciples. During my visit I saw hospitals excellent both in scientific and medical matters; we only can wish our hospitals were the same kind and quality.

    See, the account of the head of the city administration, V.I. Likhachev, cited in Belogolovyi, *op. cit.* (note 10), 57– 60. Botkin's idea to build free hospitals from the funds of physicians was popularised through the medical gazette *Vrach,* edited by V.A. Manasein (1841–1901), Professor of Therapeutics and Pathology at the Medico-Surgical Academy, and former student of Botkin. Manasein carried out a vigorous campagn for organisation of local societies of physicians, and building hospitals on the funds of these societies. By 1882, the societies and their hospitals had already been in operation in all Russian industrial centres. On the history of the hospital system in St Petersburg, see T. Grekova and I. Golikov, *Meditsinskii Peterburg* (St Petersburg: Folio–Press, 2001), 323–58.

61. I.P. Pavlov, 'Sovremennoe ob'edinenie v eksperimente glavneishikh storon meditsiny na primere pishchevareniia', Rech, chitannaia v torzhestvennom zasedanii Obshchestva russkikh vrachei v pamiat' S.P. Botkina, 1899', in Pavlov, *Izbrannye trudy, op. cit.* (note 21), 292.

11

The New Discipline of Russian Physiology: Sechenov's Laboratory

By the late 1850s, German experimental physiology was an established discipline with spectacular developments, a growing body of publications, an expanding community of physiologists, and increasing prestige of its institutions and laboratories. Studying in Germany at that time, Sechenov witnessed the advances physical physiology had made, led by the brilliant quartet of Emil du Bois-Reymond in Berlin, Hermann Helmholtz in Heidelberg, and Carl Ludwig and Ernst Brücke in Vienna. What Sechenov found in their laboratories was a new science, which applied the methods of physics, optics, acoustics, mechanics, and advanced mathematics to trace, record, and explain, with unprecedented exactitude and elegance, the complex bodily functions. In Russia, there was no tradition in physiological and medical research comparable to that of Germany and France, where experimental medical science was represented by the brilliant François Magendie and Claude Bernard. The Russian tradition in physiology started with Sechenov and his original research on both blood gases and the centres of inhibition in the brain. Sechenov virtually built physical physiology in Russia from scratch. Just as Ludwig was, in the words of du Bois-Reymond, the 'Fahnenträger der Schule', the standard bearer of a new school of physiology in Germany,[1] Sechenov became an ardent proponent and founder of the new discipline in Russia. However, his scientific fame in Russia was exceeded by the overwhelming success of his popular writings. An accomplished laboratory scientist, skilled in precise methods of research, Sechenov used to rely on facts carefully worked out in experiment. His methodology in the study of blood gases and central nervous inhibition won Sechenov esteem among the German physico-physiologists. On the other hand, his *Reflexes of the Brain*, which attempted to reduce psychological phenomena to physiological functions in response to external stimuli, simultaneously gave Sechenov a reputation among Russian radical intellectuals, mainly dilettantes in experimental physiology, for whom science was merely an omnipotent tool for attacking official ideology. *Reflexes of the Brain*, made physiology an enormously significant topic of debate in Russian society, and its author a landmark figure in the intellectual and liberal upsurge of the 1860s.

The beginning of a physiologist's career

Sechenov began to think about an academic career when studying in Germany. In 1857, he applied for the chair of physiology at both Moscow and Kazan Universities. In his application, he strategically pointed to his major assets, namely his training in German laboratories and a background in physics and mathematics from the Engineering School. In his view, laboratory training was crucial for the development of research skills and the independent critical thinking, while knowledge of higher mathematics had already become an indispensable tool in physical physiology. Sechenov also made it clear that his intention was to set up a laboratory for teaching and research, and equip it with the necessary instruments and apparatus.[2] His rhetoric was persuasive, but the University Medical Councils in Moscow and Kazan were not yet ready to undertake the introduction of the laboratory and a new program in practical training in physiology. The major reason was lack of back-up from the Ministry of Education to support such ambitious and costly innovations, which had to wait yet another decade to slowly materialise at the university medical faculties.

Sechenov found powerful supporters for his plan in the reformist administration of the Medico-Surgical Academy. Glebov, Vice-president of the Medico-Surgical Academy needed a physiologist capable of running the laboratory and conducting independent research. He encouraged Sechenov, who was finishing his studies in Heidelberg, to submit his dissertation for defence to the Medico-Surgical Academy.[3] Upon returning to St Petersburg in 1860, Sechenov, with Glebov's assistance, published his doctoral thesis in *Voenno-meditsinskii zhurnal*.[4] After the defence of his thesis, Sechenov was appointed adjunct professor. Glebov was shrewd enough to recognise the potential of the young physiologist to represent such an important medical subject and to introduce an innovative approach to its practical teaching. Years later, Sechenov reflected on his first appointment:

> Thinking now on whether I deserved a chair of experimental science at that time, I say that I knew less than our assistants now. Even those of our assistants who have not studied abroad are more familiar with physiological practice in very diverse directions.... I was able to use only frogs. It is true that I saw numerous experiments in Ludwig's laboratory, and sometimes I assisted Ludwig, but I was skilled only in those methods which I used in my project.... They [the Conference] took me because I, even with my limited knowledge, was nevertheless the first of the Russians who had partaken of Western science with such leading figures as my teachers in Germany.[5]

True, Sechenov's training in physiology was not very diverse. Inviting him to the Medico-Surgical Academy, Glebov understood, as Sechenov himself did, that electrophysiology was at the cutting edge of physical physiology, and the methods it exploited required knowledge and skills beyond conventional medical training.

In the spring of 1860, Sechenov gave his first course in electrophysiology. The subject, unknown in Russia at that time, had fascinated him during his studies with du Bois-Reymond. The central novelty of the course was the demonstration of basic electrophysiological experiments, such as the phenomena of negative variation in nerves and muscle using a special arrangement of du Bois-Reymond's devices, which Sechenov had brought with him from Berlin.[6] In the autumn, Sechenov gave a course on general physiology, which dealt in detail with respiratory function and blood as a physico-chemical system, the subject of Sechenov's own investigations at Ludwig's laboratory in Vienna. Sechenov sought to convey to his students new facts that had been carefully worked out in the laboratory, and methods and techniques by which they had been obtained and tested. Thus, to determine the mass of red blood cells, the method of Prévost and Dumas[7] and the more recent tenchniques of Hoppe-Seyler and Schmidt[8] were used. Sechenov also presented the latest results of the organo-chemical analysis of fibrin and albumen in the blood, introduced by Hoppe-Seyler and published in his *Handbuch* in 1858.[9] For the microscopic analysis of the blood, a device for counting and measuring the volume of red blood cells developed by Karl Vierordt was used.[10] To measure blood velocity, several methods and devices were applied, those by Vierordt, Alfred W. Volkmann,[11] and Ludwig. Years later, Sechenov remarked ironically about his first course in physiology:

> It turns out that I was not able to distinguish the important from the secondary in all cases, I could not express exactly in words various concepts.... There were some *naïvetés,* but the German textbooks saved me from gross mistakes.[12]

Sechenov acquired premises for his laboratory in a wing of the hospital building, next to the old anatomy theatre, consisting of two large rooms that had once served as a chemical laboratory and still had a few instruments. The Conference allotted 2,500 roubles for further equipment and instruments. Sechenov purchased from Berlin and Vienna additional electrophysiological apparatus, including Grove elements,[13] tripods with various supports, Störer switchboard, a polar switch, a magnetic interrupter of current, the rheochord for changes in the amount of constant current, magnets, Meyer absorptiometers, a Bunsen absorptiometer, blood-gas apparatus, Bunsen apparatus for extracting electrolytic hydrogen and detonating gas, Meyer forceps, glassware, and rubber tubes.[14] Another 1,000 roubles were granted to purchase a Ludwig kymograph, a 100-litre volume gasometer, a Pflüger myograph, and twenty-eight microscopes. The list of instruments is suggestive that Sechenov was planning to limit his lecture demonstrations to basic experiments in nerve and muscle physiology and, as far as his own research was concerned, he had the necessary equipment to continue his electrophysiological and absorptiometric studies.

The monthly budget of the laboratory was thirty roubles, part of which was spent on animals and on payment for taking care of them. Mainly frogs were bought, at five kopecks per hundred, but also some rats, rabbits, guinea pigs, fish, and lampreys. Sechenov received additional funding, '120 roubles for practical studies in physiological chemistry, and 200 roubles for physiological experiments'.[15] Officially, the laboratory did not have an assistant, but a graduate specialising in physiology from the Institute of Physicians usually assisted Sechenov in preparing for lecture demonstrations. Later, Sechenov recalled:

> Everybody who happened to set up a new laboratory could agree that even an experienced researcher spends years on training two or three independent assistants. We, in the 1860s, had even more difficulties as everything was so new for us and we had to do everything ourselves from scratch.[16]

Despite obvious limitations in space, equipment, and skilled staff, Sechenov's laboratory began to offer first-hand instruction in basic physiological techniques to a group of advanced students and graduates of the Institute of Physicians. They also acquired practical skills in handling various apparatus, sometimes devising and adjusting instruments to adapt them to the special requirements of particular experiments.

In 1862, Sechenov published his *O zhivotnom elektrichestve* [*Lectures on Animal Electricity*],[17] the first Russian textbook dealing with the concepts and methodology of electrophysiological studies. As with du Bois-Reymond's *Untersuchungen*, Sechenov's *Lectures* promoted and consolidated the introduction of methods and instruments in medical training using a short and dynamic history of electrophysiology.[18] Sechenov's text is a clear and comprehensive review of what had been done in the field, starting with famous experiments by Luigi Galvani, Alessandro Volta, and Leopoldo Nobili. Unlike du Bois-Reymond, who in his *Untersuchungen* sharply criticised Carlo Matteucci's experimental results, Sechenov appreciated the contributions of the Italian physicist, who was the first to demonstrate the electrical negativity of a cross-section of the muscle and described the electric oscillations in a tetanised muscle.

All successes in the field Sechenov ascribed to advancements in instrumentation. The breakthrough was du Bois-Reymond's modification of Nobili's galvanometer, a more sensitive instrument capable of detecting and amplifying weak bioelectric currents and their direction. Sechenov praised the 'sparkling wit and talent' of the German scientist, and pointed out that du Bois-Reymond's theory raised questions about the molecular structure of nerves and muscles for the first time.[19] The *Lectures on Animal Electricity* was nominated for the prestigious Demidov Prize at the St Petersburg Academy of Sciences, awarded in 1863. It seems that Zinin saw enough reason to introduce Sechenov to Karl Baer, the famous embryologist, as a possible successor at the chair of anatomy and physiology at the Academy of Sciences. Suspecting a concealed intrigue between the so-called German and Russian factions in the Academy of Sciences, in which he was assigned to play some role, Sechenov 'flatly declined' the offer to apply for membership at the Academy of Sciences. He felt that he did not deserve 'such a high honour'.[20] This episode is typical of Sechenov, he would always avoid any official responsibilities as well as any formal ceremonies held in his honour.[21]

In 1862, Sechenov requested from the Conference one-year's leave for studies in Paris at Claude Bernard's laboratory. Bernard was held in high esteem in Russian scientific and medical circles. In 1860, in appreciation of his contributions to medical science, Bernard was elected a foreign member of the St Petersburg Academy of Sciences. The older generation of Russian medical professors, like Glebov, were admirers of the French school of physiology, and its brilliant representatives, Pierre Flourens,[22] François Magendie, and Bernard himself. Their major works were translated into Russian. When Sechenov had been a student, physiology at Moscow University was taught mainly according to the French textbooks.[23] In 1862, while in Bernard's laboratory, Sechenov performed his famous study on central nervous inhibition, which would powerfully influence his research career for the following decade and would remain one of his major interests for the rest of his life. Sechenov thought of Bernard as 'the most skilful vivisectionist in Europe, a very keen observer and a sober philosopher'.[24] Indeed, Bernard's charisma, his wide physiological outlook and superior abilities as an experimenter, attracted Russian medical scientists and physiologists and, at different times, Sergei P. Botkin, Elie de Cyon, Filip V. Ovsiannikov, Ivan R. Tarkhanov, and Ivan P. Pavlov,[25] worked in his famous laboratory at the Collège de France.

According to Sechenov's description, Bernard's laboratory in 1862 occupied a small room, with just a vivisection table and several cupboards with vessels and basic instruments. The laboratory was not equipped with any electrophysiological devices or a kymograph, the key instruments at most German laboratories of the time. Apparently Bernard's research, which was rarely aimed at achieving quantitative results, did not require appropriate instrumentation. His physiology was based mainly on surgical techniques and basic chemical tests, involving organ sectioning, interruption of nervous pathways, injection of certain drugs, and observation of how a controlled intervention affected the entire living organism. Sechenov was permitted to observe Bernard's vivisectional experiments, although it was of little use as for him as he adhered to a physico-chemical approach. Sechenov worked in the auditorium, adjacent to Bernard's study, at the professor's table in front of the audience benches. There were no other researchers at the laboratory, and so he worked alone, sometimes in the presence of the old army doctor Rancheval, Bernard's unofficial assistant. For the several months that Sechenov stayed in the laboratory, Bernard only spoke occasionally to him enquiring about:

> [H]ow the Germans look at certain problems, which were of interest to him… he was not a teacher like the Germans. He worked out the topics, which arose in his mind, with his own hands, not leaving his study room… that is why it was impossible for someone as myself, who came to him for a short period of time, to learn anything in his laboratory.[26]

The project devised by Sechenov 'had direct connection to the problem of the ability of the human organism to depress impulses to movements and to withstand in general any temptation for various actions'. Basically, it was a study of inhibitory influences on the spinal chord of the frog, which presumably originated in the cerebrum, using techniques for measuring reflexes introduced by the Viennese physiologist Ludwig Turk (1810–68).[27] Apparently Bernard, with his thorough practical approach, was neither interested in the problem nor the methods of Sechenov's study and 'treated it with complete indifference'.[28]

Sechenov, however, became increasingly interested in the problems related to how 'the will is capable to evoke and suppress movements' during his sojourn in Germany. The problem of the influence of the nervous system on motor function came into focus in 1845 in connection with a famous experiment by Eduard and Ernst Weber. The experiment demonstrated that the heatbeat accelerates after cutting the vagus nerve, and after excitation of the distal end of the cut nerve, the heartbeat slows down to a full diastolic stop. The Webers correctly concluded that normally weak excitation from the cerebrum must travel continuously along the vagus moderating cardiac activity. Some twelve years later, in 1857, Eduard Pflüger showed the influence of the great cranial nerve on the movements of the intestine. These studies suggested the existence of inhibitory and stimulating influences on the inner organs, and that these influences are controlled by special areas in the brain. Describing the background of the problem, Sechenov pointed out that studies on cerebral nervous centres had been lacking for at least two reasons. Numerous experiments on the cerebrum and its function in locomotion performed by the leading French researchers in the field, Flourence, Magendie, and Longet, had yielded contradictory and confusing results. Sechenov referred to Ludwig's comment on the crude methods used in such experiments: 'It is equivalent to studying the mechanism of a watch by shooting at it with a gun.'[29]

> In contrast, a series of brilliant experiments on isolated nerves performed by du Bois-Reymond and Helmholtz in the late 1840s and early 1850s gave accurate and reliable results, and since then research in nerve–muscle physiology in Germany had been focused on more easily accessible isolated nerves.[30]

What Sechenov found interesting in Weber's studies was a hypothesis that normally weak inhibitory influences on the reflex activity of the spinal cord derive from the cerebrum. Sechenov decided to challenge the problem in 1862 in Paris. The design of his experiments was quite simple. He sectioned the cerebrum from the front, measured the strength of reflexes with a metronome – the time from the stimulation to the motor response – after each section, then applied chemical stimulation to the exposed cross section of the cerebrum, and measured the reflexes again. The experiments demonstrated that 'centres modérateurs de l'action réflexe', or 'Hemmungscentra' did exist in the frog brain, and inhibitory influences on reflex activity originated from these centres.[31] Prior to Sechenov's studies it was known that reflex reactions are intensified in the decapitated frog. Moreover, Ludwig also attempted to study a depressing action of thalami optici in the brain on reflex activity. Sechenov, however, independently discovered brain centres that depress spinal reflexes, and formulated a general physiological approach to the newly observed phenomenon of central nervous inhibition. The core of his theory was the conclusion that central nervous inhibition is the basis of co-ordination of reflex activity, and that it is a process that preceeds increased excitability of the corresponding neural elements. Sechenov's discovery was immediately recognised. Bernard wrote an introduction to Sechenov's publication in France.[32] En route to St Petersburg, Sechenov went to Vienna and Berlin, to demonstrate his experiments to Ludwig, Brücke, and du Bois-Reymond. Soon Ludwig informed Sechenov that he had succeeded in reproducing the 'beautiful experiments' at one of his lectures and thus confirmed general recognition of Sechenov's discovery.[33] Sechenov dedicated his work to Ludwig.

Reflexes of the Brain

Sechenov returned to St Petersburg in the spring of 1863, and that summer wrote his first and most famous essay, *Refleksy golovnogo mozga* [or *Reflexes of the Brain*], aimed at making a closer connection between psychic phenomena and functions of the nervous system.[34] He speculated that 'all acts of conscious and unconscious life are reflexes with regard to their origin'. In a novel style, and with many examples, he attempted to prove the formative influence of external factors on nervous processes.

> [I]n psychic life, wishes and desires are the determinants of actions… passions are rooted, directly or indirectly, in the so-called sensory systems of the man, capable of increasing to the degree of strong desires (hunger, self-preservation, sexual feeling, etc.) and manifested by very abrupt actions or deeds; therefore, they can be treated in the category of the reflexes with intensified results.[35]

Sechenov based his 'hypothetical propositions' on his discovery of the centres of inhibition in the central nervous system and suggested that his experiments demonstrating the existence of such centres 'had a direct relation to acts of consciousness and will'.[36] Why Sechenov decided to write on such a provocative topic in a genre of popular prose is an intriguing question. Ivan Petrovich Pavlov suggested an explanation which pointed to personal rather than political circumstances so much emphasised by Soviet historiography.[37] In a letter to M.N. Shaternikov, Pavlov enquired about Sechenov's relationship with Maria Alexandrovna Bokova at the time, when Sechenov was writing the essay, explaining his interest in Sechenov's private life 'for the sake of a better understanding and an adequate appraisal of the scientific image of Sechenov.' Pavlov linked the appearance of *Reflexes* with a kind of 'personal passion'. For Pavlov, the essay, 'a stroke of genius in Sechenov's thought', bore evidence of 'a strong emotional upheaval'. He was convinced that Sechenov 'was possessed by emotions of love, while writing *Reflexes of the Brain*'.[38]

Sechenov first met Maria Alexandrovna Bokova (1839–1929, née Obrucheva) in 1861 at the Medico-Surgical Academy, when female students were allowed into the Academy to listen. A young and attractive lady from a wealthy family, Bokova had received an excellent home education, including languages and music. She studied at Sechenov's laboratory and performed her first research work on colour vision under Sechenov's supervision. By that time, Maria was married to Petr Bokov, a young and very capable physician. It was a fictitious marriage arranged to free Maria from the guardianship of her parents. Like Sofia Kovalevskaia, the future brilliant mathematician, Maria belonged to a generation of young ladies, who did not follow the customary rules of the class to which they belonged.[39] They dreamed of leading independent lives, to be educated and bring benefit to society. Maria continued formal studies in Vienna and obtained her MD from Zürich University in 1870. In 1871, at the end of the Franco–Prussian War she went to work as a Sister of Mercy with the remnants of the French army near Belfort. Many years later, Sechenov wrote about the woman of his life:

> She possessed an exceptional ability to devote herself to people whom she loved and to those who needed her.... At her family estate, she treated the peasants' infirmities so skilfully and generously that she earned the confidence of the population and appreciation of the district council. For those close to her she was constantly a careful nurse, it was almost the main trait of her character.... She inherited honour to the point of punctiliousness from her father, a strict and very educated general.[40]

The early 1860s was an exciting and happy time for Sechenov both in his private life and in his research. He was inspired by recognition of his discovery of the central nervous inhibition in Germany, and zealously pursued further investigations in the direction he had himself defined. He had everything he needed for research and teaching, his laboratory, adequate equipment, a group of capable students, and the topic at hand was so excitingly new and promising. His academic career too, went smoothly. In 1864, he was promoted to ordinary professor with the chair of physiology, the only chair at the Medico-Surgical Academy, which had two ordinary professors, Sechenov and the histologist Nikolai Martynovich Iakubovich (1817–79). It seems that even troubles with the publication of *Reflexes of the Brain* did not upset Sechenov too much. In the autumn of 1863, he submitted the manuscript to *Sovremennik*, the popular liberal journal, which devoted considerable space to natural sciences and to controversial philosophical problems. Sechenov admitted that although his work was based on some valid observations and experimental data, its propositions, however, were hypothetical and conclusions were speculative and controversial – acceptable for popular writing but not for scholarly publication. For the censorship department, however, an essay written by a recognised physiologist was suspicious as it dealt with problems related to consciousness and free will, problems conventionally considered to be beyond the laws of natural history. The censors admitted that the essay, which masterfully resorted to the language of facts, scientific methods and concepts, should be referred to as scientific prose. They requested that the essay be published in a specialised journal, the *Meditsinskii Vestnik*, and insisted on changing its original title, which had been *Popytka ob'iasnit' psykhologicheskie protsessy na osnove fiziologii* [*An Attempt to Establish the Physiological Basis of Psychological Processes*].[41]

Apparently Sechenov informed Ludwig about his troubles with the censorship department. Ludwig expressed anxiety about the official reaction to the publication of *Reflexes of the Brain*, which was 'under special surveillance'. Involvement in political activities during the revolution of 1848 in Marburg had dire consequences for Ludwig's academic career in Prussia, and he had to leave Marburg for Zürich. Ludwig believed that for Sechenov, a good 'Vaterlandsfreund', a loyal patriotic subject, political activity was a career-damaging diversion. He advised Sechenov to continue his investigations on central nervous inhibition, which would reward him with more discoveries.[42] Sechenov's thoughts, however, were again averted from laboratory research. He was deeply affected by the decision of the War Ministry, in 1864, to ban the admission of women to the Medico-Surgical Academy. This prevented Bokova from entering the Academy as a student. To Sechenov's great disappointment, his petition to the Conference of the Medico-Surgical Academy was futile. He was ready to leave the Academy and go with Maria to Vienna, where she could study obstetrics at one of Viennese hospitals. Ludwig was upset by Sechenov's argument with the Academy authorities regarding women's education that gave him the reputation of a troublemaker. In Ludwig's opinion, Sechenov's resignation from the Medico-Surgical Academy would ruin his career. 'The Medico-Surgical Academy', he wrote to Sechenov, 'is so essential for your scientific work and you should concentrate your energy and talent there.'[43] Du Bois-Reymond also was convinced that Sechenov's engagement in such trivialities as women's education was useless. German colleagues had no sympathy for this cause.[44] Interestingly, many distinguished Russian professors, including Sechenov's colleagues at the Medico-Surgical Academy, Borodin and Botkin, were champions of women's education, and gave free lectures to support the women's courses. In Sechenov's case, however, the political was also personal.

Despite these diversions, Sechenov continued his study on the nervous centres and published twenty-five articles in German journals during that period, as well as preparing a monograph on the physiology of the nervous system.[45] He spent all his time at the laboratory, a habit he maintained throughout his life. He lived very close to the laboratory, in a small apartment in the grounds of the Medico-Surgical Academy, and usually worked at his laboratory until late at night with a group of five to ten students.[46] Obviously, after his troubles with the censorship department, Sechenov acquired a reputation as a radical liberal who influenced the minds of the students. One of his students recalled that he never sought popularity among his students with fine words or liberal views. He was readily available to anyone interested in scientific problems, however students never gathered at his home – only his closest associates from the laboratory.

Sechenov's course in physiology enjoyed a great popularity amongst the students of the Academy. The large lecture hall was always full during his lectures, which often lasted two hours instead of one and a half. The students came early to take their seats, and often expressed their gratitude with an ovation to the professor:

> [H]is major merit as a lecturer was his aspiration to confirm all that he said by masterfully executed demonstrations… he formulated ideas from separate facts, and connected various physical, chemical, and anatomical phenomena in one whole, which came to be a biological regularity. He explained complex concepts in such a clear, logical and simple way so that what he said sounded a complete truth to the listeners.[47]

A vivid picture of Sechenov as a lecturer is presented by A.F. Samoilov (1867–1930), Sechenov's collaborator at Moscow University's laboratory in the 1890s.[48] A fine musician, and a friend of the composer Sergei V. Rakhmaninov (1873–1943), Samoilov referred to Sechenov's voice as a wonderful, clear, slightly sharp, high baritone:

> Beautiful diction was combined with masterful speech. He had perfect command of German and French, knew Italian, and his Russian, expressive and accurate was a model of scientific language. Old-fashioned expressions that he sometimes used added a peculiar charm to his speech.[49]

Sechenov's lectures were usually accompanied by demonstrations related mainly to electrophysiology and physiology of the spinal cord. He was especially fond of demonstrating his famous experiment, inhibition of spinal reflexes by stimulating the optic nerve in the frog's brain with sodium chloride – the so-called Türck–Sechenov method.[50] Sechenov never incorporated vivisectional demonstrations on warm-blooded animals into his lectures, and referred to the authority of du Bois-Reymond, Helmholtz, and Brücke, who never used such demonstrations in their physiology courses. To Sechenov, vivisectional demonstrations were intolerable. He believed that the only place for such bloody experiments was the laboratory.[51] The techniques of vivisection and its associated instrumentation were totally absent from Sechenov's experimental practice, which depended primarily upon results from electrophysiology and its methods. His unwillingness to use warm-blooded animals was known among his students and colleagues. Tarkhanov commented:

> Sechenov could not bear yelps and screams of animals during vivisections performed by his students. He used to reproach the students for improper

> anaesthetising of the animals. Sometimes he held his ears or just ran out of the laboratory not to hear the sufferings of the animals. He ultimately could not share the viewpoint of his famous teacher Claude Bernard, who once said that during the experiment he saw neither blood nor sufferings of the animals as he gave himself entirely to the leading idea of the experiment.[52]

Obviously Sechenov was acquainted with basic vivisectional methods but was reluctant to master the necessary operative techniques to use animals other than frogs – which did not scream and yelp during the experiment. Aside from his sympathy towards animals, there is another explanation for the total absence of methods of vivisection in Sechenov's physiology. Like du Bois-Reymond and Helmholtz, Sechenov was a physical physiologist *par excellence*. The problems they studied, such as electrical phenomena in nerve and muscle, acoustic and optical mechanisms of sensory organs, mechanics of circulation, and absorptive characteristics of the blood, required methods employed by physics and mechanics. Methods of vivisection were irrelevant here, and an ideal animal in the experiment was a frog, 'the old martyr of science' in the words of Helmholtz.[53] The classical experiment of nineteenth-century physical physiology was the gastrocnemius twitch, in which an isolated muscle connected to a length of sciatic nerve, was used to study muscle contraction and the transmission of the nerve impulse. The experiment used a frog, and the current that flowed between connected skin (high potential) and muscle (lower potential) was simply known as 'Froschstrom', literally 'frog stream'.

Similar to vivisectional techniques, methods of physiological chemistry were not employed in Sechenov's research. Even his later investigations in the field of physical chemistry required physical, not chemical methods for their solution.[54] Although some of his students, like Pashutin, became interested in problems related to physiological chemistry, Sechenov was reluctant to supervise such projects:

> Unfortunately, experimental research in Sechenov's laboratory was focused exclusively on physiology of the nervous system. The experiments were performed, with rare exception, on frogs. Training in physiological chemistry was poor. Sechenov's supervision in the study of chemical problems could not give us necessary help. To use the advice and the laboratory of other professors seemed inappropriate.[55]

Apparently, Sechenov realised the lack of diversity in experimental methods his laboratory offered to the students. His laboratory was designed and equipped for 'physical physiology'. Such specialised orientation in laboratory training could not be explained by the lack of funds for expansion and diversity. In his experimental research, Sechenov simply did not deal with problems related to other areas of physiology. To acquaint the students with most recent techniques in physiological chemistry, Sechenov undertook the translation of *Lehrbuch der physiologischen Chemie*, one of the first textbooks in the field published by Willy Kühne.[56] Sechenov had studied with Kühne at Ludwig's laboratory in Vienna. A capable physiologist with a thorough training in chemistry, Kühne worked with Ernst Brücke, du Bois-Reymond, and Bernard, and pursued investigations in three different areas – vision, muscles, and digestion. In all of these he applied chemico-physiological methods, and Sechenov was particularly interested in the application of these methods to the study of muscular contraction and innervation.

Sechenov's research interests did not shift during the decade he spent at the Medico-Surgical Academy, so his students' research was concentrated on nerve and muscle physiology. From the annual account of the laboratory, we know the scope of the projects carried out under Sechenov's supervision:

> V. Pashutin: digestion in the intestine; K. Voroshilov: nitrogen exchange in the body; nutritive equivalents of meat and leguminous plants; P. Spiro: physiological–topographical research on the spinal cord; innervation of the respiratory centre in the frog; I. Tarkhanov: summation of the electrical impulses by the nervous centres; heat effect on the sensible nerves, the spinal cord and the brain in a frog; A. Tyshetskii: effect of the direct electrical stimulation of the spinal cord on the locomotion of the frog; M. Litvinov: absorption of carbon dioxide by colloids.[57]

Despite the narrow research orientation Sechenov gave to his students, the main result was that the students acquired adequate experimental skills by assisting and following Sechenov's investigation and pursuing their own research projects. Even more important was 'the method' which comprised of the following: the ability to determine a solvable problem; to adapt or, if necessary, to devise the instrument or its part to study the problem; to develop the protocol of the experiment and to ensure that it yielded reliable and reproducible results; to interpret these results, and arrive at conclusion. Although, during the decade of Sechenov's stay at the Academy, laboratory training in physiology overall still played an insignificant role in the curriculum, the laboratory did become the key component for research and teaching. The most successful of Sechenov's students began their academic careers with defined research objectives and adequate experimental skills. In most cases, these objectives were patterned according to their mentor's experimental approach and methods. I.R. Tarkhanov, V.K. Voroshilov, and P.A. Spiro continued investigative paths in neurophysiology and achieved important results in the field. The research interests of V.V. Pashutin

(1845–1901), the most capable of Sechenov's students, shifted towards physiological chemistry, and Pashutin became a distinguished pathologist.[58]

The discovery of central nervous inhibition, in 1862, ultimately determined Sechenov's area of study for nearly a decade. Only briefly did he return to his absorptiometric studies related to the improvements of the blood-gas pump, which he had devised in Ludwig's laboratory. In the modified version, Sechenov used a thin metal tube instead of a rubber one to connect the receiver for the absorbing liquid to the manometer. A similar connection was used in the air thermometer by the French physicist Henri Victor Regnault (1810–78), whose lecture demonstrations at the Collège de France were attended by Sechenov during his stay with Bernard.[59] For the four subsequent years, Sechenov worked 'exclusively on the nervous system of the frog',[60] and, in 1866, published results of his investigations in his major monograph of the period, *Fiziologiia nervnoi systemy* [*Physiology of the Nervous System*].

Sechenov's *Physiology of the Nervous System* was the first textbook in nerve–muscle physiology written in Russian, and even for the physiologists of the later generation remained a model of the 'clear statement, strict critical approach, and the manner of posing problems'.[61] Basically, Sechenov's monograph was a critical review of the voluminous work done in the field. Chapters 2 and 3, however, incorporated results of Sechenov's own experimental work, aimed at further development of the theory of central nervous inhibition, an important issue that came into focus after his discovery of specific inhibitory centres in the brain.

For Sechenov, as for a number of electrophysiologists of the time, investigation of electrical phenomena in the nerves and muscle was of crucial importance for understanding the nature of nerve excitation and muscle contraction. The major goal was 'to define every possible physical and chemical property of nerve and muscle at rest and changes of these properties, when nerve and muscle are active.'[62] In the early 1840s, du Bois-Reymond, almost simultaneously with Carlo Matteucci, discovered that isolated striated muscle under certain conditions exhibited pronounced and regular electromotive activity. This opened up the vast field of electrophysiology. In his famous *Untersuchungen,* du Bois-Reymond put forward a physical approach, which basically reduced nerve excitation and muscle contraction to electromotive forces. Du Bois-Reymond's adherence to the so-called theory of identity, which likened nerve excitation to an electrical wave, later drew the most criticism. This theory was challenged in the early 1850s by Helmholtz's investigations, completed with a myograph of his modification. Helmholtz established that the speed of transmission in nerves is considerably less than the speed of electrical transmission in a normal conductor. Helmholtz's set of experiments, devised to demonstrate the generation of heat in an isolated muscle during contraction, suggested that chemical changes also took place in the muscle during contraction and nerve excitation. Although Helmholtz otherwise supported du Bois-Reymond's electromotive theory of transmission of nerve impulses, he believed that the underlying mechanisms were chemical in essence.[63]

A decade later, in the early 1860s, du Bois-Reymond's physical reductionist theory was seriously undermined by his student Ludimar Hermann (1838–1914). Hermann proposed the so-called alteration theory, which incorporated all that was known about chemical composition of nerves and muscles in the resting state, in activity, and in rigor mortis. Hermann's theory referred all electromotive activities of living tissue to chemical changes or alterations of the substance without regard to its molecular structure. The alteration theory was widely adopted and further developed by physiologists such as Eduard Pflüger (1829–1910) and particularly by Julius Bernstein (1839–1917), another of du Bois-Reymond's students.[64] Sechenov adopted Hermann's theory, and translated and published Hermann's *Handbuch der Physiologie* (1864) in St Petersburg , which first expounded the alteration theory to a Russian scholarly audience. Sechenov appreciated Hermann's application of a mathematical approach to physiological problems and, in particular, his chemical approach to the study of nervous processes.[65]

By the mid-1860s, two main approaches in nerve–muscle physiological physiology had been formed, the physical theory of classical electrophysiology by du Bois-Reymond, and the the theory of chemical changes by Hermann. Du Bois-Reymond remained convinced of the electrical nature of nerve-muscle excitation. In the mid-1870s, however, he had to admit that that a 'secretion' of stimulating substances occurred at the borders of the contracting elements – 'contractile substances':

> Of the known natural processes which could mediate excitation, as far as I can see, only two come under consideration. Either stimulating secretion should happen at the border of the contractile substances in the form of a thin layer of ammonium or lactic acid, or some other substances excite muscles intensely. Or the effect must be electrical in nature.[66]

Du Bois-Reymond, however, throughout his life remained convinced of the electrical nature of nerve excitation. Already in 1862, Sechenov, in his *Lectures on Animal Electricity*, held that spontaneous chemical processes were the basis of nervous activity.[67] In *Physiology of the Nervous System*, he pointed out that 'transition of the nerve from rest to excitation could serve as an indication of chemical transition inside the nerve when it is active.'[68] Investigating the problem of nerve excitation, Sechenov determined the conditions under which excitation was possible, the influx of arterial blood to the nerve, the connection of the nerve with the nerve centres, and the change from rest to excitation in the nerve. He and his students performed numerous experiments with an impaired blood supply to the nerve and presented convincing proof that a change in the metabolism of the nerve first of all resulted in disturbances in its excitability, and in the corresponding physiological indices of such excitability:

> If the blood supply [to the nerve] is terminated, for a short of time, 5–10 minutes, and then resumed, the nerve restores its excitability; otherwise excitability disappears and the nerve dies. This fact clearly indicates the close link between the chemical activity of the nervous tissue and the physiological property under study [excitability].[69]

Sechenov showed that similar disturbances in excitability of the nerve occurred upon severance of the nerve from the nerve centre. He believed that the ganglia represented 'a kind of nutritive centre, whose activity maintains anatomical, chemical and physiological integrity of the nerve fibre [sensory fibre].'[70] In the living tissue, a synthesis of special substances, 'unstable chemical compounds', was responsible for both muscular contraction and conduction of nervous excitation, providing that a constant influx of nutrients and oxygen was ensured.[71] This idea was further developed and elaborated by Hermann in the third edition of his *Grundriss der Physiologie des Menschen,* published in 1874, where he suggested a cyclical chemical pathway in which, during the restitutive or synthetic phase, glycogen and oxygen were combined with some, as yet unknown, protein. Hermann called this an 'krafterzeugende' ['energy generating'] or 'inorganic' substance and assumed that it was highly unstable.[72]

In the mid-1860s, when Sechenov was writing *Physiology*, experimental data on the chemical properties of nerve fibre were limited to basic facts – nervous tissue at rest had a neutral reaction and acidic during activity:

> Of course it is impossible to draw any important conclusions from such a meagre data. But the change in nerve reaction during transition from rest to excitation is suggestive that the state of excitation is related to some unknown chemical transformations within the nerve. New facts are needed to confirm that these processes are bound to oxidation.[73]

Experimental proof of chemical processes taking place during excitation had to wait until the 1920s. Although the inadequacy of the physical approach to the process of excitation was evident to some German physiologists in the 1860s, the reverse was markedly expressed by Max Verworn (1863–1923) at the end of the nineteenth century. Verworn,

Pflüger's successor at the University of Bonn, was an anti-reductionist, especially interested in physiology of the nervous system and its cellular mechanisms. He took his chemical stand on colloid chemistry as the peculiar essence of life, a kind of vitalism. Verworn particularly objected to the reductionism implied in the analogy of the nerve as a wire conducting an electrical current. He was convinced that 'the assumption that the process of propagation of excitation in the nerve is a strictly physical phenomenon was not only false but even harmfully misleading.'[74] Sechenov, however, remained convinced throughout his life that the 'electromotive theory' of du Bois-Reymond, physical in essence, was a crucial step in understanding nerve muscle excitation on a molecular level.[75]

Concurrent with publishing the *Physiology of the Nervous System,* Sechenov published a revised version of his *Reflexes of the Brain* essay in book form in 1866. The time for publication of such a provocative book was unfortunate. That year, a Moscow student Karakozov made an attempt on the life of Tsar Alexander II, and the Government began to consolidate its power to suppress the growing liberal opposition particularly among university students. The Government also began to tighten the freedom of the liberal press. Under these circumstances the fate of the book was doomed. The St Petersburg Censorship Committee forbade sale of the book and requested that the edition 'should be confiscated and destroyed'. The grounds for such a severe verdict was that:

> [T]he book undermines the moral foundation of society.... Being an expression of extreme materialistic views expounded in such a manner as to seem based on the authority of science, the book is indictable as dangerous reading for people without established convictions.[76]

Moreover, the Censorship Committee summoned the Attorney-General of the St Petersburg High Court of Justice to institute an action against Sechenov. The case, however, was decided in favour of Sechenov in the official correspondence between the head of the Ministry of Justice, Prince S.N. Urusov, Attorney-General P.A. Tisenhausen and the Office of the Ministry for the Interior: 'The teaching of Professor Sechenov, if erroneous, must be decided upon by means of scientific discussion, and not through a legal process.'[77] Sechenov could not refute in the press the accusations of the Censorship Committee of the immorality of his book, but the addendum he wrote on the copy presented to Maria Bokova is telling:

> The teaching, which I have expounded, does not destroy the value of human virtue and morals: the foundations of our love for one another is eternal; in the same way, man will always value a good machine better and will prefer it to a bad one, when he has a choice. But in addition to this negative side of

> my teaching, let me point to its positive merit: only my viewpoint explains how man can acquire the greatest of all human virtues, the all-forgiving love, which is the ultimate understanding of the weakness of his neighbour as a human being.[78]

Obviously, the controversy over the publication of *Reflexes of the Brain*, which dragged on for nearly a year, caused Sechenov much trouble and distress, ultimately resulting in a bout of physical and mental exhaustion. During the last three years, his teaching duties, publication of the *Physiology of the Nervous System*, and translation and editing of the three voluminous German textbooks, required a great deal of work and effort. Besides, the laboratory was situated in the old, damp part of the hospital building and these unhealthy conditions added to Sechenov's state of fatigue, weakness, and anxiety. On Botkin's advice, Sechenov took a year's leave and went to Karlsbad for medical treatment.

Half of the year's leave Sechenov spent in Graz in Austria at Alexander Rollet's laboratory.[79] Rollet (1834–1903) was a good friend of Sechenov from the time both had studied in Vienna, and Sechenov valued Rollet as a sincere, honest man and as a thorough and devoted researcher. Recommended by Ernst Brücke, his own mentor, Rollet was appointed ordinary professor at the newly founded physiological institute at the University of Graz in 1863. He was actively engaged in blood research, and his studies on haemoglobin, haemolysis, and blood analysis earned him the nickname 'Blut-Rollet'.[80] He was a fine histologist too, and Sechenov used this opportunity to perform some microscopic studies of the nervous structures in the sympathetic trunk and discussed his observations with him. Sechenov's primary studies in Rollet's laboratory concerned experimental proof of reflex regulation of motor acts, one of the central propositions of *Reflexes of the Brain.* Investigating inhibitory phenomena in motor reflexes and rhythmic work of the lymph heart, Sechenov found an interrelationship between the central and peripheral portions of the nervous system. He seemed quite happy with his work, and his letters to Bokova reporting his experiments were filled with delight and excitement: 'These experiments in their importance are not inferior to my Paris experiments, and in grace of their execution, they exceed anything known to me in the area of physiology of the nervous system'.[81] In an experiment involving stimulation by induction current of the thalami optici in the frog's brain and the exposed sciatic nerve, Sechenov demonstrated the phenomena of the 'summation of separate stimuli into co-ordinated movement by the nervous centres.' He wrote to Bokova:

> What I dreamed when I was writing *Reflexes* is coming true. I am on the right track to finding the missing links of experimental analysis, which were necessary to recreate the true picture of the mechanism of the reflex regulation of motor functions.[82]

Sechenov was especially fascinated by the experiment, which he romantically called 'The White Lady' since he listened to a piece of music of the same name by Boieldieu at Graz Opera that evening. This experiment, as he wrote to Bokova, demonstrated several important findings, such as the basic mechanism of regulation of co-ordinated movement, and the existence of inhibitory centres in the spinal cord. Most interestingly, these experiments suggested to Sechenov a similarity between central inhibition of cerebro–spinal reflexes and mechanisms responsible for inhibition of cardiac activity:

> You know about the inhibitory apparatus of the heart: the centres of the vagus nerve in the medulla oblongata, then the fibres of the vagus, and then the endings of the vagus in the venous sac have an inhibitory effect on the motor elements of the heart. In the same manner, an inhibitor centre in the middle portion of the brain, then some conductors, which descend along the spinal cord and terminate in inhibitory mechanisms are linked to the reflex mechanisms.[83]

Sechenov's hypothesis, which emphasised the identity of the pathways, through which inhibitory impulses travel from the centres to the periphery, required experimental proof.

Although Sechenov correctly supposed that inhibitory influences from the middle portion of the brain were distributed through the sympathetic nervous system – marginal sympathetic trunk – he was not able to prove it experimentally.[84] Despite the obvious limitations and lack of experimental proof, Sechenov ascribed particular importance to the research at Graz, which he believed confirmed and further developed the major propositions of his theory of central nervous inhibition. He published the results in several articles and as a separate brochure,[85] and demonstrated his experiments to Rollet, Brücke, and Ludwig. Ilia Il'ich Mechnikov (1845–1916), en route from Naples, where he performed embryological studies, visited Sechenov in Graz. Mechnikov was so fascinated with the experiment on the inhibitory effect of the vagus on the heart, that as Sechenov informed Bokova, he 'began to dream of becoming a physiologist and to study in my laboratory'.[86] Sechenov would return to these experiments at Novorossiisk University in collaboration with Mechnikov.

Resignation from the Medico-Surgical Academy

In 1868, Sechenov left Graz for St Petersburg and resumed his work at the Medico-Surgical Academy. After a year's absence, he 'felt the change of the tone in the upper echelons of the Academy.'[87] By that time the reformist 'triumvirate' had disintegrated. Zinin had left for the St Petersburg Academy of Sciences, and Dubovitskii had assumed the position of Chief Medical Inspector at the War Ministry. Dubovitskii continued for a short time to protect the Academy, but after his sudden death in 1868, the presidency was abolished, and the War Ministry appointed N.I. Kozlov (1814–89) to head the Academy. Kozlov had an excellent service record. After medical studies in Kazan, Dorpat, Vienna, Paris, and Zürich, he had worked as Professor of Anatomy at Kiev University for twelve years. Later during his service in the War Ministry's Medical Department, Kozlov had earned a reputation as an experienced administrator, but it turned out that he appeared absolutely unfit for his new post at the Medico-Surgical Academy. Borodin wrote:

> He [Kozlov] does not know our order, better to say, our disorder, besides he is extremely tactless and over zealous. In a short time he managed to turn against himself not only all the students but also the majority of the professors.... [T]he students are agitated to the extreme, and I expect they will make a row one of these days.... Such a scandal may bring repressive measures, which would affect our established rules.[88]

Indeed, in March 1869, due to student unrest, the Medico-Surgical Academy was temporarily closed. Kozlov was transferred back to the War Ministry in the position of Chief Medical Inspector. But for his two-year stay at the Academy, Kozlov contrived to become its 'evil genius':

> The Academy, like a complex mechanism, adjusted and renovated by Dubovitskii, went astray. Under these circumstances and because the Conference was enlarged with new members, all graduates of the Institute of Physicians, the meetings of the Conference became an arena of arguments that were far beyond the scope of scientific and administrative matters.[89]

Not surprising then, that Sechenov's open criticism of the policy of appointment, which gave privilege to the graduates of the Institute over university graduates, met opposition from the new members of the Conference. In Sechenov's opinion, among five newly appointed professors, all graduates of the Institute, only one was a capable man – we do not know whom Sechenov meant – but the other four were 'people who knew their work, but in no way contributed to science and renown of the Academy.' Sechenov believed that the existing policy of appointment was ineffectual and even harmful to the Academy. He referred to the tradition of German universities, where scholarly merit was the prime criterion for academic position. 'I did not make a secret of my opinion about the Institute of Physicians. Therefore my colleagues, with the exception, of course, of Botkin and Gruber, had no sympathy for me.'[90] Kozlov, following the well-tested method of divide-and-conquer, skillfully used the growing tension and discord among the members of the Conference to manage the Academy his own way.

The Conference meeting of November 1869, which discussed the candidates for the two chairs of zoology and histology, is particularly telling in this respect. Sechenov had nominated two candidates, Alexander E. Golubev and Mechnikov. Golubev, a Privatdozent from Kazan University, had trained as a medical histologist in the laboratories of Rollet and Brücke. Both mentors valued Golubev as a skilful and capable researcher.[91] The young Mechnikov, in Sechenov's opinion, was an exceptionally gifted scientist with an established scholarly reputation in zoology and embryology. He had graduated from Kharkov University in just two years. In 1865, he went to a research sojourn in Naples supported by the Ministry of Education, where he began a systematic study of germ layers in invertebrate embryos. In 1867, after finishing his formal studies in Göttingen and Munich, Mechnikov returned to St Petersburg to receive his doctorate in zoology from St Petersburg University and to share the prestigious von Baer Prize of the St Petersburg Academy of Sciences with another young talented zoologist, Alexander O. Kovalevskii. The same year, at the age of twenty-two, Mechnikov received a position at the newly founded Novorossiisk University. To Sechenov, both Golubev and, particularly, Mechnikov were the most deserving candidates, whereas the men proposed by the Conference were of quite ordinary, if not dubious, scientific ability and it was even ridiculous to compare them with his protégés.[92] The Conference, nevertheless, voted down Golubev in favour of F.N. Zavarykin, a graduate of the Institute of Physicians, disregarding Sechenov's opinion of Zavarykin as a mediocre researcher.

The vote on Mechnikov was even more painful for Sechenov. Kozlov's treacherous remark, that Mechnikov deserved a position of no less than ordinary professor, but the Academy may not need an ordinary professor for 'a secondary' chair of zoology, had its desired effect. Now the clinicians stood in opposition to Sechenov, and the Conference vetoed Mechnikov. Another graduate of the Institute of Physicians, E. Brandt, attained the position of extraordinary professor. This failure showed Sechenov that it was no longer possible to win a position at the Academy by merit, only by intrigue and patronage. In a letter to Mechnikov, Sechenov pointed to a hypocritical atmosphere at the meeting and to the improper role of the head of the

Academy in the voting procedure.[93] We find a different picture of this meeting in Alexander Borodin's letter:

> [W]e had a long and hot meeting. Sechenov suffered severe defeat in connection with both of the candidates he had suggested.... I became convinced that he is far from being honest, and even not so experienced as the choristers of nihilism are trumpeting about him. Superficial effect and popularity are his ideals.... Previously, much in his ill behaviour I ascribed to his passionate temperament and some foolishness, now I am convinced that many of his deeds are deliberately dishonest.[94]

Borodin's harsh reproaches of incompetence are equally unfair as the groundless accusations of dishonesty and popularity seeking. Sechenov was known for his somewhat strange avoidance of any administrative responsibilities and publicity. Setting aside the notorious popularity among nihilists, which he never sought, Sechenov deserved esteem for his personal qualities. Ilia Mechnikov recalled, that 'everywhere Sechenov was met with a sympathetic even enthusiastic attitude. His wit and his decency inspired respect, and his influence was felt everywhere.'[95] Indeed, as we know from Alexander Samoilov, Sechenov's hot temper, 'inflammability' and overanxiousness sometimes 'took the place of his natural kindness, and he became stern and overwhelmed with anger when confronted with an unfairness.'[96] It is fair to assume that this was the case at the meeting of the Conference, and Sechenov gave way to his indignation. Most likely, Borodin took Sechenov's open and well-grounded critique of the privileges of the Institute's graduates personally. Although Borodin was not a graduate of the Institute, he had gone through a similar procedure as an Institute graduate, and Zinin had engineered his appointment.[97] Criticising inefficacy of the existing policy of appointments, Sechenov unintentionally – or possibly intentionally – questioned the academic merit of those professors who were graduates of the Institute. Apparently, Sechenov's superfluous straightforwardness and his unwillingness to accommodate himself to the situation created an insuperable barrier between him and the majority of professors of the Conference.

Frustrated by the open and offensive disregard of his opinion, Sechenov submitted his resignation. His decision created a furore at the Academy and obviously irritated the members of the Conference. These events are reported by Borodin in his letters of 1870:

> Sechenov is leaving the Academy because as he says, his opinion is not respected, that is to say, the Conference does not want to absolutely obey his stupid willfulness.[98]

Next day [after Sechenov had sent in his resignation] the deputation from the Conference came to Sechenov requesting that he stay at the Academy. Although the result of this attempt was evident to everyone, it was necessary as a statement from the Conference that it was not the Conference that became the reason for Sechenov's resignation. The rumours are circulating everywhere that Sechenov is persecuted and victimised for the truth, and that he is a martyr for his convictions. In this case the Academy is to be blamed. Sechenov shows off and drapes himself in civil virtue.... Even Sechenov's thorough friends, Botkin and Gruber, say that he acts stupidly and makes a gross mistake leaving the Academy.[99]

The episode with the voting gave a strong impetus to Sechenov's decision to leave the Academy. The major reason, as Sechenov admitted later, was that the academic years of 1868 and 1869 were the most unproductive in his life. His former illness returned with its general weakness and dizziness, not disposing him to either activity or a cheerful mood, and he regarded the new situation at the Academy more gloomily than it was in reality.[100] He seemed to have lost his usual enthusiastic interest in his laboratory. There is no record of purchasing any new equipment and instruments for the laboratory during these years.[101] Furthermore, he never expressed any interest in the new Anatomo–Physiological Institute, with its spacious physiological laboratory, which was near completion in 1870. Although Sechenov never mentioned his personal arguments with Kozlov, a total lack of understanding and co-operation between the two men was obvious. Sechenov did nothing to adjust to the new administration, and seemed to have been increasingly marginalised and excluded from the circle of those whose voices were powerful at the Academy. The situation was even more regretful for Sechenov as he never had any disagreement with the members of the Conference under the skilful and liberal rule of Dubovitskii, whose support was pivotal in Sechenov's early career at the Academy. The influential and benevolent Dubovitskii[102] seemed to ignore the biased opinion of the Minister for the Interior, Count P.A. Valuev, of Sechenov as a thorough materialist. Quite tolerant to Sechenov's somewhat eccentric and independent way of thinking, Dubovitskii valued Sechenov as a devoted scientist and as a decent and honest man.

Soon after Sechenov had sent in his resignation, he had a meeting with the 'high authorities' – we do not know their names. This episode, once again, proves Sechenov's straightforwardness and independence:

[The high authorities] asked me about the German professors with whom I had studied; they remarked in passing that I had published *Reflexes of the Brain* to no purpose, to which I answered (the conversation was in German):

> 'Man muss doch die Courage haben seiner Überzeugungen ausdrücken.' ['One must nevertheless have the courage to express his convictions'].... Of course, these people of high standing used to receive people who come to them with requests and petitions, if not in their hands, then in their mind.... But I never have a thought to ask them for anything.[103]

Leaving the Academy, Sechenov recommended Ilia Cyon as his successor and the most deserving candidate for the chair of physiology.[104] Cyon, like Mechnikov and Golubev, was from outside the Medico-Surgical Academy. He had an excellent reputation as a researcher and teacher, who contributed much to the organisation and operation of St Petersburg University's laboratory. Sechenov deliberately did not recommend either of his most capable students, Pashutin or Tarkhanov, as his possible successor. Both men would become notable scientists. Tarkhanov would succeed Cyon at the chair of physiology, and Pashutin would head the Academy in the 1890s. In 1870, however, Sechenov rightfully believed that neither of them could rival Cyon in scholarly merits. Moreover, knowing too well the customs of the Academy, Sechenov openly disregarded the fact that his nomination of Cyon would be faced with opposition from the Conference. For Sechenov, the most important imperative was that his laboratory would benefit from the research, teaching, and managerial talents of Cyon.

Notes

1. Du Bois-Reymond to Ludwig, letter dated January 9, 1853, Berlin, in P.F. Cranefield (ed.), S. Lichtner-Ayèd (trans.) *Two Great Scientists of the Nineteenth Century: Correspondence of E. du Bois-Reymond and C. Ludwig* (Baltimore: Johns Hopkins University Press, 1982), 78.
2. Sechenov's letters to the medical faculties of Moscow and Kazan Universities are published in K.S. Koshtoyants, *et al.* (eds), *Nauchnoe nasledstvo: I.M. Sechenov: neopublikovannye raboty, perepiska i dokumenty* (Moscow: Akademiia Nauk, 1956), 19–21.
3. Sechenov received this offer in 1859 through Botkin, who was personally connected with Glebov, see I.M. Sechenov, *Autobiographical Notes,* D.B. Lindsley (ed.), K. Hanes (trans.), (Washington: American Institute of Biological Sciences, 1965), 88; see also Ludwig to Sechenov, letter dated 14 May 1859, Vienna, in H. Schröer, *Carl Ludwig* (Stuttgart: Wissenschaftliche Verlagsgesellschaft M.B.H., 1967), 249.
4. I.M. Sechenov, *Dannye dlia budushchei fiziologii alkogol'nogo op'ianeniia* (St Petersburg: Tipografiia Treia, 1860), 3. Published in K. Koshtoyants (ed.), *I.M. Sechenov, Izbrannye proizvedeniia* (Moscow: Izd-vo Akademii Nauk, 1952-1956), 2 vols, Vol II, 35–98. A physiologist, Sechenov declared in his thesis, is 'a physico–chemist, who deals with the living organism.' This

statement referred to a controversy between the two competing schools in physiology, physico–chemical, and morphological, and demonstrated passionate devotion of the young Sechenov, a student of du Bois-Reymond and Helmholtz, to a physico–chemical trend in physiology, see N.N. Vvedenskii, 'Ivan Mikhailovich Sechenov: nekrolog, in *Trudy S. Peterburgskogo obshchestva estestvoispytatelei,* xxxvi (1906), 1–44: 6, in K.M. Bykiov (ed.) *I.M. Sechenov, I.P. Pavlov, N.E. Vvedenskii: Fiziologiia nervnoi sistemy: Izbrannye proizvedeniia* (Moscow: Medgiz, 1952), 59–81: 64.

5. Sechenov, *op. cit.* (note 3), 99.
6. *Ibid.*, 99.
7. Jean-Baptiste A. Dumas (1800–84), a leading French chemist, taught organic chemistry at the École Polytechnique and at the École de Médécine. He succeeded Louis Joseph Gay-Lussac (1778–1850) at the Sorbonne and lectured for several years at the Collège de France. Together with Jean L. Prévost (1790–1850), a physiologist from Geneva, he published on the size and shape of blood corpuscles, 'Examen du sang et de son action dans les divers phénomènes de la vie', *Annales de chimie et de physique,* xviii (1821), 280–96. In the second part of the paper, published two years later, they described the formation of the blood clot and the method of preparing fibrin for quantitative analysis, 'Examen de sang et de son action dans les divers phénomènes de la vie', *Annales de chimie et de physique,* xxiii (1823), 50–68. In his discussion of fibrin in the blood, Sechenov drew on Johannes Müller's important studies on the separation of the components of the blood, and Müller's famous filtration experiment, which had shown that fibrin was not inside the red cells. These were reported in Müller's *Handbuch der Physiologie,* well known to Sechenov. On Müller's research on the blood, see P. Mazumdar, 'Johannes Müller on the Blood, the Lymph, and the Chyle', *Isis,* xxxiii (1975), 242–53.
8. Alexander Schmidt (1831–1914) carried out basic studies on blood clotting with Felix Hoppe-Seyler in Berlin and also studied gas exchange with Ludwig in Vienna and then in Leipzig, see K.E. Rothschuh, *History of Physiology,* G.B. Risse (trans. and ed.), (Huntington: R.E. Krieger, 1971), 284–7.
9. F. Hoppe-Seyler, *Handbuch der physiologish- und pathologisch-chemischen Analyse* (Berlin: Hirschwald, 1858).
10. Karl von Vierordt (1818–84), Professor of Physiology in Tübingen, was a student of Johannes Müller. His research interests lay in the area of measurable, quantitative physiological phenomena, especially the gas exchange in respiration and specifically the elimination of carbonic acid. His sphygmograph was the first pulse-recording device. He also investigated blood velocity and the application of spectrophotometry in physiology and chemistry, see Rothschuh, *op. cit.* (note 8), 242.

11. Alfred W. Volkmann (1800–77), Professor of Anatomy and Physiology at Halle, was active in the field of haemodynamics, in which he exchanged ideas with his teacher Ernst H. Weber, as well as Carl Ludwig; see A. Volkmann, *Die Haemodynamik nach Versuchen* (Leipzig: Veit, 1850); see also Rothschuh, *op. cit.* (note 8), 176, 179.
12. Sechenov, *op. cit.* (note 3), 102.
13. 'Gas Voltaic battery', the first long-life battery capable of generating a constant current, was constructed in 1843. It consisted of platinum electrodes in two jars of sulphuric acid with a layer of oxygen above one, hydrogen above the other. Sir William Robert Grove (1811–96), an English lawyer, for several decades of his life was engaged in experimental natural science.
14. Koshtoyants *et al., op. cit.* (note 2), 25–8.
15. L. Popel'skii, *Istoricheskii ocherk kafedry fiziologii v imperatorskoi voenno-meditsinskoi akademii za 100 let (1798–1898)* (St Petersburg, 1899), 54–6.
16. Sechenov, *op. cit.* (note 3), 101.
17. I.M. Sechenov, *O zhivotnom elektrichestve* (St Petersburg: Izd-vo Voenno-meditsinskogo departamenta, tipografiia Treia, 1862).
18. Sechenov's appeal to the history of electrophysiology in his *Lectures* supports the claim that during the same period such discipline builders as du Bois-Reymond, Virchow, and Bernard, all used the historical treatment of physiology and allied sciences to promote and legitimise the introduction of experimental methods into medicine. See, N. Jardine, 'The Laboratory Revolution in Medicine as Rhetorical and Aesthetic Accomplishment', in A. Cunningham and P. Williams (eds), *The Laboratory Revolution in Medicine* (Cambridge: Cambridge University Press, 1992), 310–21; see also J. Bleker, *Die Naturhistorische Schule, 1825–1845: ein Beitrag zur Geschichte der klinischen Medizin in Deutschland* (Stuttgart: G. Fischer, 1981), Ch. 6; on du Bois-Reymond as an historian of the natural sciences, see H. Boruttau, 'Emil du Bois-Reymond als Physiologe und Historiker der Naturwissenschaften', *Berliner klinische Wochenschrift,* xxvi (1919), 926–8.
19. Sechenov, *op. cit.* (note 17), 23–5.
20. Koshtoyants, *op. cit.* (note 2), 32.
21. Vvedenskii, *op. cit.* (note 4), 58.
22. On Flourens' research on the functions of the cerebellum and its role in voluntary movements and their co-ordination, see E. Clarke, *The Human Brain and Spinal Cord: A Historical Study Illustrated by Writings from Antiquity to the Twentieth Century* (San Francisco: Norman Publications, 1996).
23. Sechenov, *op. cit.* (note 3), 47.
24. *Ibid.*, 104–5.

25. On Bernard's influence on Pavlov and the relation of Bernard's scientific vision to that of Pavlov, see D. Todes, *Pavlov's Physiology Factory. Experiment, Interpretation, Laboratory Expertise* (Baltimore: Johns Hopkins University Press, 2002), 45–6; 199–200; 353–4.
26. Sechenov, *op. cit.* (note 3), 106–7.
27. Ludwig Türck (1810–68) studied the effect of a partial section through the spinal column in 1857. He also discovered the laws concerning the secondary degeneration in the nervous system in 1849. In 1861, together with Johannes N. Czermak (1828–73), he received the Montyon Prize of the Paris Academy of Sciences for the invention of the laryngoscope. See E. Lesky, 'Ludwig Türck: Neuroanatom und Neurophysiologe', in K.E. Rothschuh (ed.), *Von Boerhaave bis Berger: Entwicklung der kontinentalen Physiologie im 18. und 19. Jahrhundert mit besonderer Berücksichtigung der Neurophysiologie* (Stuttgart: Fischer, 1964), 129–33.
28. Sechenov, *op. cit.* (note 3), 106.
29. *Ibid.*, 107–8.
30. *Ibid.*, 107–8.
31. I.M. Sechenov, *Physiologische Studien über die Hemmungmechnismen für die Reflextatigkeit des Rückenmarkes im Gehirne des Frosches* (Berlin: Hirschwald, 1863). In his *Autobiographical Notes,* Sechenov gives an overview of the current state of knowledge on the problem of inhibition, as well as desribes his famous experiment. A comprehensive discussion on Sechenov's discovery of central nervous inhibition is found in K.S. Koshtoyants, *Essays on the History of Physiology in Russia,* D.P. Boder *et al.* (trans.), (Washington: American Institute of Biological Sciences, 1964), 126–41.
32. I. Sechenov, 'Sur les centres modérateurs de mouvements réflexes dans le cerveau de la grenouille', *Comptes rendus hebdomadaires des séances de l'Académie des Sciences,* 56 (1863), 50–3; 185–7.
33. Ludwig to Sechenov, letter 25 November 1862, Vienna, in M.N. Shaternikov, 'Life of I.M. Sechenov', in Shaternikov (ed.), I.M. Sechenov, *Selected Works* (Moscow: Izd-vo Vsesouznogo instituta eksperimental'noi meditsiny, 1935), ii–xxxiii: xix, see also Sechenov, *op. cit.* (note 3), 108.
34. I.M. Sechenov, 'Refleksy golovnogo mozga', first published in *Meditsinskii vestnik,* 47 (1863), 461–84; and 48 (1863), 493–512; for an English-language edition, see I.M. Sechenov, *Reflexes of the Brain,* K. Koshtoyants (ed.), S. Belsky (trans.), (Cambridge: MIT Press, 1965).
35. Sechenov, *op. cit.* (note 3), 110
36. *Ibid.,* 109.
37. Following the logic of Soviet historiography, Sechenov was very keen to overthrow the official philosophical and religious dogmas on free will, and in doing so he laid the groundwork for the school, which later flourished as the Soviet school of psychology. See M.G. Iaroshevskii's biography, *I.M. Sechenov*

(Leningrad: Nauka, 1968), 75–80. Another conventional view is that *Reflexes of the Brain*, which Sechenov intended to publish in *Sovremennik* [*The Contemporary*], was written to support and promote materialistic ideas on the nature of consciousness exposed by N.G. Chernyshevskii, a popular writer, editor of the *Sovremennik*, and hero of revolutionary democrats. See Koshtoyants, *op. cit.* (note 31), 142–7.

38. I.P. Pavlov to M.N. Shaternikov, letter dated 23 September 1929, St Petersburg; Shaternikov to Pavlov, letter dated 27 September 1929, Moscow; Pavlov to Shaternikov, letter dated 3 October 1929, St Petersburg. The originals are preserved in the Archive of the Russian Academy of Sciences, first published in Koshtoyants, *op. cit.* (note 31), 151–2. Shaternikov was Sechenov's assistant at Moscow University in the 1890s, a close friend and the first biographer of Sechenov.
39. Sofia Kovalevskaia, a daughter of General V. Korvin-Krukovskoi, arranged a fictitious marriage to a notable biologist Vladimir Kovalevskii. Both Vladimir and Sofia were close friends of Maria and Sechenov in the early 1860s. On Kovalevskii biography, see A.H. Koblitz, *A Convergence of Lives: Sofia Kovalevskaia – Scientist, Writer, Revolutionary* (Cambridge: Birkhäuser Boston, 1983).
40. Sechenov, *op. cit.* (note 3), 118. A devoted couple throughout life, Sechenov and Maria Bokova would only officially marry in 1888, after Maria's divorce from Bokov, and had a wedding ceremony in the church, see P.G. Kostiuk, S.P. Mikulinskii and M.G. Iaroshevskii (eds) *I.M. Sechenov: K 150- letiiu so dnia rozhdeniia* (Moscow: Nauka, 1980), 555.
41. On Sechenov's troubles with the censorship department in 1863 and 1866, whe *Reflexes of the Brain* appeared in a book form, see Konievetz, 'Censorship and Science', *Russkie vedomosti,* 116 (1905), cited in Shaternikov, *op. cit.* (note 33), xxi. All facts mentioned in this article are taken from official sources, xx–xxv. Soviet historiography paid much attention to the social response to Sechenov's popular works; as a typical example, see Koshtoyants, *op. cit.* (note 31), 142–52, 157–65. The majority of the historiographical works of the 1980s do not contain any new information or interpretation of the topic. An English language source on Sechenov's *Reflexes of the Brain* is D. Todes, *From Radicalism to Scientific Convention* (unpublished PhD thesis, University of Pennsylvania, 1981), 249–66.
42. Ludwig to Sechenov, letter dated 15 November 1863, Vienna, in Shaternikov, *op. cit.* (note 33), xx–xxi.
43. Ludwig to Sechenov, letter dated 2 November 1864, Vienna, in *ibid.,* xxvi.
44. Sechenov, *op. cit.* (note 3), 109.
45. *Ibid.,* 119.

46. A.S. Stal', *Perezhitoe i peredumannoe studentom, vrachem i professorom* (St Petersburg, 1908), 22.
47. *Ibid.,* 24–5.
48. On Samoilov, see N.A. Grigor'yan, *A.F. Samoilov* (Moscow: Izd-vo Akademii Nauk, 1963); on Samoilov and Pavlov, see A. Vucinich, *Science in Russian Culture* (Stanford: Stanford University Press, 1970), Vol. II, 315–16.
49. A.F. Samoilov, 'Sechenov i ego mysli o roli myshtsi v nashem poznanii prirody', in K.S. Koshtoyants (ed.), *Izbrannye stat'i i rechi* (Leningrad: Izd-vo Akademii Nauk, 1946), 43–69: 44.
50. Popel'skii, *op. cit.* (note 15), 55.
51. I.M. Sechenov, 'Zapiska o prepodavanii fiziologii cheloveka i vysshykh zhivotnykh na vysshykh zhenskikh kursakh', 1894, see Koshtoyants *et al., op. cit.* (note 2), 180.
52. I.R. Tarkhanov, 'Pamiati professora I.M. Sechenova', *Trudy obshchestva russkikh vrachei,* 73 (1906), 69–75: 73.
53. Cited in F.L. Holmes, 'The Old Martyr of Science: The Frog in the Experimental Physiology', *Journal of the History of Biology,* xxvi (1993), 311–28: 326.
54. Sechenov to Liapunov, letter dated November 1892, Moscow, in Koshtoyants *et al., op. cit.* (note 2), 225. A.M. Liapunov (1857–1918), Sechenov's brother-in-law, was an outstanding mathematician at the St Petersburg mathematical school of P.L. Chebyshev.
55. V.V. Pashutin, *Avtobiografiia* (St Petersburg: K.P. Rikker, 1899), 4.
56. W. Kühne, *Lehrbuch der physiologischen Chemie* (Leipzig: Engelmann, 1866–8); I.M. Sechenov (ed. and trans.), V. Kune, *Uchebnik fiziologicheskoi khimii* (St Petersburg: Izd-vo V. Kovalevskogo, 1866–68). *Lehrbuch* was a course of lectures Kühne taucht at the Berlin Pathological Institute. On Kühne, see Sechenov, *op. cit.* (note 3), 110; Rothschuh, *op. cit.* (note 8), 239–40; J.S. Fruton, *Contrasts in Scientific Style: Research Groups in the Chemical and Biological Sciences* (Philadelphia: American Philosophical Society, 1990), 80–2.
57. Koshtoyants *et al., op. cit.* (note 2), 50–1.
58. I.R. Tarkhanov and V.V. Pashutin became professors at the St Petersburg Medico-Surgical Academy, V.K. Voroshilov became professor at Kazan University, and P.A. Spiro at Novorossiisk University, see N.A. Grigor'yan, 'O pervoi russkoi fiziologicheskoi shkole', *Istoriia biologii,* v (1975), 137–51; see also A.A. Mozzhukhin, 'Fiziologicheskie shkoly v Voenno-meditsinskoi akademii', in Mozzhukhin (ed.), *Shkoly v nauke* (Moscow: Izd-vo Akademii Nauk, 1977), 443–7.
59. I.M. Sechenov, 'Pneumatologische Notizen', *Zeitschrift für rationelle Medizin,* iii, 10 (1861), 285–92; and 'Neuer Apparat zur Gewinnung der Gase aus dem Blut', *Zeitschrift für rationelle Medizin,* iii, 23 (1865), 16–20. Henri

Victor Regnault was a professor of physics and from 1845 to 1871, Director of the Royal Porcelain Factory in Sèvres. Regnault was interested in the physical state of gases, as well as in respiratory gas exchange, see Sechenov, *op. cit.* (note 3), 107.

60. Sechenov, *op. cit.* (note 3), 119.
61. Vvedenskii, *op. cit.* (note 4), 72.
62. I.M. Sechenov, *Fiziologiia nervnoi sistemy* (St Petersburg: Izd-vo A. Golovacheva, 1866), 62.
63. *Ibid.,* 64.
64. On the chemical approach and on Bernstein's electrochemical theory, see T. Lenoir, 'Models and Instruments in the Development of Electrophysiology, 1845–1912', in *Historical Studies in the Physical and Biological Sciences* xvii, 1 (1986), 1–54: 20.
65. Sechenov translated the second edition of Hermann's, *Grundriss der Physiologie des Menschen,* of 1864. In turn, Hermann was interested in Sechenov's absorptiometric investigation. The chapter on blood gases contains a detailed account on Sechenov's method and results, see L. Hermann, *Grundriss der Physiologie des Menschen* (Berlin: Hirschwald, 1867), 46–9; *idem., Grundriss der Physiologie des Menschen* (Berlin: Hirschwald, 1874), 42–7.
66. E. du Bois-Reymond, *Gesammelte Abhandlungen z. Muskel- und Nervenphysik* (Leipzig: Vogel, 1875–7), 2 vols, Vol. I, 36.
67. Sechenov, *op. cit.* (note 17), 194.
68. Sechenov, *op. cit.* (note 62), 21; 36.
69. *Ibid.*, 21.
70. *Ibid.,* 77.
71. *Ibid.,* 79.
72. Hermann, *Grundriss* of 1874, *op. cit.* (note 65), 229–30, see also Lenoir, *op. cit.* (note 64), 21.
73. Sechenov, *op. cit.* (note 62), 23.
74. M. Verworn, *Allgemeine Physiologie: ein Grundriss der Lehre vom Leben* (Jena: G. Fischer, 1895), 59. Verworn's monographs *Vitalismus und Neovitalismus,* and *Allgemeine Physiologie* reflected his keen intellect and tendency to discern the general and theoretical issues behind the phenomena. On Verworn, see Rothschuh, *op. cit.* (note 8), 302–3.
75. I.M. Sechenov, 'O deiatel'nosti Gal'vani i du Bua-Reimona v oblasti zhivotnogo elektrichestva', in *Trudy fiziologicheskogo instituta Moskovskogo universiteta,* iii, 5 (1899), 1–8: 4.
76. Cited in Shaternikov, *op. cit.* (note 33), xxii, xxiii.
77. Cited in *ibid.,* xxiv.
78. Cited in *ibid.*, xxii.
79. Sechenov, *op. cit.* (note 3), 122.

80. On Rollet, see Sechenov, *op. cit.* (note 3), 84; and Rothschuh, *op. cit.* (note 8), 236–7. On Rollet's biography see, O. Zoth, 'Alexander Rollet', *Pflügers Archiv,* 101 (1904), 103–11.
81. Sechenov to Bokova, letter dated 12 October 1867, Graz, cited in Koshtoyants, *op. cit.* (note 31), 130.
82. Sechenov to Bokova, letter dated 30 January 1868, Graz, *ibid.,* 132.
83. Sechenov to Bokova, letter dated 29 February 1868, Graz, *ibid.,* 133–4.
84. Sechenov to Bokova, letter dated 19 October 1867, Graz, *ibid.,* 135.
85. I.M. Sechenov, *Über die elektrische und chemische Reizung der sensiblen Rückenmarksnerven des Frosches* (Graz: Leuschner und Lubenskii, 1868). Cited in Sechenov, *Über die elektrische und chemische Reizung,* in Sechenov, *Selected Works, op. cit.* (note 33), 177–211: 191–2.
86. Sechenov to Bokova, letter dated 17 March 1868, Graz, cited in Koshtoyants, *op. cit.* (note 31), 134.
87. Sechenov, *op. cit.* (note 3), 126.
88. Borodin to E.S. Borodina, letter dated 24 October 1868, St Petersburg, in S.A. Dianin (ed.), *Pis'ma A.P. Borodina, 1857–71* (Moscow: Gos. izd-vo, muzykal'nyi sector, 1927), Vol. I, 131.
89. N.I. Ivanovskii (ed.), *Istoriia Imperatorskoi voenno-meditsinskoi (byvshei Medico-khirurgicheskoi) akademii za 100 let. 1798-1898* (St Petersburg: Tipografiia Ministerstva vnutrennikh del, 1898), 602.
90. Sechenov, *op. cit.* (note 3), 126.
91. *Ibid.,* 127.
92. See Sechenov's 'Zapiska' of 1869, regarding the scientific merits of Zavarykin versus Golubev, and of Mechnikov versus Brandt, in Koshtoyants *et al., op. cit.* (note 2), 186–93; see also Sechenov, *op. cit.* (note 3), 127–8; and K.K. Kekcheev and N.A. Shustin, 'K ukhodu I.M. Sechenova iz Mediko-khirurgicheskoi akademii v 1870 g.', in *Fiziologicheskii zhurnal SSSR,* i, 21 (1936), 2–14.
93. Sechenov to Mechnikov, letter dated 16 November 1869, St Petersburg, in S.I. Shtraikh (ed.), *Bor'ba za nauku v tsarskoi Rossii: neizdannye pis'ma I.M. Sechenova, I.I. Mechnikova i dr.* (Moscow–Leningrad: Gossotsgiz, 1931), 57–8.
94. Borodin to E.S. Borodina, letter dated 16 November 1869, St Petersburg, in Dianin, *op. cit.* (note 88), Vol. I, 169.
95. I.I. Mechnikov, 'Vospominania o Sechenove', in I.I., Mechnikov, *Stranitsy vospominanii* (Moscow: Izd-vo Akademii Nauk SSSR, 1946), 45–64: 47.
96. Samoilov, *op. cit.* (note 49), 47.
97. Apparently Borodin, who respected Zinin as a father, was aware of strained relations between Sechenov and Zinin. Sechenov once mentioned his disappointment in Zinin in connection with some dishonest deed of the latter (Sechenov did not specify which), see Sechenov to Mendeleev, letter

dated 30 December 1860, St Petersburg, in Koshtoyants *et al., op. cit.* (note 2), 210. However, their 'business' relations at the Academy were quite agreeable.

98. Borodin to E.S. Borodina, letter dated 24 September 1870, St Petersburg, in Dianin, *op. cit.* (note 88), Vol. I, 237.
99. Borodin to E.S. Borodina, letter dated 30 September 1870, St Petersburg, in *ibid.*, 240–1.
100. Sechenov, *op. cit.* (note 3), 125.
101. *Rossiiskii gosudarstvennyi voenno-istoricheskii arkhiv*, Moscow, fond 1, for 1860–70.
102. Sechenov, *op. cit.* (note 3), 126.
103. *Ibid.*, 129.
104. Sechenov to Mechnikov, letter dated 19 April 1870, St Petersburg, in Shtraikh, *op. cit.* (note 93), 59.

12

A Few Steps Further: The Operation of the Physiological Laboratory under Cyon

Elie de Cyon's remarkably short stay at the Medico-Surgical Academy, from the time of his scandalous appointment in 1872, until the time of his no-less scandalous resignation in 1875, marked a period of transition. The Academy progressed from a small laboratory, which was only exclusively orientated towards physical physiology, to an adequately equipped and spacious physiological institute. During these three years, Russian physiological literature, which had started with Sechenov's textbook on the physiology of the nervous system, was enriched by Cyon's original textbook on general physiology and a monograph on electrotherapy. In many ways, this period represents the founding stage in the development of Russian experimental physiology. Cyon's activities as a builder of discipline reflected, in a curious way, the ideological contradictions that became particularly prominent during a tormented decade of the post-reform era. Therefore, any picture of the introduction of the laboratory into Russian medicine is incomplete without Cyon.

Among nineteenth-century Russian physiologists, Ilia Fadeevich Tsion (1842–1912) – later known as Elie de Cyon – was a controversial and somewhat obscure figure. He was, however, quite an extraordinary personality. His experimental talent and scientific productivity won him early success and recognition both in Europe and at home, while his quarrelsome disposition, arrogance, intolerance, and severe conservative viewpoints made him a notorious figure in Russia, and later in France. Cyon's contemporaries had polarised opinions about him, either admiration or scarcely concealed enmity. On one side, we find that physiologist A.A. Ukhtomskii (1875–1924) described Cyon as 'a brilliant teacher and researcher',[1] summing up the general opinion held by physiologists and students at St Petersburg University. Furthermore, Sechenov, Botkin, and Ovsiannikov, the major figures in the field, facilitated Cyon's career at the University and the Medico-Surgical Academy. However, on the other side, Mechnikov pointed out that Cyon's personal qualities frequently overshadowed his scholarly merits: 'Many who knew him, including me, did not like him greatly for his wicked character and inability to take a somewhat moral viewpoint.'[2]

Image 12.1

Ilia F. Cyon.
Courtesy of the Karl Sudhoff Institute for the History of Medicine and Natural Science, Leipzig University.

The same pattern is seen in any comments on Cyon in the Soviet historiography. Physiologist and historian Khachatur S. Koshtoyants, whilst admitting that Cyon was a brilliant and important physiologist, has contrasted the progressive values of Sechenov's activities with Cyon's task 'to eradicate the materialistic outlook from the minds of the youth'.[3] The comparison was inevitable and wanting. The two leading physiologists seemed to represent the two opposing camps – revolutionary democrats and reactionary monarchists – and in such treatment Cyon's outstanding scientific accomplishments were of secondary, if any, importance. With all differences in their background, character, reputation, and political views, Cyon and Sechenov did have a great deal in common. Both men were accomplished experimenters and the first Russian physiologists acknowledged in Germany and France. Both were strong and respected advocates of the laboratory in two leading Russian institutions, the St Petersburg Medico-Surgical Academy and the University. Both trained a number of researchers in their laboratories, who later had successful academic careers. Finally, and most importantly, both men made major contributions to the rise of experimental physiology and greatly influenced further development of independent physiological thought and expertise in Russia.

Little is known about Cyon's early years and his family. He was born in 1842, in a small village of the Kovno district in the Baltic province of the Russian Empire – now Lithuania – to a Jewish family of the lower middle-class – or 'meshchane'. The family, however, was keen on educating their children. In 1858, Il'ia finished gymnasium in Chernigov – now Ukraine – and entered the Warsaw Medico-Surgical Academy. He did not stay long. Poland was torn by political unrest at the end of the 1850s, and its educational institutions were under continuous re-organisation. Cyon tried Kiev University, which also appeared unpromising in scientific study opportunities. In 1859, he moved to Berlin, attracted by its renowned medical faculty. Here, Cyon studied under du Bois-Reymond, Virchow, and the neuropathologist Robert Remak (1815–65). Remak, Cyon's principal mentor, maintained a close connection with the Jewish–Polish culture to which Cyon belonged. Remak supervised Cyon's doctoral thesis on impairments of the nerve–muscle apparatus in chorea and the relation of chorea to rheumatic conditions of the joints (polyarthritis), pericarditis, and endocarditis. Soon after his defence, Cyon translated his thesis from Latin into German and published it in the annual Viennese medical journal *Wiener medizinische Wochenschrift*.[4]

Cyon's thesis was essentially clinical. At the same time, it contained some important elements of his future studies on the innervation of the heart. Undoubtedly, Remak was the first to inspire Cyon's interest in experimental research, as well as in pursuing an academic career. Remak's early and remarkable success while a student of Johannes Müller was the discovery of the marrowless nerve fibres in the sympathetic nervous system which, as Remark confirmed, originated in the ganglion cells.[5] Doubted by the eminent microscopists Purkyne and Müller, and criticised by such authorities as Gabriel Valentin, Theodor Bischoff, Alfred Volkmann, Rudolph Wagner, and Jacob Henle, Remak's findings were later proved correct. Remak was not defeated by all these controversial opinions and polemics, and continued his research. His other important discovery of ganglion cells in the heart suggested a neurogenic explanation of the relatively autonomous action of the heartbeat, which was known to be independent of the central nervous system.[6] Remak's histo–physiological studies on the heart and function of the sympathetic nervous system played a major role in turning Cyon's attention to the innervation of the heart, the foremost topic in Cyon's research career. Cyon's later work on electrotherapy, which won him the Gold Medal of the French Academy of Sciences in 1870, also bore testimony to Remak's influence.[7] As early as 1858, Remak published a collection of his works on electrotherapy, including his earlier study on galvanotherapy, which had brought him recognition among the Parisian academicians, clinicians Gabriel Andral, Pierre-François-Olive Rayer, Alfred Armand Velpeau, and the physiologist Claude Bernard.[8]

Although Remak's manifold interests and important contributions to histophysiology, comparative and microscopical anatomy, embryology, and cytology made him very well suited for academic teaching, his career was not successful. Supported by Müller and, particularly, by Alexander von Humboldt, Remak attempted several times to get a position at Berlin University and at Vilna University in the Russian Empire. Non-baptised Jews were barred from having teaching careers at Prussian and Russian universities. Remak believed that changing religion to advance his career was not acceptable. Only in 1859, the year Cyon entered Berlin University, was Remak made an extraordinary professor due to the influential recommendations of the departing Lucas Schönlein, in whose clinical laboratory Remak worked as an assistant.[9] Obviously, Cyon became aware of the thorny academic path of his mentor, and it is fair to suggest that Cyon's decision to convert to the Russian Orthodox Church received its first impulse in Berlin.[10]

In 1865, Cyon left Berlin for St Petersburg. He defended his Berlin doctoral thesis on chorea at the Medico-Surgical Academy, a condition to qualify for an academic appointment in Russia. Upon his defence, Cyon obtained a position as assistant at the Medico-Surgical Academy's department of nervous diseases and mental disorders, headed by a well-known psychiatrist, Ivan Mikhailovich Balinskii (1827–1902). The same year, Cyon seized the opportunity to go back to Germany to specialise in mental and nervous diseases for three years. He soon completed a research project assigned to him by Balinskii, and published two articles on the care of mental patients in the asylum in German and Austrian journals, as well as two articles and a monograph on the tabes dorsalis.[11] Cyon devoted the rest of his time during his sojourn to experimental physiology, a promising and always fascinating field, which obviously attracted him more than studies in mental disorders. An eager and ambitious researcher, he intended to pursue experimental studies on the innervation of the heart, a project of his own devising. A strong background in histology made him quite well-equipped for such an undertaking. Together with his brother Moisei, also a graduate of the Berlin medical faculty, Cyon began his studies in the laboratory of du Bois-Reymond. Even though du Bois-Reymond was not very interested in

Cyon's project, his laboratory was the best choice for mastering electrophysiological methods – an ideal career boost for any experimental physiologist. At different times Eduard Pflüger (1829–1910) and Albert von Bezold (1836–68) studied the phenomenon of electrotonus – the alteration in excitability and conductivity of a nerve or muscle during electrical stimulation – at du Bois-Reymond's laboratory. Later, both physiologists performed their classical investigations on organ innervation, Pflüger, on innervation of the intestine in 1855, and Bezold on the innervation of the heart in 1863.[12]

From Berlin, Cyon moved to Leipzig, attracted by Ludwig's fine Physiological Institute. Ludwig was known for his support of any feasible research project by his students, and his new Institute became particularly popular among young scientists from Russia. Cyon's project on innervation of the heart was not entirely within the range of Ludwig's diverse research interests, although Ludwig's work of 1864 on the sympathetic innervation of the vascular muscle layers suggests that he became interested in Cyon's studies. Ludwig seemed to appreciate the originality of the project and Cyon's enthusiasm. Skilful in various techniques, including vivisection, and in devising ingenious instruments, Ludwig became a valuable advisor and collaborator for Cyon in studying the innervation of the heart. It was with Ludwig that Cyon acquired basic vivisectional skills to work with isolated organs and a quantitative physico–chemical approach, which required the use of a variety of instruments and devices. For his studies on the influence of temperature on the heartbeat – first discovered by E.H. Weber in 1846 – and on the rate of contractions in an isolated segment of the heart, Cyon devised an ingenious instrument which allowed him to trace the effect of heat in the isolated heart of a frog.[13]

The most remarkable success Cyon achieved in collaboration with Ludwig was the discovery of the 'cardio–sensory' or depressor nerve. A series of experiments imitated the condition of elevated blood pressure in the heart and aorta. Electrical stimulation of the central end of the depressor nerve, which conveyed the sensory impulses from the heart and aorta to the brain, elicited dilatation of the blood vessels and hence a sharp drop in blood pressure. These experiments suggested to Cyon and Ludwig the existence of a self-regulatory mechanism, by which the heart can be relieved of too great a load. This mechanism can be seen as analogous to the action of the valves in a steam boiler.[14] The explanation of autoregulation of the systemic blood pressure was precisely mechanistic, in accordance with the leading German trend. On the other hand, the relationship between the heart and the blood vessels, interpreted in terms of the reflex from the sensory nerves in the heart and aorta (interoceptors) to the motor nerves of the blood vessels, was the first important landmark in studying reflex regulation of the cardiovascular system.

Cyon at Bernard's laboratory: the first time

A series of Cyon's works on the innervation of the heart, including the collaborative work on the 'depressor nerve', was awarded the Montyon Prize of the Paris Academy of Sciences in 1867. Claude Bernard, chairman of the awarding committee, took a special interest in Cyon's work. In 1858, Bernard discovered the active vasodilator reflex,[15] which drove his earlier studies of 1852 on the problem of nervous control of the blood vessels and the temperature change related to it, and discovery of the vasoconstrictor nerves.[16] Bernard's explanation of the function of vasomotor nerves established the concept of a physiological equilibrium of the two antagonistic innervations. Apparently, Bernard recognised the originality of Cyon's research on the innervation of the heart and invited him to work at the laboratory of the Collège de France. Here, Cyon performed a series of experiments showing the existence of sympathetic cardiac nerves that originate from the spinal cord. The cardiac nerves originating in the brain had already been known as vagi (parasympathetic). Cyon convincingly demonstrated that sympathetic cardiac nerves have a marked positive chronotropic action – an increase in heart rate – so they have the opposite effect on the heart than that of the vagi. His explanation of the function of the cardiac nerves was simple and elegant. If organs need an intensive blood supply, which provided by the enhanced heart activity, impulses from the brain to the heart are conducted through the sympathetic cardiac nerves, but while at rest when the demand of the organism in blood supply is low, the brain sends signals to the heart through the vagi (parasympathetic) nerves thus securing the inhibitory action on the heart.[17] Cyon's investigation also clarified the important problem concerning the autonomic rhythmic activity of the heart. He correctly stated that impulses from the central nervous system, conveyed through sympathetic or parasympathetic nerves, regulate the heart rhythm, either increasing or decreasing it and thus adapting it to various conditions.[18]

In Bernard's laboratory, Cyon also performed investigations on the influence of carbon dioxide and oxygen on heart function,[19] a project apparently suggested by Bernard. In 1855–6, studying the mechanism of carbon monoxide toxicity in animals, Bernard found that carbon monoxide replaces oxygen in the red blood cells. He developed an original method to measure carbon dioxide after absorption in a base, and carbon monoxide after its transformation to carbon dioxide by means of electric sparks.[20] These studies finally confirmed Bernard's theory of organic combustion, which disproved the influential theory of Antoine-Laurent Lavoisier (1743–94), who had attributed the origin of animal heat to direct oxidation taking place exclusively in the lungs.[21] Bernard stated that organic combustion is an indirect oxidation aided by special ferments, and using cardiac catheterisation, in 1853, demonstrated that this vital combustion takes place in all tissues. This study, as did the majority of his other studies in various areas of physiology, reflected Bernard's pre-occupation with the idea of 'milieu intérieur', or internal environment of an organism, and regulatory mechanisms that maintain the stability, or homeostatis, of this environment.

Bernard's influence on Cyon is undeniable, but the choice of the major subject of Cyon's research career, innervation of the heart, rested primarily with his original mentor, Remak. However, Bernard and Ludwig crucially influenced his approach and methods employed in his research. To Bernard, Cyon owed his superb surgical skills and ability to work on certain organs, the heart in particular. Like his French mentor, Cyon extensively exploited virtuoso vivisectional techniques using ligatures, sections, stimulation of living animals, and basic chemical tests, none of which required many instruments. At the same time, he also worked on isolated organs. He was very keen on devising and modifying various instruments and applying them to achieve quantitative conclusions, the cornerstone of the physico–chemical approach of Cyon's German teacher, Ludwig.

Studies in Paris and Leipzig were highly rewarding for Cyon. He gained support from both leading physiologists, and published about thirty research papers in German, French, and Russian.[22] Such successful and versatile productivity stood in sharp contrast to the results of the research sojourn of Fedor Zavarykin, graduate of the Institute of Physicians, who studied in Ludwig's laboratory at the same time as Cyon. In a reference letter on Zavarykin's research at Ludwig's laboratory, Sechenov pointed out that 'Dr Zavarykin was engaged solely in practical exercises in medical chemistry for which there was no need to stay abroad as all necessary means for such studies are available at home.' Sechenov was also aware of Ludwig's quite unfavourable opinion of Zavarykin.[23] Moreover, Zavarykin presented to the Conference of the Medico-Surgical Academy his collaborative work with Ludwig, 'Zur Anatomie der Niere', as his original research. Sechenov's verdict was harsh and straightforward: 'Mr Zavarykin in this work was a skilful performer of Professor Ludwig's instructions.'[24] It was on those grounds that Sechenov was against Zavarykin's appointment to the chair of histology in 1869, and it was Zavarykin who headed the campaign against Cyon's appointment after Sechenov's resignation in 1870.

Cyon in St Petersburg: rise and fall

Cyon returned to St Petersburg in 1868 to teach a course on nervous diseases at the Medico-Surgical Academy. Luckily for Cyon, F.V. Ovsiannikov (1827–1906), Professor of Anatomy and Histology at both the St Petersburg Academy of Sciences and at the University, invited Cyon to the University's laboratory.[25] A fine histologist, Ovsiannikov was keen on physiological research and the University's laboratory was his own creation. He had studied with Bernard and later performed some important neurophysiological investigations in Ludwig's laboratory in Leipzig.[26] Ovsiannikov's organisational abilities, his influence and connections were pivotal both for setting up the new laboratory to work, and for Cyon's successful career at the University. Supported by Ovsiannikov, Cyon adequately equipped the laboratory at a cost of 3,124 roubles, which made it possible for him to give lectures with demonstrations and to organise practical training for advanced students in laboratory techniques, including vivisection, organ isolation, and the application of basic physiological devices.[27] For his own research, however, Cyon had to resort to laboratories of Bernard and Ludwig in Paris and Leipzig respectively, where he used to go for short visits. In Ludwig's laboratory, Cyon performed studies on the nerves of the vessels and of the peritoneum, which clarified the role of the afferent nerves of the viscera.[28] Another important study, which Cyon carried out in Bernard's laboratory, was on the synthesis of urea in the liver and its secretion in the blood.[29]

In 1868, Cyon started his teaching career at St Petersburg University as assistant and Privatdozent and two years later, in 1870, was promoted to extraordinary professor. Ovsiannikov continued to teach histology, and N.N. Bakst offered a special course on the physiology of sensory organs.[30] Cyon's lectures on general physiology were always accompanied by vivisectional demonstrations and were extremely popular, like theatrical performances played to a full house. In the best traditions of scientific public demonstrations, Cyon's lectures displayed the importance and power of physiological knowledge and skills to deal with complex functions of a living organism. Ivan P. Pavlov, Cyon's student at the University recalled:

> Professor I.F. Cyon made a tremendous impression on us physiologists. We were simply astonished by his masterfully simple account of the most complex physiological questions and his truly artistic ability to perform experiments. You cannot possibly forget such a teacher.[31]

Indeed, Cyon was said to be such an adroit vivisectionist that sometimes coming into the laboratory on his way to the theatre, elegantly dressed in his frock coat and starched white shirt, he could perform experiments involving vivisection without covering his clothes, for there was no need.[32]

Cyon always welcomed interested students to his laboratory to assist him in his research. Pavlov, then a student at the St Petersburg University, carried out his first research under Cyon's guidance. The connection between the spinal cord and the heart through sympathetic nerves still remained unclear and controversial. Studying the afferent nerves of the heart, Pavlov and V.N. Velikii (1851–1911), showed that sympathetic nerve fibres innervating the heart are not part of the pharyngeal nerves. Pavlov and Velikii demonstrated that nerves accelerating the heartbeat extend from the spinal cord, not the brain, through the stellate ganglion and thus confirmed the hypothesis of Cyon and the German physiologist, Bezold.[33] Such experiments, which involved finding and isolating nerves from the tissue, their ligation, and connection of the electrodes to the central end of the nerve fibres, required solid operative skills. Obviously, Cyon helped the young researchers and it was under his supervision that Pavlov acquired his admirable operative skills. In Cyon's laboratory at the Medico-Surgical Academy, Pavlov, in collaboration with M.I. Afanas'ev (1850–1910), performed experimental work on the innervation of the pancreatic gland, one of the most complex organs to study. The young investigators were awarded a gold medal at the St Petersburg University physico–mathematical faculty. Published in the *Pflügers Archiv* in 1877 with acknowledgements to Cyon,[34] this work became an overture to a series of famous Pavlovian works on the innervation of the digestive glands. That Pavlov never subscribed to the opinion of Cyon's enemies, and never forgot his teacher, is best expressed in Pavlov's letters to Cyon from 1897 to 1905. For Pavlov, his research at Cyon's laboratory and his mentor's 'incomparable lectures' remained 'the best recollections of my youth… the time when you were so passionately teaching us and we were so passionately studying… and this connection remained throughout life.' At the peak of his fame, in 1904, when Pavlov, the powerful Director of the Imperial Institute for the Experimental Medicine, was awarded the Nobel Prize, he sent a telegram to his old, ill, and abandoned teacher in Paris with an expression of sincere gratitude and admiration to his mentor, who 'made possible my triumph'.[35]

In 1870, Sechenov left the Medico-Surgical Academy and recommended Cyon as his successor. In a note to the Conference, on Sechenov's request, Botkin presented Cyon as an accomplished scholar and teacher deserving the chair of physiology.[36] The Conference, however, suggested another candidate, A. Shkliarevskii, professor of medical physics from Kiev. A special commission of seven professors, headed by F.N. Zavarykin, Cyon's most fierce and unappeasable opponent, was to make judgments on the scientific merits of both candidates. The 'work' of the Commission dragged on for almost a year. The report, which ran to a hundred and ten pages, is an astonishing document. Full of sarcastic critiques of all twenty-two of Cyon's scientific publications, it accused the author of 'scientific and literary plagiarism, conceit, rude polemic means and vulgarity.' The 'analysis' of Cyon's work sounds even more unfair and, at times, incompetent since none of the judges was a physiologist.[37] The conclusion ran: 'In our opinion, it is a sheer impossibility to entrust scientific supervision of young people entering our Academy to Mr Cyon.'[38] Professor Balinskii did not sign the report, instead presenting a ninety-one-page balanced account of Cyon's scientific accomplishments. In Balinskii's opinion, the works of Shkliarevskii were insignificant and he could not even qualify for a position.[39] Despite the support of Balinskii and Botkin, and the recommendations of Ovsiannikov and Sechenov, Cyon was voted down. In view of this extraordinary situation, N.I. Kozlov, by then Chief Medical Inspector at the War Ministry, appealed to War Minister D.A. Miliutin. Disregarding the results of the voting, Miliutin appointed Cyon ordinary professor in 1872. Miliutin's unprecedented decision was based on the recommendations of the leading European physiologists, Bernard, Ludwig, and Pflüger. Moreover, Miliutin, in a note to the commission, pointed out that the arguments used against Cyon in the report were unacceptable in scholarly disputes.[40]

Cyon's appointment was by no measure a sinecure. Cyon was a remarkable laboratory leader and fulfiled his teaching responsibilities both at the Academy and University to the utmost of his abilities. His experience and managerial skills were pivotal in setting up the laboratory at the new Anatomo-Physiological Institute, and in creating a well-defined training programme in experimental physiology at the Academy. Nevertheless, the appointment through the War Ministry had dire consequences for Cyon. The radical press responded fiercely to the interference of Miliutin and his appointment. The 'Cyon affair', as it became known, agitated a considerable number of radical intellectuals, for whom it was a convenient motive to demonstrate their unappeasable rejection of any action of the government and its high officials. Cyon's monarchist views, staunch adherence to Russian orthodox dogmas, and his expressed enmity to nihilism only added flavour to the criticism of Miliutin's decision. Cyon's distinguished professional qualities and research abilities seemed of no interest and importance for radical intellectuals.[41] It was not surprising, then, that Cyon's first formal speech at the Medico-Surgical Academy entitled 'Serdtse i mozg' ['The Heart and the Brain'], the field in which Cyon was a recognised authority was, in view of such intellectuals, a mere demonstration of Cyon's anti-materialistic position. As commented by Cyon:

> In the large auditorium, where I delivered my speech, there was a group of students, who had been already poisoned by the teaching of my predecessor, a patriarch of nihilism [Cyon meant Sechenov]. While I was talking about pure science, they listened to me with careful attention. But when I, finishing the talk, concluded that human knowledge has its limits beyond which everything is dark, and would remain dark forever, a storm of indignation exploded among these students.[42]

Interestingly, Cyon made a point that such a view was widely accepted among the leading European physiologists. Apparently, Cyon had in mind the famous speech by Emil du Bois-Reymond 'Über die Grenzen des Naturkennens' ['On the Limits of Knowledge'] delivered at the meeting of the German Society of Natural Scientists and Physicians in 1872. For du Bois-Reymond, the origin of matter and energy, and the problems of free will, perception, sense, feeling, and consciousness were 'Welträtsel' ['world mysteries'] that the human mind could never comprehend, *ignorabimus* ['we will never know'].[43] Cyon published 'The Heart and the Brain' as a separate brochure in both Russia and France.[44]

During his ensuing years at the Medico-Surgical Academy, Cyon acquired a notorious reputation as reactionary and conservative. He openly expressed provocative statements demonstrating his negative attitude towards Darwinism and towards Sechenov's 'nihilistic treatise'.[45] On the other hand, like any other conservative of the time, Cyon was against anything associated with radicalism and nihilism, which would soon grow into anarchist and terrorist movements and throw Russia into the chaos and blood of the 1880s, ultimately manifesting itself in the assassination of some high officials and Tsar Alexander II.

The atmosphere at the Medico-Surgical Academy during Cyon's tenure remained tense and increasingly hostile against Cyon, both on the part of the radical students and from the Conference. It is hard to imagine, however, that the professors of the Conference shared the views of radical students. Mild liberal Sechenov, who had harshly criticised the policy of appointment, and monarchist Cyon, who dared break unwritten rules of this policy, equally experienced covert and sometimes open opposition from the Conference. Cyon's scholarly merits were not recognised, and his industrious and successful activities at the Academy were simply not appreciated, or welcomed, by the Conference. Moreover, routine work and meagre, if any, research contributions of the majority of the members of the Conference contrasted with the diligence and energy of Cyon. Under these circumstances, Cyon displayed his quarrelsome and unpleasant character, and intolerant and disdainful attitude towards his opponents. In a foreword to the collection of works performed in his laboratory for the year 1873,

Cyon wrote: 'The best answer to the attacks directed against my teaching activities at the Academy is to publish an account of the research carried out by my students under my guidance in a short period of time.' In the second part of the collection entitled 'K moim kritikam' ['To My Critics'], Cyon did not spare any disdain towards those members of the commission whose negative opinion about his scholarly merits had been completely ignored: 'I am proud that, in spite of having been blackballed, I was appointed on the grounds of the references of the first luminaries of European science.'[46]

Though short and unbelievably controversial, Cyon's tenure marked a new period in teaching experimental physiology at the Medico-Surgical Academy. In 1871, the physiological laboratory was moved from the First Land Forces Hospital to the spacious new building of the Anatomo-Physiological Institute. Within two years, Cyon managed to equip the laboratory with the best apparatus and instruments available. The list of equipment Cyon purchased during his visits to Europe includes more than forty items: electrophysiological devices of different modifications, acoustic, vocal, optical instruments, all kinds of recording apparatus such as a myograph, sphygmograph, cardiograph, kymograph, and others. Most illustrations in Cyon's famous *Atlas zur Methodik der physiologischen Experimente und Vivisectionen* [*Atlas of Methodology for Physiological Experiments and Vivisection*] were pictures of the instruments and apparatus from the laboratory of the Medico-Surgical Academy.[47] The spacious premises consisted of several rooms – two rooms were equipped for vivisection work, two for electrophysiological studies and studies of sensory organs, one room for blood analysis, and one for physiological chemistry. The equipment and maintenance of such a facility, as well as a new programme in practical studies for an increasing number of students, required a substantial infusion of money. In the Chief Medical Inspector, Kozlov, Cyon had an influential and powerful supporter at the War Ministry to obtain the necessary funding.[48] Cyon possessed excellent managerial and financial abilities, and certain skills to become a trustworthy person with high officials. Co-operation between the two men resulted in the creation of one of the best physiological institutions in Europe. In appreciation of Kozlov's key role in his career at the Academy, Cyon dedicated his two-volume textbook, published in 1873–4, to Kozlov.[49]

Cyon's book was the first original textbook on general physiology in Russian, and a clear and comprehensive statement of the theory and practice of the discipline. It was basically his lectures on general physiology delivered at the Academy in the first year of his teaching. The first lecture was a vivid account of the history of the main trends in contemporary anatomy, physiology, physical physiology, and vivisection. The lectures on blood circulation, respiration, digestion, nerve and muscle physiology and sensory organs contained detailed and clear explanations of a wide variety of methods, techniques, and instrumentation, applied to the study of bodily functions. One of Cyon's major innovations in teaching physiology was the introduction of graphics and the application of self-recording devices in his lecture demonstrations. The use of graphics visualised the phenomenon under investigation and facilitated its evaluation and analysis. From the time of his collaboration with Ludwig, Cyon had a special interest in graphic methods which permitted a particular kind of measurement in his studies on the nerves of the heart and their vasomotor function. The method and the analysis of mechanically recorded graphs, introduced by Ludwig in the late 1840s and later widely used for various physiological purposes, became a symbol of modern physiology and an embodiment of precision and exactitude in its experimental practices.

Cyon's lectures and practical course enjoyed great popularity among the students of the Medico-Surgical Academy. Masterful in vivisectional technique, he accompanied his lectures with elegantly executed experiments on warm-blooded animals. The large lecture hall for three hundred students was always full, and for practical classes Cyon divided the students into three or four groups. These practical classes were also accompanied by demonstrations of the principal physiological experiments, and were usually held in the evening and sometimes lasted until late at night. Cyon devoted much of his time and energy to the practical course, and never cancelled it even when there were only two or three students in the laboratory. He also had enough energy to organise and carry out a private practical course for physicians in the physiology of respiration and blood circulation. This course was also held in the evening, and many influential members of the St Petersburg medical establishment attended it.[50] As he did at the University, Cyon gathered around him a group of students interested in physiological research at the Medico-Surgical Academy. The students' projects were mainly related to Cyon's primary subject, the innervation of the heart and other organs. He was very keen to demonstrate the first achievements of his students. A collection of works, published in 1874, included two articles by Cyon and five by his students – 'Innervation of the Spleen' and 'Effect of Temperature Changes on the Central Ends of the Nerves of the Heart' by I.R. Tarkhanov, 'Innervation of the Uterus' by M. Shershevskii, 'Investigation on Vasomotor Nerves' by M. Polkov, and 'Functions of the Semilunar Ducts' by I. Solukha.[51]

Cyon also pursued active research and publication. He took part in the competition announced by the French Academy of Sciences. The academic commission, headed by Claude Bernard and the physicist Antoine-Henry Becquerel (1852–1908), awarded a gold medal for Cyon's work, 'The basic physical and physiological principles of the application of electricity in medicine'. Cyon revised the work and in 1873 published it under the title *Principes d'électrothérapie* [*Principles of Electrotherapy*] in France, and the following year in Russia.[52] At the same time, Cyon was working out a theoretical basis for his interference theory of inhibition. Cyon never accepted Sechenov's theory of central nervous inhibition. Ludwig and Cyon's studies on the depressor nerve suggested a new reflex mechanism, different to Sechenov's concept of the reflex as an end process, based on the notion of the reflex arc. For Cyon, reflex was a circular or self-regulating mechanism. The impulse coming from the nerve centre causes certain changes in organ function, and the nerve centre having received information about these changes – through the reciprocal feedback mechanism – sent additional impulses to the organ. Cyon performed studies on the speed of transmission of excitation along the spinal cord in a normal animal. For this study, he devised a special device, since Helmholtz's instrument allowed measurement of the speed of transmission only in the peripheral nerves isolated from the animal. Cyon's instrument allowed accurate measurement of the speed of transmission of the excitation, and reflex impulses in the brain.[53] He demonstrated that the speed of transmission of excitatory impulses along the spinal cord was very slow. He gave this fact particular importance since it allowed, as Cyon believed, refutation of a phenomenon known as 'Sechenov's inhibition'. The Sechenov–Turk method of the inhibition of reflexes had demonstrated that the speed of spreading of impulses through the spinal cord was considerably inhibited. He attributed inhibition of reflexes to low conductivity of the impulses through the spinal cord but not to a special process of inhibition, which, as he believed, did not exist. He held that only two states, rest or excitation are possible in the elements of the nervous system. He interpreted inhibition in terms of interference of the waves of excitation. Thus, for Cyon, inhibition or rest occurs when there is a collision or interference of the waves of excitation, and waves of opposite phases extinguish each other.[54] This supposition seemed plausible since the German physiologist Julius Bernstein had suggested wavelike movement of the excitation along the nerve fibre. Rudolf Heidenhein, in 1870, was the first to dispute Cyon's controversial experimental data and unconvincing conclusions.[55] Sechenov also published two papers criticising Cyon's theory of interference for the lack of experimental proof and unsoundness of its postulates. Although Cyon's theory of interference was not accepted, its discussion focused the attention of some physiologists, in particular Nikolai Vvedenskii, onto his theoretical postulates. Vvedenskii showed that interaction of waves of excitation caused depression of the nerve activity; however, he attributed it to the resultant state of inhibition. Interestingly, in 1874, Cyon's paper on the theory of interference appeared in a collection of articles published in honour of Ludwig by his students. This collection, however, was lacking a contribution from Sechenov, one of Ludwig's first and most recognisable students, and a close friend.[56]

In the spring of 1874, Cyon had to interrupt his service to the Medico-Surgical Academy. The conventional version of the circumstances that forced him to leave the Academy, confirmed by Cyon himself, is the continuing hostility of a group of radical students who provoked unrest at his lectures.[57] On Cyon's request, the authorities of the Academy had to demand 'zhandarmy' – or gendarmes – to rebuff a group of students who interrupted his lectures. The arrest of ten students, suspected in organising disruption, caused even more unrest among the students of the Medico-Surgical Academy, who stopped attending classes and in the evenings gathered at meetings to protest against the repressive measures. Student unrest spread to the University, Technological Institute, and Mining Institute.[58] In the mid-1870s, student riots agitated by radicals and politically active students and severe repressive measures against them were frequent events in St Petersburg and other big university cities.

The reactionary position of one of the professors could be, and actually was, convenient grounds for students to manifest their political demands and discontent with the actions of the Government. However, the physiologist Leon A. Orbeli (1882–1958) pointed to another reason for student disorder at Cyon's lectures. Orbeli reported, from the words of Pavlov, that on the day of the exam, a group of students came to Cyon and demanded satisfactory marks for the course without examination. Cyon firmly rejected their demand. The next day, students not having met the requirements of the exam failed to pass. A row flared up, and student discontent manifested itself by unrest at Cyon's lectures.[59] Cyon took leave, suggested by the War Ministry, and left for Paris to work at Bernard's laboratory. The following autumn, Cyon returned to St Petersburg and tried to resume teaching, but student unrest at the Medico-Surgical Academy broke out once more, and he was forced to leave again for Paris. In 1875, Cyon resigned from both the Medico-Surgical Academy and the University. Pavlov, by then a student at the Academy, was bitterly disappointed. It was obvious to Pavlov that Cyon's successor, Ivan R. Tarkhanov, a recent graduate of the Academy and three years younger than himself, could not replace a remarkable mentor and scientist.[60]

Cyon's return to Paris

In Paris, back in Bernard's laboratory, Cyon found a quiet harbour after all his stormy misfortunes in St Petersburg. Bernard's favourable disposition, respect of his colleagues, and the possibility to continue experimental research, atoned for the absence of an official position for him within the laboratory. The place was quite modest, especially in comparison with the research facilities in Leipzig, St Petersburg, and Berlin. Throughout the late 1860s and 1870s, Bernard, together with other distinguished scholars such as the chemist Adolf Wurtz, was involved in a campaign aimed at laboratory construction and the improvement of research facilities. Through communication with Cyon, Bernard knew in detail about the laboratory facilities at the Medico-Surgical Academy, as well as about Ludwig's famous Institute in Leipzig. In his introductory lecture at the Museum of Natural History in 1877, Bernard referred to 'the splendid physiological institute in St Petersburg', and 'spacious institutions, which house physiology and experimental medicine in Germany and Russia.'[61] Victor Duruy, the like-minded French Minister of Public Instruction,[62] in proposing a new budget to help create research and teaching laboratories in Paris similar to those in Germany and St Petersburg, also referred to the example of 'the new physiological laboratory just completed in St Petersburg, at a cost equivalent to three million francs'.[63]

During three years of his stay with Bernard, Cyon, in his usual manner, continued active research and publication. He performed several original experimental studies, in nerve–muscle physiology, and published an article on the application of the telephone – then cutting-edge technology – in physiological research.[64] Interestingly, Cyon also started studies on the effect of barometric pressure on the functions of the organism[65] – the premier topic for Paul Bert (1833–86), professor of physiology at the Sorbonne, Bernard's faithful student, and Cyon's future competitor for Bernard's chair. Cyon's most important work at that time, however, was the richly illustrated *Methodik der physiologischen Experimente und Vivisektionen, mit Atlas.* [*Methodology of Experimental Physiology and Vivisection, with Atlas*].[66] The *Atlas* was specifically designed as a source book for both researchers and physicians. In a preface, Cyon pointed to the crucial importance of the application of experimental methods, techniques, and instrumentation in the physiological research and in medical practice of the scientifically trained physician. In the *Atlas*, methods and techniques used in all areas of experimental physiology are skilfully presented and accompanied by comprehensive guidelines for their application. Superb depiction of the instruments and apparatus, and detailed information on their inventors and manufacturers, made the book a unique catalogue of the material culture of nineteenth-century experimental physiology. Cyon dedicated the work to Ludwig, 'a teacher and a friend', in appreciation of his most valuable contribution to the development of physiological instrumentation upon which his famous physico–chemical approach rested.

In 1878, after Bernard's death, Cyon applied for the chair of medicine at the Collège de France, a position which Bernard had held since 1855, when he succeeded his celebrated teacher, François Magendie (1783–1855). To be considered for a position, Cyon had to obtain a doctoral degree conferred in France, which he did easily in a short period of time, once again demonstrating his diligence and unflinching capacity for work. His doctoral thesis, his third during his academic career, was on the function of semi-lunar canals and the formation of the sense of space, the topic of his research at the Medico-Surgical Academy.[67] The head of the committee, which conferred Cyon's doctoral degree, was the notable pathologist Edmé Félix Alfred Vulpian (1826–87). In Vulpian, Cyon had a powerful supporter, who had presented his works to the Paris Academy of Sciences and favourably regarded his achievements.[68] Nevertheless, Cyon lost, and Paul Bert received the chair at the Collège de France.

Bert was the most celebrated student of Bernard and was depicted as such in the famous picture of Bernard at work by Léon L'Hermite. From 1863, Bert became, in his own words, Bernard's first assistant to share 'tanière obscure et humide' at the laboratory of the Collège de France. In 1869, he was awarded the chair at the Sorbonne. Bert dedicated his first important research, *On Respiration* published in 1870, to his mentor. Studies on respiration led Bert to other important research, the first to interpret clearly the effect of changes in barometric pressure on the living organism. The barometric studies won him, in 1875, the 'Grand prix biennual de l'Institute', one of highest scientific honours in France. In 1878, the year of Bernard's death, he published a monumental volume *La pression barométrique,* which included a substantial and detailed historical section followed by records and analysis of Bert's own 668 experiments. Faithful to his mentor, Bert was among a select few at the death bed of Bernard, abandoned as he was by his pious family.[69]

Cyon's decision to compete with Bert is quite a damning example of his conceit, over ambitions, and tremendous arrogance. Cyon left the laboratory in chagrin, blaming Bert for intrigues against him,[70] and his research and teaching career had come to an end. Cyon never forgave Bert his last lost chance to continue an academic career, and used every opportunity to defame him politically and personally. As a deputy, Bert fought for compulsory, lay education, free of interference from the Catholic Church, especially the economically powerful Jesuits. Bert's monograph *La morale des Jésuits,* published in 1880, sharply criticised religious, ethical, and political foundations of clericalism. As a response to his old foe, Cyon published an essay, 'La guerre à Dieu et la morale laique' ['The War with the God and Secular Morality'] in the newspaper *Le Gaulois.* He fiercely attacked Bert and the suggested measures for securisation of education, which, in Cyon's opinion, would degrade and destroy moral and cultural foundations of the society. Despite opposition, Bert became Minister of Education (1881–2) in the Government of Léon Gambetta.[71]

Cyon's new career, now in journalism, was supported by his old friend Katkov. In 1877–8, Cyon published in the Russian newspapers *Moskovskie vedomosti, Russkii vestnik,* and *Golos*, edited by Katkov, about two hundred popular articles on the newest scientific achievements, as well as essays on history of science and short biographies of scientists. Later, these articles were published as a book, *Nauchnye besedy* [*Scientific Discussions*].[72] Cyon appeared a skilful polemist and productive popular writer. However, some of his critical articles such as the 'Descending of Man According to Häckel', which disputed evolutionary theory, lacked sound arguments and were bluntly reactionary.[73] From 1880, Cyon's journalistic interests shifted more to political and financial problems, ranging from social studies on nihilism and anarchism to Russo–French political and financial relations and various aspects of Russian economic and military policy.[74]

By the mid-1880s, Cyon became a successful conservative journalist both in Russia and in France. He was a leading author, director, and co-owner of the French journal *Le Gaulois* and headed the influential French journal *La nouvelle revue.* Journalist activities were useful for Cyon's relationship with some notable Russians, such as the writer Ivan S. Turgenev, who used to live in Paris. Typically for Cyon, his quarrelsome and unbalanced character provoked distrust and an aversion in Turgenev, who dealt with him in connection with the organisation of an art exhibition.[75] While in Paris, Cyon continued to contribute to the Russian conservative newspapers and journals edited by Katkov. Cyon's articles 'Piatnadtsat' let Respubliki' ['Fifteen Years of the Republic'] and 'Nigilisty i nigilizm' ['Nihilists and Nihilism'], published in Katkov's *Russkii vestnik* in 1885–6, caused a fierce reaction and indignation from the democratic press in Russia and among Russian political émigrés in Paris, who saw in Cyon an unappeasable enemy. Indeed, in the first article, Cyon defamed the French Third Republic, skilfully using arguments supporting a monarchistic and clerical claim on the evident degradation of political and financial spheres, as well as the morals and ethics of society in France during the Republic. The second article was even more scandalous, full of wrath, unmasking, and attacks on the nihilistic movement – although some of these attacks were quite fair. In particular, Cyon presented a comprehensive and skilful critical analysis of the writing published in Paris and Geneva by the ideologue of anarchism, Petr Kropotkin, and by the leaders of the Narodnaia volia [People's Will], an organisation responsible for a number of terrorist acts and the assassination of Alexander II.[76]

In Paris, Cyon, keen on money matters, became actively engaged in financial operations on the stock exchange. These activities allowed him to get acquainted with some important French financiers and bankers. Recommended by Katkov, Cyon became a confidential agent of the Russian

Ministry of Finance, headed by I.A. Vyshnegradskii (1831–95) from 1887–92, and was soon granted the high rank of Active Councillor of State carrying the entitlement to nobility.[77] Cyon played quite an important role in attaining the Franco–Russian financial agreement of 1888. This was the first large loan from France – the Russian government traditionally tended to deal with Germany, Netherlands and England in financial matters. Cyon's financial career, so successful at the start, ended ungracefully. He was charged with illegitimate financial deeds and in 1895 was degraded. Count Sergei I. Vitte (1849–1915), one of the most significant political figures of the time (Minister of Finance in 1892, and head of the Government in 1905–6), in his memoirs gave an extremely negative portrait of Cyon. Vitte pointed to Cyon's personal detestable qualities, his talent for an abominable attitude whether to his wives or colleagues at the Medico-Surgical Academy, and his loathsome methods in fabricating and publishing accusations against any of his opponents. Most disgusting, in Vitte's opinion, was that Cyon, under his extremely conservative political views, concealed a sheer lust for personal profit.[78] Cyon indeed published a series of malicious pamphlets against Vyshnegradskii and Vitte, and their financial policy.[79]

Both Cyon and Katkov were actively engaged in the political and financial events of the time. Ardent proponents of Franco–Russian alliance, their publications in both France and Russia in the late 1880s were intended to promote and support the emerging political and financial rapprochement between the two countries. After the conclusion of the Franco–Russian alliance of 1891, aimed at counterbalancing the trinity alliance and growing German hegemony, Cyon published a voluminous monograph on the history of Franco–Russian relations in 1886–94, which included numerous documents and reminiscences.[80]

From the late 1880s, until his death in 1912, Cyon again became loyal to physiology. He contributed to the *Pflügers Archiv*, and maintained close contact with its editor, Eduard Friedrich Wilhelm Pflüger (1829–1910), one of the leading physiologists of the time. Cyon also preserved connections with Ludwig and sometimes visited the laboratory in Leipzig. He conducted several investigations on secretory glands, the thyroid gland, hypophysis, and epiphysis cerebri in his private laboratory in Paris. He published his preliminary results in several articles including one dedicated to Ivan Pavlov,[81] and in two monographs, one on the physiology of the thyroid gland, of 1889, and another on the glandular vessels as regulatory protective organs of the central nervous system, in 1910. Most of Cyon's works published in Germany and France during that period were the results of the previously performed investigations and some of his revised works, usually richly illustrated and well produced. During this period he published a collection of his physiological works in 1888 and several monographs: on the anatomy and physiology of the cardiac nerves, in 1905; on anatomy and physiology of the heart, in 1907; and on the labyrinth of the ear as an organ for mathematical sense of space and time, in 1908. Cyon's last monograph, on the ear as an organ of orientation in space and time, was published in Paris in 1911, a year before his death.[82]

For Russian physiology and medicine, Cyon remains one of the first and most remarkable laboratory leaders of the time. His major contribution was the introduction and development of an integrative approach to physiological teaching and research, which embraced vivisectional techniques, graphical methods in representation of physiological phenomena, and the use of instruments to study bodily function. Important innovations, which Cyon introduced at both the Medico-Surgical Academy and the University, were vivisectional demonstrations, systematic practical training for the students in a variety of experimental methods, and supervision of student research in a wide range of projects. In Cyon's time, physiology as a well-defined and established discipline at the Medico-Surgical Academy was practised in a spacious, excellently equipped laboratory in a new building. It signified an important change from a small, poorly equipped laboratory for a few students, to a large laboratory for systematic training and research available for a large number of students. Due to Cyon's efforts in setting up the new physiological institute and promotion of the research performed during his professorship, the Academy's laboratory was ranked among the best physiological institutions in Europe. Controversial and versatile, Cyon's activities were a peculiar mix of talent and contradictory – even disgraceful – deeds. His extremely conservative political views, and notorious hatred of nihilism, reflected the complex and tragic period Russia was entering at the dawn of the new century.

Notes

1. A.A. Ukhtomskii, 'I.M. Sechenov v Peterburgskom-Leningradskom universitete', *Fiziologicheskii zhurnal SSSR,* 40, 5 (1954), 527–39: 529.
2. I.I. Mechnikov, 'Vospominaniia o Sechenove', in *idem, Stranitsy vospominanii* (Moscow: Izd-vo Akademii Nauk SSSR, 1946), 10.
3. K.S. Koshtoyants, *Essays on the History of Physiology in Russia,* D.P. Boder *et. al.* (trans.) (Washington: American Institute of Biological Sciences, 1964), 158–60; 291.
4. E. Cyon, *De choreae indole sede et nexu cum rheumatismo articulari, peri et endocarditide* (Berlin: Berolini, 1864); *idem,* 'Die Chorea und ihr Zusammenhang mit Gelenk-Rheumatismus, Peri- und Endocarditis', *Wiener medizinische Wochenschrift,* ii (1865), 115–31.

5. R. Remak, 'Vorläufige Mitteilung microscopischer Beobachtungen über die Entwicklung ihrer Formelemente', *Müllers Archiv für Anatomie, Physiologie und wissenschaftliche Medizin* , iv (1836), 145–61.
6. R. Remak, 'Über die Ganglien der Herznerven des Menschen und deren physiologische Bedeutung', *Wochenschrift für die gesamte Heilkunde,* xiv (1839), 9–54.
7. E. Cyon, *Principes d'electrothérapie* (Paris: J. B. Baillière, 1873); E. Cyon, *Osnovy elektroterapii* (St Petersburg: K. Ricker, 1874).
8. R. Remak, *Galvanotherapie der Nerven- und Muskelkrankheiten* (Berlin: Hirschwald, 1858). The work was dedicated to the eighty-ninth birthday of Alexander von Humboldt.
9. B. Kisch, 'Robert Remak', in *Transactions of the American Philosophical Society,* 44 (1954) 227–96; see also E. Hintzsche, 'Robert Remak', in *Dictionary of Scientific Biography* (New York: Charles Scribner, 1970–present), Vol. 11, 367–70.
10. Cyon was converted in 1865, see *Rossiiskii gosudarstvennyi voenno-istoricheskii arkhiv,* Moscow, fond 1, opis' 1, 1865, listy 1– 10, lichnoe delo I.F. Tsiona.
11. E. Cyon, *Die Lehre von der Tabes dorsalis, kritisch und experimentell erläutert* (Berlin: Librecht , 1866); *idem.*, 'Über Irrenpflege und Irrenanstalten', *Virchows Archiv für pathologische Anatomie und Physiologie und für klinische Medizin,* 42 (1867), 149–90.
12. On Bezold and his work, see R. Herrlinger and J. Krupp, *Albert von Bezold, 1836–1868: Ein Pioneer der Kardiologie* (Stuttgart: Wissenschaftliche Verlagsgesellschaft, 1964); on Pflüger and his discovery of the inhibitory action of the nerve splanchnici on the intestine, and on the debates concerning studies of the reflex mechanism in the spinal cord, see E. Cyon, 'Eduard Pflüger: ein Nachruf', in *Pflügers Archiv für die gesamte Physiologie,* 132 (1910), 1–19.
13. E. Cyon, 'Über den Einfluss der Temperaturveränderungen auf Zahl, Dauer und Stärke der Herzschläge', in *Arbeiten aus der physiologischen Anstalt zu Leipzig* (Leipzig: Veit, 1866), 77–85.
14. E. Cyon and C. Ludwig, 'Die Reflexe eines der sensiblen Nerven des Herzens auf die motorischen der Blutgefässe', *Berichte über die Verhandlungen der königlich-sächsischen Gesellschaft der Wissenschaften zu Leipzig,* xviii (1866), 128–49. On the discovery of the nerve depressor, see H. Schröer, *Carl Ludwig* (Stuttgart: Wissenschaftliche Verlagsgesellschaft M.B.H., 1967), 160–2; see also V.O. Samoilov and A.S. Mozzhukhin, *Pavlov v Peterburge–Petrograde–Leningrade* (Leningrad: Lenizdat, 1989), 39–41.
15. C. Bernard, 'De l'influence de deux ordes de nerfs qui determinant les variations de couleur du sang veineux dans les organs glandulaires', *Comptes rendus hebdomadaires de l'Académie des sciences,* 48 (1858), 245–53.

16. C. Bernard, 'De l'influence du système nerveux grand sympathique sur la chaleur animale', *Comptes rendus hebdomadaires des séances de l'Académie des Sciences,* 34 (1852), 472–5; *idem,* 'Recherches expérimentales sur le grand sympathique et spécialement sur l'influence que la section de ce nerf exerce sur la chaleur animale', *Comptes Rendus des Séances et Mémoires de la Société de Biologie,* v (1853), 77–107. On Bernard's research on vasomotor nerves, see J. Olmsted, *Claude Bernard and the Experimental Method in Medicine* (New York: Collier, 1961), 81–5.
17. E. Cyon and M. Cyon, 'Über die Innervation des Herzens vom Rückenmarke aus', *Centralblatt für die medicinischen Wissenschaften, 51* (1866), 389–416; E. Cyon and M. Cyon, 'Sur l'innervation du Coeur', *Comptes rendus hebdomadaires des séances de l'Académie des Sciences,* 64 (1867), 670–4.
18. Samoilov and Mozzhukhin, *op. cit.* (note 14), 35–40.
19. E. Cyon, 'De l'influence de l'acide carbonique et de l'oxygène sur le coeur', *Comptes rendus hebdomadaires des séances de l'Académie des Sciences,* 64 (1867), 1049–53.
20. C. Bernard, 'Sur la quantité d'oxygène que contient le sang veineux', *Comptes rendus séances de l'académie des Sciences,* 47 (1858), 393.
21. M. Grmek, 'Claude Bernard', in *Dictionary of Scientific Biography, op. cit.* (note 9), 24–34: 30.
22. For a full list of Cyon's publications, see E. Cyon, *Die Gefässdrüsen als regulatorische Schutzorgane des Zentral-Nervensystem* (Berlin: Julius Springer, 1910), 359–71; see also N. Artemov, *Ilia Faddeevich Tsion* (Nizhnii Novgorod: Izdatel'stvo Nizhegorodskogo universiteta, 1996), 60–73.
23. I.M. Sechenov, *Autobiographical Notes,* D.B. Lindsley (ed.), K. Hanes (trans.) (Washington: American Institute of Biological Sciences, 1965), 127.
24. I.M. Sechenov, 'Note on Zavarykin's research abroad, 1866', in K.S. Koshtoyants *et al.* (eds), *Nauchnoe nasledstvo: I.M. Sechenov: Neopublikovannye raboty, perepiska i dokumenty* (Moscow: Akademiia Nauk, 1956), 181–2.
25. V.V. Grigor'ev, *Imperatorskii S. Peterburgskii Universitet v techenie pervykh piatidesiati let ego sushchestvovaniia* (St Petersburg, 1870), 391–4.
26. Ovsiannikov worked in Bernard's laboratory in 1860. In Ludwig's laboratory in Leipzig, he studied the action of the vasomotor centres and demonstrated the presence of nerve centres regulating blood pressure in the vessels. His other important research was on the reflex functions of the medulla and spinal cord in the rabbit. See P. Owsjannikow, 'Die tonischen und reflektorischen Centren der Gefäßnerven', *Berichte über die Verhandlungen der königlich-sächsischen Gesellschaft der Wissenschaften zu Leipzig,* xxiii (1871), 21–33; *idem,* 'Über einen Unterschied in den reflektorischen Leistungen des verlängersten und des Rückenmarkes der Kaninchen',

Berichte über die Verhandlungen der königlich-sächsischen Gesellschaft der Wissenschaften zu Leipzig, xxvi (1874), 308–18.

27. Grigor'ev, *op. cit.* (note 25), 393.
28. E. Cyon, 'Über die Wurzeln, durch welche das Rückenmark die Gefässnerven für die Vorderpfote aussendet', *Berichte über die Verhandlungen der königlich-sächsischen Gesellschaft der Wissenschaften zu Leipzig,* xx (1868), 104–12; *idem,* 'Über die Nerven des Peritoneum', *Berichte über die Verhandlungen der königlich-sächsischen Gesellschaft der Wissenschaften zu Leipzig,* xx (1868), 119–27.
29. E. Cyon, 'Über die Harnstoffbildung in der Leber: vorläufige Mittheilung', *Zentralblatt für medizinische Wissenschaften,* viii, 37 (1870), 580–1.
30. Nikolai Nikolaevich Bakst (1843–1904) studied, under Helmholtz, the problem of transmission in the motor nerve in human, see N.N. Bakst, *O skorosti peredachi razdrazhenii po dvigatel'nym nervam cheloveka* (St Petersburg: Tip. St Petersburgskogo Universiteta, 1867).
31. I.P. Pavlov, 'Avtobiografiia', in I.V. Natochin *et al.* (eds), Pavlov, *Izbrannye trudy* (Mosow: Meditsina, 1999), 24–6: 24; on Cyon and Pavlov, see D. Todes, *Pavlov's Physiology Factory: Experiment, Interpretation, Laboratory Expertise* (Baltimore: Johns Hopkins University Press, 2002), 51–2, 55–7.
32. W.H. Gantt, *Russian Medicine* (New York: Hoeber, 1937), 111. Gantt was an American collaborator in Pavlov's laboratory in 1925–29.
33. I.P. Pavlov, *Polnoe sobranie sochineni* (Moscow-Leningrad: Izd-vo Akademii Nauk SSSR, 1940), Vol. 1, 35, see also Samoilov and Mozzhukhin, *op. cit.* (note 14), 32–4.
34. Russian translation from the article in *Pflügers Archiv für die gesamte Physiologie,* xvi (1877), 173–89, in I.P. Pavlov, *Polnoe sobranie trudov* (Moscow–Leningrad: Izd-vo Akademii Nauk SSSR, 1946), Vol. 2, 173–88. Later, Afanis'ev became a notable pathologist, and Velikii, a professor of physiology at Tomsk University.
35. E.M. Kreps (ed.), *Perepiska I.P. Pavlova* (Leningrad: Nauka, 1971), 61–2.
36. *Rossiiskii gosudarstvennyi voenno-istoricheskii arkhiv,* Moscow, fond 316, opis' 38, delo 351, list 4–5 January 1871.
37. The commission included a chemist N.N. Zinin, a pharmacologist I.V. Zabelin (1834–75), a toxicologist I.M. Sorokin (1833–1917), a physicist P.A. Khelbnikov (1829–1902), an anatomist V.L. Gruber (1814–1890), and a psychiatrist, I.M. Balinskii (1827–1902).
38. F.N. Zavarykina, *et al.,* 'Otchet akademicheskoi komissii: ob uchenykh trudakh Tsiona i Shkliarevskogo', in *Protokoly zasedanii Konferentsii Imperatorskoi mediko-khirurgicheskoi akademii za 1872 god* (St Petersburg, 1873), 40–150: 122.
39. I.M. Balinskii, 'Otchet ob uchenykh trudakh Tsiona i Shkliarevskogo', in *ibid.,* 151–216.

40. L. Popel'skii, *Istoricheskii ocherk kafedry fiziologiiv Imperatorskoi voenno-meditsinskoi akademii za 100 let (1798-1898)* (St Petersburg, 1899), 70–4. It might be that the above-mentioned reference letters are preserved in Cyon's archive somewhere in Paris.
41. Articles against Cyon appeared in the journals, *Znanie,* xi (1873), *Otechestvennye zapiski,* vii (1874), and in the journal published abroad, *Vpered,* cited in Koshtoyants, *op. cit.* (note 3), 158.
42. E. Cyon, 'Nigilisty i nigilizm', *Russkii vestnik*, vii (1886), 262–98: 296.
43. E. du Bois-Reymond, 'Über die Grenzen des Naturkennens', Rede in der zweiten allgemeinen Sitzung der 45. Versammlung deutscher Naturforscher und Ärzte zu Leipzig, 1872, in E. du Bois-Reymond (ed.), *Reden von Emil du Bois-Reymond* (Leipzig: Verlag von Veit, 1912), Vol. I, 441–74: 462.
44. E. Cyon, 'Serdtse i mozg', speech given at the St Petersburg Medico-Surgical Academy, 21 January 1873, in *Protokoly zasedanii Konferentsii Imperatorskoi medico-khirurgicheskoi akademii za 1873* (St Petersburg, 1874); also I.F. Cyon, *Serdtse i mozg* (St Petersburg: Izd-vo Anrepa, 1874); in French, E. Cyon, 'Le coeur et le cerveau', *Revue scientifique de la France et de l'etranger,* v (1873), 1–33.
45. E. Budilova, *Bor'ba materializma i idealizma v russkoi psikhologicheskoi nauke* (Moscow: Izd-vo Akademii Nauk SSSR, 1960), 35–6; see also, Koshtoyants, *op. cit.* (note 3), 160–2.
46. I.F. Cyon, 'K moim kritikam', in *idem, Raboty, sdelannye v fiziologicheskoi laboratorii Imperatorskoi medico-khirurgicheskoi akademii za 1873 god s prilozheniem kriticheskikh statei professora Tsiona* (St Petersburg: K. Ricker, 1874), 87–120: 87.
47. Popel'skii, *op. cit.* (note 40), 81–2.
48. On Kozlov's career at the War Ministry, see *Russkii biograficheskii slovar'* (St Petersburg: Izd-vo Voennogo ministerstva, 1903), 25 vols, Vol. IV, 53–4.
49. E. Cyon, *Kurs fiziologii* (St Petersburg: K. Ricker, 1873– 4), 2 vols.
50. Popel'skii, *op. cit.* (note 40), 80.
51. Cyon, *Raboty, op. cit.* (note 45), 40–74.
52. E. Cyon, *Principes d'électrothérapie* (Paris: Baillière, 1873); in Russian, E. Cyon, *Osnovy elektroterapii* (St Petersburg: Izd-vo Anrepa, 1874).
53. E. Cyon, 'O skorosti provedeniia vozbuzhdeniia po spinnomu mozgu', in Cyon, *Raboty, op. cit.* (note 46), 75– 80; E. Cyon, 'Über die Fortpflanzungsgeschwindigkeit der Erregung in Rückenmarke, *Bulletin de la classe physico-mathématique de l'Académie Impériale des Sciences de St Pétersbourg,* iv, 19 (1872), 394–400.
54. E. Cyon, 'K teorii reflektornoi deiatel'nosti', in Cyon, *Raboty, op. cit.* (note 45), 81–4; *idem,* 'Zur Lehre von der reflektorischen Erregung der Gefässnerven', *Pflügers Archiv für die gesamte Physiologie,* viii (1874), 327–40.

55. R. Heidenhain, 'Über eine neuer Theorie von Cyon', *Pflügers Archiv für die gesamte Physiologie,* iii (1870), 551– 8.
56. E. Cyon, 'Zur Hemmungstheorie des reflectorischen Erregungen', in *Beiträge zur Anatomie und Physiologie: Als Festgabe Carl Ludwig zum 15 October 1874 gewidmet von seiner Schülern* (Leipzig: Vögel, 1874–5), 196.
57. Koshtoyants, *op. cit.* (note 3), 158–60.
58. N.N. Strakhov to L.N. Tolstoy, letter dated 8 November 1874, St Petersburg, in *Perepiska L.N. Tolstogo s N.N. Strakhovym* (St Petersburg: Izd-vo Muzeia Tolstogo, 1914), 53–4; see also Popel'skii, *op. cit.* (note 40), 78–9.
59. L.A. Orbeli, *Vospominaniia* (Moscow-Leningrad: Nauka, 1966); see also Artemov, *op. cit.* (note 22), 27–8
60. Pavlov, *Avtobiografiia, op. cit.* (note 31), 25.
61. C. Bernard, transl. M.A. Antonovich, *Zhiznennye iavleniia, obshchie zhivotnym i rasteniiam: lektsii, chitannye v Muzee estestvennoi istorii v Parizhe* (St Petersburg: Izd-vo Panteleeva, 1878), 9.
62. A. Wurtz, *Les hautes études pratique dans les université allemandes* (Paris: Imprimerie impériale, 1870). The report was prepared at the request of the minister, Victor Duruy, who used it in the campaigm aimed at gaining comparable facilities for France. See, W. Coleman, 'The Cognitive Basis of the Discipline: Claude Bernard on Physiology', *Isis,* 76 (1985) 49–70: 58, note 25.
63. V. Duruy, speech, 18 April 1868, *Revue des cours scientifiques,* v (1868), 343–4 cited in A. Rocke, *Nationalizing Science: Adolphe Wurtz and the Battle for French Chemistry* (Cambridge: MIT Press, 2001), 293.
64. E. Cyon, 'La forme de la contraction musculaire produite par l'excitation des racines antérieures', *Comptes rendus hebdomadaires des séances et mémoires de la Société de Biologie,* (1876); *idem,* 'Sur la secousse musculaire par l'excitation des raciness de la moelle epinière', *Comptes rendus hebdomadaires des Séances et mémoires de la Société de Biologie,* (1877); *idem,* 'Note sur le fonctionnement physiologique du téléphone', *Comptes rendus hebdomadaires des séances et mémoires de la Société de Biologie,* xxix (1877), 458.
65. E. Cyon, 'Action des hautes pressions baromètrique sur la circulation et la respiration', *Congrès international de médecine,* 1878. Cyon published his studies on the influence of atmospheric pressure on respiration and circulation much later, in 1882.
66. E. Cyon, *Methodik der physiologischen Experimente und Vivisektionen, mit Atlas* (Giessen: J. Rickersche Buchhandlung; St Petersburg: C. Ricker, 1876). On Cyon's collection of apparatus at the Medico-Surgical Academy, see Popel'skii, *op. cit.* (note 40), 79–80.
67. E. Cyon and Dr Solucha, 'Über die Funktion der halbzirkelförmigen Kanäle', *Pflügers Archiv für die gesamte Physiologie,* viii (1874), 306–27. E. Cyon, *Recherches expérimentales sur les fonctions des canaux sémi-circulaires et*

sur leur rôle dans formation de la notion de l'espace. Thèse pour le Doctorat en médecine à la Faculté de médecine, 1 Avril 1878 (Paris: Imprimerie Emile Martinet, 1878).

68. On Cyon's thesis and its discussion at the Paris Medical Faculty, see D. Anuchin, 'Dissertatsiia i disput Tsiona', *Moskovskaia meditsinskaia gazeta,* xix (1878), 482–8; see also Artemov, *op. cit.* (note 22), 31.
69. E. Ackerknecht, 'Paul Bert's Triumph', *Bulletin of the History of Medicine,* Suppl. III (1944), 16–31: 19, 21. See also N. Mani, 'Paul Bert als Politiker, Pädagoge, und Begründer der Höhenphysiologie', *Gesnerus,* xxiii, (1966), 109–16.
70. Cyon pointed to the negative role of Paul Bert in his decision to leave the laboratory, see E. Cyon, *Die Gefässdrüsen als regulatorische Schutzorgane des Zentralnervensystems* (Berlin: Julius Springer, 1910), 364; see also Cyon's article on Paul Bert in Hirsch, *Bibliographische Lexikon der hervorragenden Ärzte aller Zeiten und Völker* (Vienna, 1884), Vol. 1, 427–8.
71. E. Cyon, 'La guerre à Dieu et la morale laique: résponse á M. Paul Bert' (Paris: *Le Gaulois,* 1881). On the ideological debate between Cyon and Paul Bert, see G.E. Feldman, *Paul Bert, 1833–86* (Moscow: Nauka, 1979), 32–4. On Gambetta's government, see C. Jones (ed.), *Cambridge Illustrated History of France* (Cambridge: Cambridge University Press, 1994), 218–22.
72. E. Cyon, *Nauchnye besedy* (St Petersburg: K. Rikker, 1880).
73. Ernst Häckel (1834–1919), the notable German biologist, was a proponent of Darvin's evolutionary theory. His most acclaimed monographs are the *General Morphology of the Organisms,* published in 1866, and the *The Riddle of the Universe,* 1901.
74. See E. Cyon, 'Historische, politische und finanzwissenschaftliche Schriften', in Cyon, *Die Gefässdrüsen, op. cit.* (note 70), 369–71. Some historians of medicine saw Cyon as the author of the notorious *Protocols of the Elders of Zion,* see S. Kagan, *Medical Leaves* (Chicago, 1942), Vol. IV, 138–45; Ackerknecht, *op. cit.* (note 69), 22. Other historians do not accept this story as being not only unproven but unlikely, see D. Joravsky, *Russian Psychology: A Critical History* (Oxford: Blackwell, 1989), 485, fn61. On Cyon's Jewish confession, see Popel'skii, *op. cit.* (note 40), 77.
75. Turgenev to D.V. Grigorovich, letter dated 16 December 1881, Paris, in I.S. Turgenev, *Polnoe sobranie sochinenii i pisem* (Moscow: Nauka, 1981), Vol. 13, 165. See also, Artemov, *op. cit.* (note 22), 39.
76. E. Cyon, 'Piatnadtsat' let Respubliki' *Russkii vestnik,* viii, 178 (1885), 461–516; *idem,* 'Nigilisty i nigilizm', *Russkii vestnik,* vi, 426– 43; vii, 262–98; viii, 772–826. In France, Cyon published a separate brochure, see E. Cyon, *Nihilisme et anarchie: Etudes sociales* (Paris: C. Lévy, 1892.). On Cyon as a journalist, see Artemov, *op. cit.* (note 22), 33–44; see also S.I. Shtraikh (ed.), *Bor'ba za nauku v tsarskoi Rossii: neizdannye pis'ma I.M.*

Sechenova, I.I. Mechnikova i dr. (Moscow–Leningrad: Gossotsgiz, 1931), 61–2.

77. Cyon signed his published works 'de Cyon' in French, and 'von Cyon' in German.
78. B.V. Anan'in (ed.), S. Vitte, *Iz arkhiva S. Iu. Vitte: vospominaniia* (St Petersburg: Institut istorii PAN), 3 vols, Vol. I, 281–6.
79. E. Cyon, *Oú la dictature de M. Witte conduit la Russie?* (Paris: Hanner, 1897).
80. E. Cyon, *Histoire de l'entente francorusse 1886–94: documents et souvenirs* (Paris: Chamerot et Renouard, 1895).
81. E. Cyon, 'K voprosy o fiziologicheskoi roli *gl. Pinealis:* predvaritel'nye opyty', *Arkhiv biologicheskikh nauk,* ii (1904), 297–308.
82. E. Cyon, *Gesammelte physiologische Arbeiten* (Berlin: Hirschwald, 1888); *idem, Beiträge zur Physiologie der Schilddrüse und des Herzens* (Bonn: Martin Hager, 1898); E. Cyon, *Les nerfs du coeur: Anatomie et physiologie* (Paris: Tipographie Félix Alcan, 1905); E. Cyon, *Das Ohrlabyrinth als Organ der mathematischen Sinne für Raum und Zeit. Den Manen von J.P.M. Flourens, E.H. Weber und K. Vierordt gewidmet* (Berlin: Julius Springer, 1908); *idem, Die Gefässdrüsen, op. cit.* (note 70).

PART III

From Physics and Chemistry of the Body to Physical Chemistry:

Sechenov's Research on Blood Gases and Salt Solutions

13

Russian Universities in the Sea of Change, 1870–1886

> [M]ais les sujets ne s'improvisent dans la science; s'ils éclatent parfois comme la lumière, dans les découvertes, c'est par des faits qu'il faut bien posément et bien consciencieusement constater, avant de s'y fier.
>
> [B]ut subjects in science are not improvised; if they explode sometimes like light in discoveries, it is through facts, which must be established steadily and conscientiously before you can rely on them.
>
> George Sand, *Valvèdre*[1]

In 1870, after ten years at the St Petersburg Medico-Surgical Academy, Sechenov had to start his career anew. He resigned from the Academy in chagrin, following a change in management that left him exposed to a group of men who did not respect his scientific judgments. They had not wished to appoint the candidates he supported, or the successor he suggested. By his own will, Sechenov left the laboratory, an established institution for experimental physiology of his own creation, and his secure position of ordinary professor and even declined promotion to the higher rank of Active Councillor of State. Sechenov realised that his situation was not an easy one. He had a reputation as a 'dangerous materialist' in the Ministry of Education, while his unwillingness to resort to his connections in the medical establishment decreased his chances of getting an academic position in St Petersburg or elsewhere. Sechenov wrote to Mechnikov about his intention to study chemistry in Mendeleev's laboratory at St Petersburg University. Sechenov's hope 'to have one more tool to struggle for existence in the rank of professor'[2] explains his otherwise unreasonable decision to abandon physiology for chemistry and to move to Mendeleev's laboratory without any position whatsoever.

For all the distress over the uncertainty about his future career, the year spent in Mendeleev's laboratory was important for turning Sechenov's attention from neurophysiology back to blood-gas research, with which he had so successfully started his career as a scientist in Vienna. Mendeleev's influence was decisive in shaping Sechenov's outlook on solutions, a subject of particular interest for Mendeleev, and a topic that would become the primary investigation area for Sechenov in the following fifteen years. An independent researcher, who had run his own laboratory for ten years, Sechenov appreciated the possibility of following the customary schedule of a laboratory scientist enjoying an environment so familiar and dear to his heart. He had his room, all the necessary equipment, and reagents for synthesis of nitrous methyl ether, a project assigned to him by Mendeleev. Sechenov liked his apprentice work, and the synthesis of nitrous methyl ether, a substance that boils at minus 12°C, fascinated him. The results of Sechenov's work were published in the *Zhurnal fiziko-khimicheskogo obshchestva* [*Journal of the Russian Physico-Chemical Society*] in 1871.[3] He seemed, however, not ready to abandon physiology completely, as he would do later at St Petersburg to commit to salt solutions. Later he wrote: 'To be a student of such a teacher as Mendeleev was both pleasant and useful, but I had already partaken of too much physiology to be disloyal to it, so I did not became a chemist.'[4] In 1871, Sechenov finally received a call to Novorossiisk University. He would devote seventeen of his most productive years to teaching and research, first at Novorossiisk, and then at St Petersburg University from 1875 to 1887.

The 1870s and 1880s were tormented years in Russian history, which witnessed contradictory results of the reforms of the 1860s and the Russo–Turkish war of 1877–8, social upheavals and political repressions, the assassination of Alexander II by anarchistic terrorists in 1881, and the new rein of Alexander III, which marked the last period of political stability for Imperial Russia. During these decades, Russian literature, music and art achieved unprecedented heights and, like a marvellous mirror, reflected all of the dramatic turns in Russian history and admirably exposed the country's life and soul. One of the landmarks of the period was a steady growth in university science and its increasing role in the rapidly developing economy. Representatives of university science served as advisors on various committees and councils, created under the Ministry of Finance to facilitate the development of industry and technology. Zinin, Russia's foremost expert in aniline was one of the creators of the dye industry. Mendeleev too, was very keen on technological questions. He initiated the exploitation of the Baku oil fields, and their large-scale expansion began in 1873 at Mendeleev's urging. Another of Mendeleev's important contributions was the modern petroleum industry in the region of the Caspian Sea. As chief scientific adviser, Mendeleev continuously argued that the country's future economic prosperity lay in a national investment in science.[5] The number of institutions of higher education increased, and by the 1880s totalled fifty-two, including nine universities, and the number of university students tripled, growing from 4,125 in 1865 to 12,432 in the late 1880s.[6] The social role of the universities rose into prominence despite the abolition of their autonomy and the repressive police measures imposed against politically active students and faculties.

A short discussion on Russian universities in the sea of change of the 1870–80s, begins with the quotation from the speech of A.S. Norov, Minister of Education, at Kazan University in autumn 1855, after Russia's bitter defeat in the Crimea: 'We have always regarded science as an essential necessity, and now we regard it as our first need. If our enemies are superior to us, they owe their advantage solely to the power of their education.'[7] This clearly states the Government's awareness of the gap in education between Russia and the West, which always powerfully determined the levels of scientific, technological, and industrial developments. The educational reform of the 1860s was aimed at modernisation to meet new economic and military demands. In Russia's critical circumstances after her defeat in the Crimean War, the Government's increasing apprehension that economic, administrative, educational, and social problems required more attention than grand diplomacy and military power, determined a course of reforms in all these spheres. With all the paradoxical turns in their course and implementation, the reforms of the 1860s appeared to be a major turning point in Russia's economic and social renovation and well-being.

An important achievement of the educational reform was the adoption of the liberal university statute in 1863, followed by a statute for high schools in 1864. The new statute incorporated the ideas of the enlightened bureaucrats, scholars, and public figures under the leadership of Minister Alexander Golovnin. It opened up higher education to all social classes, as well as granting universities corporate rights and special community status with a wide range of actions independent from the Ministry of Education. A university council of professors was now responsible for sanctioning teaching programmes and educational publications, organisation of scientific research and selection of students for studies abroad. The council acquired an important role in preparing university budgets and in setting the policy of academic appointments and promotions. The universities now had authorised rights to divide their faculties into departments, or to create new departments that proved conducive to the rapid growth of specialisation in academic training and expansion of the university curriculum, particularly at physico–mathematical faculties. The enlightened officials and educational reformers understood that modernisation of university education, and development of science, could not move ahead without considerable investment. From the early 1860s, the Government support for university science began growing slowly but steadily, so the universities began to expand their research and teaching facilities, and created new departments, institutes, and laboratories. The 1863 statute and increasing government support were crucial changes that drove the transformation of the universities into the foremost centres of scientific investigation in Russia, a position that had been previously occupied by the St Petersburg Academy of Sciences.[8] The unprecedented intellectual upsurge of the 1860s also helped the universities acquire such positions. Lust for knowledge obtained by scientific method penetrated society and was especially pronounced in a flood of publications on various social, economic, scientific, and educational issues. New periodicals, which sprang up one after another following the relaxation of the censors, widely discussed the newest developments in natural science, physiology, and scientific medicine associated with the awaited improvements in social well-being.

By the mid-1860s, however, such optimism had dissipated. Disappointment in the provision of reforms, particularly the peasant reform of 1861, began to build up in society, especially among the radical youth involved in numerous student unions and political organisations, which had mushroomed in the early 1860s. One of the sources of student unrest was inconsistency in pursuing liberal principles in education policy, which did not ensure students' corporate rights. The growing discontent and opposition movement resulted, in 1866, in the first attempt on the life of Alexander II, by D. Karakozov, a student of Moscow University and member of an extremist group.[9] An oppressive government reaction was inescapable, and universities suffered from the ensuing repressive measures more than other institutions of higher learning. The appointment of Dmitrii Andreevich Tolstoy (1823–89) as Minister of Public Education manifested an official return to educational conservatism. Chief Procurator of the Holy Synod and an Honorary Member of the St Petersburg Academy of Sciences, Tolstoy was an influential state official and one of the most staunch reactionaries. Tolstoy's conviction that university autonomy undermined autocratic ideology in great measure determined university policy for some fifteen years (1866–80). In the middle and high schools, the introduction of the 'classical system' into the curriculum, which relied upon the intensified study of Latin and ancient Greek, was aimed at deterring students from social, economic, and political problems. The so-called classical gymnasiums prepared students to enter university, while technical schools (real'nye uchilishcha) were intended for basic training in technical specialties. To keep university students and instructors away from political activities, Tolstoy strongly recommended university councils engage students in serious research. These recommendations of 1869 were conducive to the development of university science, which the Ministry of Education increasingly began to support.[10] The creation of a physiological laboratory at Novorossiisk University in 1870 is illustrative of the Ministry's attempts to expand research and teaching facilities in provincial universities.

Centralisation of Russian science was responsible for the concentration of the best research and teaching facilities in St Petersburg. The presence of the Government, the Academy of Science, and an admirable assortment of collections and libraries gave the St Petersburg educational institutions prestige and importance, and comparatively better funding than provincial universities. Small and remote provincial universities usually suffered most from continuous lack of funds, bureaucratic impediments, and petty academic intrigues.[11] Despite severe limitations, however, some of the provincial universities managed to maintain high standards in teaching and research. The chemical laboratory at Kazan University remained Russia's best centre for training in organic chemistry. V.V. Markovnikov, the world-class organic chemist, pointed out: 'As a graduate of Kazan University, I am proud that this poor University, which had no more than three hundred students always found possibilities to help its professors in their teaching and research.'[12]

Although the Government did increase funding for university science, the working space, instruments, equipment, and technical staffs of the laboratories in the Russian universities were still no match for the research facilities in Germany. The systematic expansion of university laboratories, even for such notable university scholars as Mendeleev and Sechenov in St Petersburg, and Butlerov first in Kazan and then in St Petersburg, was hampered by budgetary uncertainties and financial limitations. Even in the 1870s, with a number of established chemical and physiological laboratories in St Petersburg, Kazan, and Kharkov, Russian graduates preparing for academic careers had to study in foreign laboratories to acquire sufficient experience and skill to start independent research careers. In his report of 1874 to the council of the physico–mathematical faculty of Novorossiisk University, Sechenov recommended:

> For those who study physiology the most important laboratories are: Hoppe-Seyler's laboratory in Strasbourg, Donders' laboratory in Utrecht, and Ludwig's laboratory in Leipzig. In the first, one can get systematic training in physiological chemistry, in the second, more than anywhere else, training in physical methods of research in the animal organism. Significance of the third laboratory is known all over the world.[13]

For the most capable Russian graduates studying abroad, it opened possibilities to position themselves within the European scientific community in terms of publications, scientific connections, and recognition which, in turn, ensured successful careers at home. This was the case for scholars such as Sechenov, Mendeleev, Borodin, Botkin, and Cyon. In his popular article on Russian universities in *Vestnik Evropy*, Sechenov pointed out that half of the Russian professors engaged in teaching natural sciences in the early 1880s were those who had studied in Europe during the late 1850s and early 1860s.[14] Indeed, in those years, Russian students and researchers at Heidelberg University outnumbered those from any other country.[15] Heidelberg was, at the time, the most reputable centre for teaching and research in natural sciences with its excellent choice of laboratories including Bunsen's at the university, Helmholtz's physiological laboratory, and a number of private chemical laboratories. A decade later, in the mid-1870s, Russian researchers were the most numerous at Leipzig University's laboratories, those of the structural chemist Hermann Kolbe and of the physiologist Carl Ludwig. It was with foreign students that Ludwig and Kolbe gained much of their educational fame; their Russian students were among the most successful in achieving academic positions and in international renown for their research. Karl Rothschuh's family tree of Ludwig's students includes only prominent Russian physiologists, such as Sechenov, Cyon, and Pavlov. Among Russians, who did their postgraduate studies in Kolbe's laboratory, Alexander M. Zaitsev (1841–1910), Vladimir V. Markovnikov (1837–1904), and Nikolai A. Menshutkin (1842–1907) were all world-class chemists. They made a great impact in Kazan, St. Petersburg, and Moscow during and after the life of their teacher Alexander M. Butlerov (1828–86), the leading organic chemist in Russia.[16] The scholarly reputation and close connections of Sechenov and Butlerov with Ludwig and Kolbe were responsible for the presence and success of their students in the laboratories of Leipzig University. Table 13.1 illustrates the number of foreign students present in their laboratories.[17]

Internationally, the reputation of the Russian scholars in the first wave of the scientific move towards the West was also important for maintaining connections with the European scientific community, and for improving the image of the country in the eyes of the broader public in Europe. For Sechenov, the period from 1860 to the early 1880s represented a final phase in the Westernisation of Russia, a process that was elevated to new heights by his colleagues, primarily chemists, physiologists, and mathematicians, who had 'introduced Russia to the family of enlightened nations'.[18]

The remarkable growth of Russian science and a wide range of scientific activities at the universities in the 1870s were overshadowed by increasing student unrest and social instability. Looking back, Sechenov wrote about this time: 'In the 70s, the Government reaction against anarchistic terror reached its climax and was expressed by a whole series of extremely severe administrative and police measures against the students.'[19] On 1 March 1881, in the centre of St Petersburg, Alexander II, 'the Tsar Liberator' died at the hands of a terrorist. This was the eighth attempt on his life. The elaboration of the new university statute was accelerated in a situation in

Table 13.1

Foreign Students in the Physiological and Chemical Laboratories at Leipzig University (1860–90)

	Russia	UK	Scandinavia	US	Austria	Switzerland
Ludwig's Lab	20	5	16	13	-	-
Kolbe's Lab	21	20	-	10	3	7

which the reactionaries characterised universities as 'hotbeds of tsar-killers'. On 23 August 1884, Alexander III confirmed the new university statute, which was, in essence, thoroughly reactionary. This put an end to the partial autonomy of the universities and their activities were fully subjected to governmental bureaucracy. Characterising the activities of the Russian universities for the period from 1860 to 1885, Sechenov had to admit that, despite notable achievements, the scientific work of the 1870s did not emulate either the pioneering zeal or the theoretical depth of that produced during the 1860s.[20] For Sechenov himself, however, his university years were the most productive in his long academic career.

Notes

1. Cited in D.I. Mendeleev, *Dnevniki 1861 i 1862 godov [Diaries 1861and 1862]*, in S.I. Vavilov *et al.* (eds), *Nauchnoe nasledstvo* (Moscow: Izdatel'stvo Akademii Nauk SSSR, 1948), 109–28: 111.
2. Sechenov to Mechnikov, letter dated 24 November 1870, St Petersburg, in S.I. Shtraikh (ed.), *Bor'ba za nauku v tsarskoi Rossii. Neizdannye pis'ma I.M. Sechenova, I.I. Mechnikova, L.S. Tsenkovskogo, V.O. Kovalevskogo, i dr.* (Moscow–Leningrad: Gossotsgiz, 1931), 71.
3. I.M. Sechenov and D.I. Mendeleev, 'Syntez, fizicheskie kharateristiki i nekotorye reaktsii azotisto-metilovogo efira', *Zhurnal fiziko-khimicheskogo obshchestva*, iii, issue 7 (1871), 250.
4. I.M. Sechenov, *Autobiographical Notes,* D.B. Lindsley (ed.), K. Hanes (trans.) (Washington: American Institute of Biological Sciences, 1965), 128.
5. On Mendeleev and the rise of the oil and petroleum industry in Russia, see D.I. Mendeleev, *Arkhiv* (Leningrad: Izd-vo Leningradskogo universiteta, 1951), 31–2, cited in B.M. Kedrov, 'Kharakteristika estestvoznaniia', in S.P. Mikulinskii and A.P. Iushkevich (eds), *Razvitie estestvoznaniia v Rossii, XVIII – nachalo XX veka* (Moscow: Nauka, 1977), 248–59: 250–1; see also D.I. Mendeleev, *Problemy ekonomicheskogo razvitiia Rossii* (Moscow: Izd-vo sotsial'no-ekonomicheskoi literatury, 1960); on Zinin and aniline dye

industry, see I.I. Solov'ev, 'Khimiia', in Mikulinskii and Iushkevich, *idem*, 199–206: 202– 4.

6 P.N. Miliukov, 'Universitety v Rossii', in F.A Brokgauz and I.A. Efron (eds), *Rossiia: Entsiklopedicheskii slovar* (St Petersburg: Izd. Firma Brokgauz i Efron, 1902), v. 18, 788–800: 796.

7. On educational reforms in nineteenth-century Russia, see A. Vucinich, *Science in Russian Culture* (Stanford: Stanford University Press, Vol. I, 1963, Vol. 2, 1970), Vol. II, 35–65: 36; see also R.G. Eimontova, *Russkie universitety na grani dvukh epoch: ot Rossii krepostnoi k Rossii kapitalisticheskoi* (Moscow: Nauka, 1985), 78–92.

8. Vucinich, *op. cit.* (note 7), Vol. II, 51.

9. Karakozov belonged to the terrorist group called 'Hell', which saw its mission as committing regicide, manipulating liberals and other opposition forces, and punishing traitors in its midst, see I.K. Pantin, E. D. Plimak, and V.G. Khoros, *Revolutsionnaia traditsiia v Rossii, 1783–1883* (Moscow: Mysl', 1986), 212; see also A. Polunov, M. Shatz (trans.), *Russia in the Nineteenth Century: Autocracy, Reform, and Social Change, 1814–1914* (New York: M.E. Sharpe, 2005), 145.

10. Eimontova, *op. cit.* (note 7), 85–7.

11. I.I. Mechnikov, *Stranitsy vospominanii* (Moscow: Akademiia Nauk SSSR, 1946), 77–86.

12. V.V. Markovnikov, 'Moskovskaia rech o Butlerove. 1886', in L. Ivanova (ed.), *A.M. Butlerov: Po materialam sovremennikov* (Moscow: Nauka, 1978), 48–66: 65.

13. Cited in K.S. Koshtoyants *et al.* (eds), *Nauchnoe nasledstvo: I.M. Sechenov: Neopublikovannye raboty, perepiska i dokumenty* (Moscow: Akademiia Nauk SSSR, 1956), 113. Franciscus Cornelius Donders (1818–89), Professor of Ophthalmology and Physiology at Utrecht University, made important contributions to the chemistry of organic tissues and physiological optics. His other physiological research was concerned with the negative pressure in the intrapleural space and the velocity of nervous processes, see biography by R. J ter Laage 'F.C. Donders', in C.C. Gillispie (ed.), *Dictionary of Scientific Biography* (New York: Charles Scribner, 1970–present), Vol. IV, 162–4.

14. I.M. Sechenov, 'Nauchnaia deiatel'nost' russkikh universitetov po estestvoznaniiu za poslednee dvadtsatipiatiletie', *Vestnik Evropy*, xi (1883), 330–42: 334.

15. S.G. Svatikov, *Russische Studenten in Heidelberg: Unveröffentlichte Texte von S. G. Svatikov,* E. Wischhöfer (ed.), (Heidelberg: Heidelberg University Press, 1997), 15. Sergei Griegor'evich Svatikov (1880–1944), a Russian jurist and historian, taught in St Petersburg and at the Sorbonne. In 1940, he moved to the USA.

16. In 1858, Butlerov worked with Adolph Wurtz in Paris, and with August Kekulé (1829–96) in Heidelberg. In 1861, Butlerov proposed the term 'chemische Struktur' in place of Charles Gerhardt's 'constitution' to mean that the particular arrangement of atoms within a molecule was the *cause* of its physical and chemical properties, and assumed the tetrahedral arrangements of carbon valences. He also investigated isomeric isodibutylenes, recognising the existence of isomeric change. For Butlerov's comprehensive biography, see A.E. Arbusov, *Kazanskaia shkola khimikov* (Kazan: Tatarskoe knizhnoe izdatel'stvo, 1971), and G.V. Bykov, *A.M. Butlerov: Ocherk zhizni i deiatel'nosti* (Moscow: Izd-vo Akademii Nauk SSSR,1961). The English language sources on Butlerov are A. Rocke, *The Quiet Revolution: Hermann Kolbe and the Science of Organic Chemistry* (Berkeley: University of California Press, 1993), 257–9, 261–4; W.H. Brock, *The Norton History of Chemistry* (New York: W.W. Norton, 1993), 256–60; M. Nye, *From Chemical Philosophy to Theoretical Chemistry: Dynamics of Matter and Dynamics of Disciplines 1800–1950* (Berkeley: University of California Press, 1993), 101–2.
17. On the foreign students at Ludwig's laboratory, see K. Rothschuh, *History of Physiology*, G. Risse (trans. and ed.), (New York: Krieger Publishing, 1973), 210; on Russian students in Kolbe's laboratory, see Rocke, *op. cit.* (note 16), 319–21.
18. Sechenov, *op. cit.* (note 14), 336.
19. Sechenov, *op. cit.* (note 4), 143.
20. Sechenov, *op. cit.* (note 14), 342; see also Vucinich, *op. cit.* (note 7), Vol. II, 64 and Eimontova, *op. cit.* (note 7).

14

Sechenov at Novorossiisk University: New Laboratory, New Challenges

Sechenov's appointment to Novorossiisk University was not an easy one, and negotiations between the University Council and the Ministry of Education lasted almost a year. Founded in 1865, Novorossiisk University was situated on the Black Sea coast, in Odessa. In April 1870, Ilia Mechnikov, recently appointed Professor of Zoology, petitioned the University Council for the election of Sechenov to the chair of zoology: 'In Sechenov, the University would acquire an excellent teacher and a distinguished scientist.' Mechnikov strategically pointed out that although the department of zoology had no chair of physiology, it had both the necessary premises and funds for setting up a physiological laboratory.[1] In August 1870, the Council elected Sechenov ordinary professor in the Department of Zoology. The Minister of Education, Count D.A. Tolstoy was, however, reluctant to approve the appointment on the pretext of insufficiency of funds for a new position of Ordinary Professor. In the opinion of some high Ministry officials, Sechenov, a dangerous materialist and radical, would have a 'harmful influence on the minds of the students'.[2] Botkin tried with his connections at the Ministry of the Interior to facilitate the appointment, but to no avail.[3] By lucky chance, E.V. Pelikan, by then Head of the Medical Department at the Ministry of the Interior, intervened and the appointment went through. Pelikan was a close acquaintance of both Sechenov and Botkin from the early 1860s, and obviously had never subscribed to a biased opinion about Sechenov. The Ministry officials' fears were pointless:

> I kept very quiet. I did not lead a single student astray, did not provoke a single rebellion and did not construct barricades. I so much delighted the Trustee of the Educational District that he made me an Active Councillor of State.... Apparently, the Trustee continued to testify before the high authorities about my loyalty during the following years.[4]

Before the bureaucratic procedures associated with the appointment were concluded, Sechenov began to make arrangements for a new laboratory. Experienced in these matters, he knew what he needed – physical premises, equipment, supplies, an independent budget, and a trained assistant. He wrote to Mechnikov:

Image 14.1

Ilia I. Mechnikov.
Courtesy of the Wellcome Library Collection.

> I need to know as soon as possible, appropriation for purchasing instruments and equipment, annual budget for research purposes, and availability of a position for my assistant P.A. Spiro. His presence is absolutely necessary. With his help I can make a lot of simple devices which, while we are without appropriate instruments, are of great importance for teaching and research.... Has gas equipment been installed in the laboratory?[5]

Obviously, Sechenov understood that the financial, administrative, and technical means of a small university were no match to those of the Medico-Surgical Academy. The budget of the new laboratory, 672 roubles,[6] however, frustrated him. Such meagre funding made setting up a new laboratory, with its equipment and maintenance, a sheer impossibility:

> Suppose that I spend all money [600 roubles a year] exclusively on instruments for the laboratory, not spending anything on the lectures, my own work and work with the students, even then I would need 10 years to equip my laboratory properly, taking into account that every small thing has to be ordered from elsewhere.[7]

The University Council appeared supportive and granted Sechenov 2,000 roubles for the instruments and 500 roubles to move from St Petersburg to Odessa, which he also spent on laboratory equipment. During his summer trip to Europe, Sechenov was able to purchase basic equipment – a set of electrophysiological devices, instruments for haemodymamic and vivisectional studies, anatomical instruments, glassware, two chemical balances, three induction apparatus, porcelain and metal dishes, air pumps, and a spectroscope, all at the cost of 2,725 thalers.[8] The spacious rooms for the new laboratory were not adapted for experimental work. Technical arrangements for experiments, particularly with the 'absorptiometer', a bulky apparatus with multiple tubes connected to reservoirs of mercury, demanded a great deal of time and effort. A mechanic prepared, according to Sechenov's instructions, metal parts for the apparatus; the glass parts Sechenov made himself since there was no glassblower in Odessa.[9] Little by little, Sechenov managed to create adequate conditions for research and teaching. A standard set of electrophyiological devices was used in lecture demonstrations. Practical training was, however, not given as the enrolement to the zoology department was quite low.

At Novorossiisk University, Sechenov was lucky to be surrounded by a group of friends, young talented zoologists, Mechnikov and Alexander Kovalevskii, physicist Nikolai A. Umov, and Sechenov's assistant Petr A. Spiro (1844–93). Sechenov was particularly close to Mechnikov:

> Of all the young people I have known, I have never in my life met a more fascinating person than Ilia Ilich Mechnikov, by the liveliness of his mind, his inexhaustible wit and his well-rounded education. He was so serious and productive in science. At that time he had done a great deal in zoology and he had acquired great reputation as a scientist.[10]

They first met in 1865 in the Italian Sorrento, where Mechnikov was performing his embryological studies. Mechnikov became very keen on physiological research since his visit to Graz, where Sechenov demonstrated a complete heart block in the frog by irritating the vagus nerve. Now, at the Novorossiisk laboratory, they collaborated in performing a series of experiments on the influence of the vagus nerve on the heart, and demonstrated that the fibres of the vagus, which inhibited the action of the heart, ended in the nervous centres of the heart similarly to the afferent fibres.[11]

In the early 1870s, Mechnikov experienced a personal tragedy. His first wife, Ludmila, whom he married in 1869, died in 1873 in Madeira, where Mechnikov had brought her to ease her suffering from consumption. He was attending his dying wife compassionately, thinking to abandon science to stay with Ludmila on Madeira. Shocked and depressed by her death, Mechnikov attempted suicide. He was frail and his eyesight worsened. Returning to Odessa, Mechnikov found deep compassion in the kind Sechenov. Olga, Mechnikov's second wife, whom he married in 1875 in Odessa, later wrote that Sechenov's 'affection was a great comfort to Mechnikov at that sad time'.[12] They remained very close to each other after Sechenov left Odessa, sharing personal and scientific successes, failures and aspirations, research plans and problems in their fascinating correspondence. Their letters offer interesting insights into the intellectual milieu of nineteenth-century Russian science including observations on academic life, appointments, patronage, and characteristics of their colleagues. In *Stranitsy vospominanii*, Mechnikov left warm recollections about Sechenov, his kind and sincere disposition, his intelligence and wit, and the influence that Sechenov exerted on his younger friends.[13]

While in Odessa, Sechenov continued translating German textbooks by Otto Funke and a recently published textbook by Ludimar Hermann.[14] This choice was not accidental. Funke's textbook had attracted Sechenov's attention as early as 1857 during his studies in Funke's laboratory in Leipzig. Funke's textbook contained a detailed account on the experimental technique applied to the studies on physico–chemical properties of the blood, including Funke's methods for its crystallisation. The text was accompanied by a fine atlas with coloured pictures of various blood crystals.[15] For Sechenov, the experimental part of the textbooks was particularly useful for its technical content and methodology. In turn, Hermann's *Grundriss der Physiologie des Menschen* was a comprehensive and clear statement of the fundamental principles of experimental physiology. The *Grundriss* thoroughly reviewed blood-gas research in Germany, including a discussion on Sechenov's studies, which were widely accepted among German physiologists engaged in the problems of respiratory functions of the blood. Similarly, Felix Hoppe-Seyler and Wilhelm Kühne, the two leading physiological chemists of the time, in their textbooks, referred in detail to Sechenov's blood-gas research.[16] These studies earned Sechenov a reputation among German physiologists. There was an appreciation of the exactness of his results, as well as the refinements he established to his methodology and apparatus.

In Russia, Sechenov's contributions to experimental physiology were frequently overshadowed by some of his popular writings, which tended to treat the philosophical problems of perception, consciousness, and free will on mechanistic–materialistic grounds, as problems which required an experimental approach. The Russian physiological community was small and there were no specialised societies or journals for physiology, and the few rather small physiological laboratories were poorly staffed and funded. Nevertheless, physiology had a high public profile in Russia due to its direct association with a materialistic tradition and new developments in natural sciences. The interest in the problems related to modern trends in science, such as positivism, rationalism, determinism, and materialism was enormously high and science issues were hotly debated in the widely read literary journals. Among Russia's great writers, it was Ivan S. Turgenev who was the first to depict a representative of the new scientific worldview in Bazarov, the nihilist medical scientist in *Fathers and Sons.* Bazarov was directly associated, by contemporaries, with Sechenov, a physiologist and author of the controversial *Reflexes of the Brain.*

Sechenov's fondness for psychology, which at that time did not have clear boundaries with philosophy, and his neurophysiological studies brought him into a realm of problems that merged with the problems related to religion, politics, and all of nature. Sechenov did not, however, seem prepared to deal with such complex and ideologically controversial problems. Boris Chicherin, the notable statesman, jurist, and publicist, wrote:

> Sechenov's attempts by shallow jumps to deduce psychology from physiology, obviously did not have a scientific foundation whatsoever, and only led to introducing the logical visionary into exact investigative methods. A natural scientist not familiar with anything except his speciality, could easily fall into a one-sided interpretation of such notions as free will. Random readings in psychology, without broad education in philosophy, could confuse an unprepared mind.[17]

Publication of *Reflexes of the Brain* had brought Sechenov troubles with the Censorship Committee and a reputation as a radical. In Odessa, however, he again returned to popular writing, and this time it was a polemic with K.D. Kavelin (1818–85), the well-known liberal, historian, and publicist. In his book, *The Task of Psychology*, Kavelin criticised Sechenov's attempts to explain the spiritual activities of the man by means of material principles. As an answer, Sechenov published two articles in the widely read journal *Vestnik Evropy*, in which he argued that the task of psycho–physiology is to determine the material substratum of psychical processes, and that such studies could and should be pursued by physiologists using a repertoire of experimental methods and approaches.[18]

Sechenov's polemics with Kavelin were widely debated in society, as reflected by Leo Tolstoy in *Anna Karenina* (1874). Tolstoy discusses neither the essence of the polemics, nor the 'topic in vogue' – free will – which are of no interest to his favourite hero Konstantin Levin. A masterful episode with Stepan Oblonskii, however, gives the reader a feeling of how widely Sechenov's ideas spread throughout society and how these ideas were at times re-interpreted:

> At that moment there had happened to him what happens to most people when unexpectedly caught in some shameful act: he had not had time to assume an expression suitable to the position in which he stood toward his wife now that his guilt was discovered. Instead of taking offence, denying, making excuses, asking forgiveness, or even remaining indifferent (anything would have been better than what he did), he involuntarily ('reflexes of the brain', thought Oblonskii, who was fond of physiology) smiled his usual kindly and, therefore, silly smile.[19]

Among Russia's great writers, it was Fyodor M. Dostoyevsky who was preoccupied with the challenge science presented to faith, particularly in two of his novels, *Notes from Underground* (1864) and *The Brothers Karamazov* (1880). Master of the psychological novel, Dostoyevsky was keen on modern trends in philosophy and psychology. For the deeply Christian Dostoyevsky, free will, the cornerstone of Christian orthodoxy, could not be explained on mechanical grounds, as Sechenov attempted to do. Sechenov's polemics with Kavelin yielded a harsh response from Dostoyevsky in one of his letters:

> It is not the same in Europe; there you can meet Humboldt and Bernard and other such people with universal ideas, with tremendous education and knowledge not only in their own speciality. In our country, however, even very gifted people, for instance, Sechenov, are basically ignorant and uneducated outside of their own subject. Sechenov knows nothing about his opponents (the philosophers), and thus he does more harm than good with his scientific conclusions. As for the majority of students, male or female, they are an ignorant lot. What is the benefit in this for mankind?[20]

Later in his *Autobiographical Notes,* Sechenov would answer these reproaches, pointing out that, although he was made out to be a nihilist philosopher because of *Reflexes of the Brain,* he had not heard, during the forty years since its appearance in print, 'of a single instance, thank heavens, in which this small book had moved anyone to evil because of a false understanding of its points.'[21]

Sechenov's psychological writings seem not to have been taken seriously by many of his colleagues, physiologists, and physicians, who justly valued him as a thorough experimental scientist and teacher.[22] The attitudes of most liberal intellectuals and philosophers were both condescending and critical. Sechenov, however, persistently revised, developed, and published his psychological studies, which were largely ignored in the 1880s, both in Russia and in France, where Sechenov published his *Études psychologiques.*[23] It was in Soviet Russia that Sechenov's contributions to psychology were appreciated. He would be seen as the founder of a new school of psychology, which, despite its obvious ideological biases and constraints, became one of the most influential schools of twentieth-century psychology.[24]

Notes

1. Mechnikov's Note to Novorossiisk University Council, 1870, cited in K.S. Koshtoyants, *et al.* (eds), *Nauchnoe nasledstvo: I.M. Sechenov: neopublikovannye raboty, perepiska i dokumenty* (Moscow: Akademiia Nauk SSSR, 1956), 103–4.
2. A confidential letter by I.D. Delianov, the influential official at the Ministry of Education, to S.P. Golubtsov, Trustee of the Odessa Educational District, cited in Koshtoyants, *op. cit.* (note 1), 105–6; see also F.N. Serkov, 'Odesskii period nauchnoi i obshchestvennoi deiatel'nosti Sechenova', in P.G. Kostiuk *et al.* (eds), *Ivan Mikhailovich Sechenov: k 150-letiiu so dnia rozhdeniia* (Moscow: Nauka, 1980), 134–42.
3. I.M. Sechenov, *Autobiographical Notes,* D.B. Lindsley (ed.), K. Hanes (trans.) (Washington D.C.: American Institute of Biological Sciences, 1965), 128.
4. Sechenov, *ibid.,* 135. Active Councillor of State was a civil rank corresponding to a military rank of major-general.
5. Sechenov to Mechnikov, letter dated 26 October1870, St Petersburg , in S.I. Shtraikh (ed.), *Bor'ba za nauku v tsarskoi Rossii: neizdannye pis'ma I.M. Sechenova, I.I. Mechnikova, L.S. Tsenkovskogo, V.O. Kovalevskogo, i dr.* (Moscow–Leningrad: Gossotsgiz, 1931), 69–70.
6. Koshtoyants, *op. cit.* (note 1), 110.
7. Sechenov to Mechnikov, letter dated 12 October 1870, St Petersburg, in Shtraikh, *op. cit.* (note 5), 68.
8. Koshtoyants , *op. cit.* (note 1), 109–10. In the list of instruments Sechenov meticulously indicated prices, makers and their addresses: Sauerwald in Berlin, Gräner & Fridrichsen in Tübingen, Schortmann in Leipzig, Rüprecht, Abservorst, Meyer & Wolf, and Lenoir, all in Vienna.
9. Sechenov, *op. cit.* (note 3), 130.
10. *Ibid.,* 132–3.
11. I. Sechenov and I. Mechnikov, 'Zur Lehre über die Vaguswirkung auf das Herz', *Zentralblatt für medizinische Wissenschaften,* xi (1873), 163–71.

12. O. Mechnikov, *Life of Elie Metchnikoff 1845–1916* (London: Constable & Company, 1921), 78. Olga Mechnikova, née Belokopytova, remained a devotee and admirer, and collaborator of her husband throughout her life.
13. I.I. Mechnikov, *Stranitsy vospominanii* (Moscow: Akademiia Nauk SSSR, 1946), 45–64.
14. O. Funke, *Lehrbuch der Physiologie* (Leipzig: Leopold Voss, 1857); I.M. Sechenov (ed. and trans.), O. Funke, *Uchebnik fiziologii* (St Petersburg: Izd-vo Cherkesova, 1872–3); I.M. Sechenov (ed.), L. Hermann, *Osnovy fiziologii cheloveka* (St. Petersburg: Izd-vo Cherkesova, 1875).
15. Funke was interested in the problems of the physico–chemical properties of the blood. In 1851, he found haemoglobin in the blood of the spleen, see O. Funke, 'Über das Milzvenenblut', *Zeitschrift für rationelle Medizin,* i (1851), 172–9; *idem,* 'Neue Beobachtungen über die Kristalle des Milzvenen- und Fisch-Blutes', *Zeitschrift für rationelle Medizin,* ii (1852), 198–207.
16. L. Hermann, *Grundriss der Physiologie des Menschen* (Berlin: Hirschwald, 1874), 42–7; W. Kühne, *Lehrbuch der physiologischen Chemie* (Leipzig: Engelmann, 1875), 225–33; F. Hoppe-Seyler, *Physiologische Chemie* (Berlin: Hirschwald, 1877), 377–99.
17. B. Chicherin, *Vospominaniia: puteshestvie za graitsu* (Moscow: Sever, 1935), 89.
18. I.M. Sechenov, 'Zamechaniia na knigu g. Kavelina *Zadachi psikhologii, Vestnik Evropy,* xi (1872), 386–420; *idem,* 'Komu i kak razrabatyvat' psikhologiiu?' *Vestnik Evropy,* ii, 4 (1873), 548–628: 552.
19. L.N. Tolstoy, *Anna Karenina* (Iaroslavl: Verkhne-Volzhskoe izdatel'stvo, 1964), 5.
20. F.M. Dostoyevsky to A.F. Gerasimova, letter dated March 1877, St Petersburg, cited in K.S. Koshtoyants, *Essays on the History of Physiology in Russia,* D.P. Boder *et al.* (trans.), (Washington: American Institute of Biological Sciences, 1964), 170.
21. Sechenov, *op. cit.* (note 3), 111.
22. On Sechenov and his psychological studies, see A.A. Ukhtomskii, 'Sechenov v St Peterbugskom universitete', in A.A. Ukhtomskii, *Dominanta dushi: iz gumanitarnogo nasledstva* (Rybinsk: Rybinskoe podvor'e, 2000), 168–87: 182.
23. I. Sechenov, *Études psychologiques: traduites du russe par V. Derély: avec une introduction de M. Wyrouboff* (Paris: C. Reinwald, 1884).
24. On Sechenov and the Soviet school of psychology, see E.A. Budilova, 'Sechenov i Sovetskaia psikhologiia', in Kostiuk, *op. cit.* (note 2), 336–490.

15

A Simple Model: Transition from Blood-Gas Research to Studies on Salt Solutions

At Novorossiisk University, Sechenov continued his studies on blood gases, which he had started in Ludwig's laboratory in Vienna in 1859 using a strictly physical approach. In Vienna, Sechenov modified Meyer's apparatus and optimised the existing methods of isolating gases from the blood to quantitatively study absorption of carbon dioxide by the blood. Sechenov's predecessors in the field, Emile Fernet and Lothar Meyer, showed that in the serum and in the whole blood, the quantity of bound carbon dioxide – CO_2 remaining in the blood when placed in a vacuum – exceeded the quantity of 'free' carbon dioxide – CO_2 extracted from the blood – four-to-six-fold. Meyer also showed that carbon dioxide dissolved in the blood does not follow the Dalton–Henry law, suggesting that it is not in a state of physical binding but rather in a state of loose or reversible chemical binding.[1] Using the apparatus of his modification, Sechenov demonstrated that nearly all carbon dioxide contained in the blood could be liberated by the vacuum, suggesting that carbon dioxide contained in the blood is in the 'free' state. Although Sechenov's method rendered an accurate estimation of the quantity of carbon dioxide in the blood, his studies of 1859, as well as studies by Fernet and Meyer, still lacked convincing proof of the state of carbon dioxide in the blood, which was directly connected to the question of 'freeing the body from CO_2 through respiration.'[2]

In the early 1870s, when Sechenov returned to his studies on blood gases, the problem of the state of carbon dioxide in the blood still remained unclear. At this stage of his research, the crucial question for Sechenov was to determine whether carbon dioxide in the blood occurred in a bound state or in a free state. He started his investigations on serum, which was known to carry most of the blood CO_2.[3] Lothar Meyer and Rudolf Heidenhain, in 1863, demonstrated that solutions of sodium carbonate, Na_2CO_3 – a salt produced by weak carbonic acid, as a result of dissociation of CO_2 in water – played an important role in the absorption of CO_2 by the serum. Sechenov determined that Na_2CO_3 absorbs CO_2 in the blood differently from that in pure water. He reasoned that the uptake of CO_2 by inorganic salt and water could be regarded as a simplified model system for its uptake in the blood.

He started a new series of experiments on the absorption of CO_2 by solutions of other salts formed by weak acids, which were capable of binding CO_2 chemically. From that time on his studies broke into two separate parts, using blood, or salt solutions.

The experiments with blood went in their own sequence. Sechenov confirmed that:

> [T]the magnitude of chemical absorption of CO_2 in the serum in regard to its respiratory function is better than water, and better than an aqueous solution of alkali carbonate. It [serum] draws CO_2 from the tissues more strongly than water and gives it up in the lung cavity more easily than bicarbonate.[4]

Sechenov attributed the high absorptive characteristics of the serum to 'the presence of globulins in it'. If the globulins were precipitated by the addition of magnesium sulphate, $MgSO_4$, the liquid remaining showed only faint signs of a weak chemical absorption of carbon dioxide. Sechenov's next step was to determine the absorptive properties of the erythrocytes. He demonstrated that carbon dioxide was present in the erythrocytes not only in the state of the physical solution and in the state of a bicarbonate, but also in a state of weak chemical bonding with haemoglobin. Sechenov isolated haemoglobin by inducing it to crystallise by repeated freezing and settling blood samples. Now he could compare the absorptiometric characteristics of a suspension of corpuscles, of the liquid remaining after precipitation of haemoglobin, and of haemoglobin itself. Haemoglobin was quite a recent area of study. A decade earlier, in 1862, Felix Hoppe-Seyler, investigating 'Blutfarbstoff', [blood pigment] isolated a crystalline substance and called it haemoglobin. Applying analytical methods from physics and chemistry such as colorimetry, polarisation, and spectrum analysis, Hoppe-Seyler identified haemoglobin by means of its absorption spectrum and decomposed it into bile pigments.[5] Sechenov's results showing that haemoglobin was able to bind CO_2 chemically were so unexpected and new that Hoppe-Seyler was first sceptical about the correctness of the experiments. Earlier, Hoppe-Seyler had demonstrated that carbon dioxide did not change the spectrum of haemoglobin. Sechenov, however, was convinced of the validity of his results since 'the chemical absorption of CO_2 appeared with great clarity and in vast quantities'.[6]

Based on these observations, Sechenov made an important suggestion that erythrocytes, due to their high absorption capacity, could be considered carriers of CO_2 from the tissues to the external environment, later known as carboxyhaemoglobin. However, the elucidation of the various forms of CO_2 binding in the blood proved to be a much more difficult undertaking than unravelling of the problems surrounding the binding of oxygen. As early as

1857, Lothar Meyer demonstrated chemical binding of oxygen in the blood. Independently of Claude Bernard, Meyer found that carbon monoxide was capable of quantitatively expelling oxygen from the blood.[7] In the mid-1860s, Hoppe-Seyler demonstrated that haemoglobin within red corpuscles carried oxygen in a loose chemical bond, which could be blocked by exposing the blood to carbon monoxide gas. The method of spectroscopy enabled Hoppe-Seyler to show a weakly bound chemical compound, which he called oxyhaemoglobin, which explained the chemical binding of oxygen in the blood. Since the main mass of O_2 can bind to haemoglobin, erythrocytes were considered the carriers of O_2 from the external air into the tissues.[8] The problem of carboxyhaemoglobin and its role in the processes of gas exchange as suggested by Sechenov, was thoroughly investigated by Christian Bohr (1855–1911) at Ludwig's laboratory in the mid-1880s. In his doctoral thesis, Bohr made use of Sechenov's absorptiometric technique to find that each gram of haemoglobin would bind approximately 2.4ml CO_2 at a pCO_2 of 30mmHg.[9] It was not until 1914, however, that Oxford physiologist John Scott Haldane (1860–1936) was able to establish conclusively that carbon dioxide was more strongly bound in the non-oxygenated blood than in oxygenated blood, the so-called 'Haldane effect'. Together with Cambridge physiologist, Sir Joseph Barcroft (1872–1947), Haldane developed, improved and simplified analytical techniques, which enabled him to accurately determine the quantity of blood gases.[10]

Another important conclusion Sechenov made while studying physico–chemical conditions, in which gaseous exchange occurs in the blood, was that exchange of CO_2 between blood and alveolar air in the lungs could be interpreted in terms of physical forces – i.e. diffusion – which occurs due to the difference of carbon dioxide tensions in blood, lungs, and tissues. Sechenov's hypothesis was proved correct in 1910 by the Danish physiologist August Krogh (1874–1949). Working in Christian Bohr's laboratory in 1895–9, Krogh developed methods for determining gas tensions in the blood and demonstrated that gas exchange in the lungs can be explained by diffusion alone without the involvement of active secretory processes.[11] Sechenov published results of his studies first in the *Pflügers Archiv* in 1874,[12] and then in the monograph *Die Kohlensäure des Blutes* published in the multilingual *Mémoires de l'Académie Impériale des Sciences de St Pétersbourg* and as a separate edition in Russian in 1879. These studies by Sechenov were included in the major textbooks on physiological chemistry by Willy Kühne and Hoppe-Seyler, and on general physiology by Carl Ludwig and Ludimar Hermann.[13]

One of the most interesting points in Sechenov's treatment of blood gases in his studies of 1873–5 is a simplified analogue model, in which solution of weak salt represents the blood:

> [I]magine for a moment that water instead of blood flows through our veins. Water, with its capability to dissolve carbonic acid and give it up by means of diffusion to the air, could replace blood very well. Imagine further, blood is replaced by a weak solution of sodium carbonate not fully saturated with carbonic acid to the formation of a bicarbonate. In this case, a salt solution could also function successfully. The salt solution could draw CO_2 from the tissues to saturation and give surplus up into the lungs, since a bicarbonate solution exposed to air loses CO_2.[14]

Sechenov was well acquainted with the methodology and standard of the absorptive coefficients for water developed by Bunsen – these had been published by Bunsen in his famous *Gasometrische Methoden* in 1857. Sechenov's model was a modification of a model once suggested by Bunsen, in which water reproduced certain characteristics of the blood. Sechenov used solutions of weak salts as a prototype for blood, since the substances which absorb CO_2 in the blood are similar to certain compounds in solutions of weak salts. Obviously, Sechenov understood that his model was not intended to be a true description of the prototype, reproducing only certain salient features of the blood, a sophisticated solution, until then very poorly understood. Sechenov believed, however, that his simple model enabled him to refine absorptiometric techniques and to establish exact quantitative parameters of the absorption of CO_2 by the salt solution that would elucidate various forms of the binding in the blood.[15]

The experiments with salt solutions grew in number and complexity, and Sechenov began increasingly to realise the complexity of his seemingly simple model:

> For almost five years, I studied the problem of the state of carbon dioxide in the blood. This, apparently, quite simple question required for its solution not only experiments with all principal elements of the blood separately and in various combinations with each other, but moreover, with a long series of salt solutions.[16]

Difficulties arising in explaining and interpreting his results, as well as a lack of adequate facilities and means for his studies pushed Sechenov to seek the possibility of returning to St Petersburg. He wrote to Mendeleev:

> [T]aking into account the type of my present research it is extremely important for me to be near you and in St Petersburg, where the means and conditions for my work are incomparably better than here.... With your help I could probably work successfully now in physiological chemistry as I have already many important problems in hand.[17]

After Cyon's resignation in 1875, the chair at St Petersburg University remained vacant. Mendeleev and Ovsiannikov on Sechenov's behalf interceded before the Council of St Petersburg University. Fortunately for Sechenov, Count Tolstoy, Minister of Education, visited Novorossiisk University in 1875 and expressed support for the wilful professor. This time, Sechenov's appointment to St Petersburg University went smoothly without any delays or bureaucratic impediments.

The period in Novorossiisk University marked an important turn in Sechenov's research career – his transition from studies of respiratory function of the blood to investigation of the nature of salt solutions, one of the most important problems in nineteenth-century physical chemistry.

Notes

1. I. Sechenov, 'Beiträge zur Pneumatologie des Blutes', in I.M. Sechenov, *Selected Works,* M.N. Shaternikov (ed.) (Moscow: Izd-vo Vsesouznogo instituta eksperimental'noi meditsiny, 1935), 3–24: 3.
2. *Ibid.*, 4.
3. On carbon dioxide in the serum, see A. Schäffer, 'Über die Kohlensäure des Blutes und ihre Ausscheidung mittels der Lunge', *Sitzungsberichte der mathematisch-naturwissenschaftlichen Klasse der Kaiserlichen Akademie der Wissenschaften,* 41(1860), 589–94, cited in Sechenov, *op. cit.* (note 1), 5.
4. Sechenov, *op. cit.* (note 1), 11–12.
5. F. Hoppe-Seyler, 'Über das Verhalten des Blutfarbstoffes im Spectrum des Sonnenlichtes', *Virchows Archiv für pathologische Anatomie und Physiologie und für klinische Medizin,* 23 (1862), 446–9.
6. I. Sechenov, 'Die Kohlensäure des Blutes', *Mémoires de l'Académie Impériale des Sciences de St Pétersbourg,* sér. 7, vol. 36, no 13, (1879), 1–62: 26; see also I.M. Sechenov, *Autobiographical Notes.* D.B. Lindsley (ed.), K. Hanes (trans.) (Washington: American Institute of Biological Sciences, 1965), 140.
7. L. Meyer, *Die Gase des Blutes* (Göttingen: Dieterich, 1857), 85.
8. F. Hoppe-Seyler, *Handbuch der physiologisch- und pathologisch-chemischen Analyse für Ärzte und Studierende* (Berlin: Hirschwald, 1865), 203–5; 207–8.
9. C. Bohr, *Experimentalle Untersuchungen über die Sauerstoffaufnahme des Blutfarbstoffes* (Copenhagen: O.C. Olsen & Co Buchdrukerei, 1885). On Bohr's research on elaboration of the techniques of physical measurement of processes underlying physiological functions during his professorship at Copenhagen University, see P. Astrup and J. Severinghaus, *The History of Blood Gases, Acid and Bases* (Copenhagen: Munksgaard, 1986), 160.
10. In 1900, John S. Haldane suggested a very fast and accurate method using ferricyanide for determining the oxygen content in 20ml blood samples with high precision, to 0.2 per cent accuracy. Next year, Haldane, together with Joseph Barcroft, described a modified ferricyanide method for determining

oxygen and carbon dioxide in the same sample, see Astrup and Severinghaus, *op. cit.* (note 9), 146, 161.

11. On Krogh research, see *ibid.*, 109–12; on the oxygen secretion controversy, see B. Schmidt-Nielsen, *August and Marie Krogh: Lives in Science* (New York: American Physiological Society, 1995), 78–94.
12. I. Sechenov, 'Über die Absorptionmetrie in ihrer Anwendung auf die Zustande der Kohlensäure im Blute', *Pflügers Archiv für die gesamte Physiologie*, viii (1874), 1–39.
13. On Sechenov's research on blood gases, see L. Hermann, *Grundriss der Physiologie des Menschen* (Berlin: Hirschwald, 1874), 42–7; W. Kühne, *Lehrbuch der physiologischen Chemie* (Leipzig: Engelmann, 1875), 225–33; F. Hoppe-Seyler, *Physiologische Chemie* (Berlin: Hirschwald, 1877), 377–99.
14. Sechenov, *op. cit.* (note 6), 138.
15. On models in nineteenth-century electrophysiology see, T. Lenoir, 'Models and Instruments', in *Hisorical Studies in the Physical and Biological Sciences*, xvii (1986), 1–54: 3.
16. Sechenov, *op. cit.* (note 6), 138.
17. Sechenov to Mendeleev, letter dated March 1875, Odessa, in K.S. Koshtoyants *et al.* (eds), *Nauchnoe nasledstvo: I.M. Sechenov: neopublikovannye raboty, perepiska i dokumenty* (Moscow: Akademiia Nauk SSSR, 1956), 220–1.

16

Sechenov at St Petersburg: 'Galvanic studies' – A Final Proof

In the autumn of 1876, Sechenov started teaching at St Petersburg University. He once remarked, not without humour, that it took him five years to move from the Vyborg side – where the Medico-Surgical Academy was situated – to the Vasil'evskii Island. Here, in one of the nicest areas of St Petersburg, the University occupied, as it still does, the building of the Twelve Boards (Ministries), a magnificent architectural monument of the first half of the eighteenth century, designed by Domenico Trezzini (1670–1734). The University had four faculties, History and Philology, Philosophy and Law, Physics and Mathematics, and Oriental Languages. Established as an Imperial University in 1819 by decree of Alexander I, the University, by the 1870s, had become one of the leading educational and research centres in the country. Its physico–mathematical faculty continued the admirable traditions of the St Petersburg Academy of Sciences in mathematical science. At the University, Pafnutii L. Chebyshev (1821–94) created a prolific school of mathematics famous for its pure and applied research. The University also housed laboratories in organic and analytical chemistry headed up by renowned chemists, Alexander Butlerov and Dmitrii Mendeleev.

The physiological laboratory, founded by Ovsiannikov and Cyon in 1866, was a subdivision of the zoology department at the physico–mathematical faculty. After Cyon's resignation, the laboratory was neglected for almost two years, and Ovsiannikov, busy in his laboratory at the Academy of Sciences, was happy that Sechenov took over the University's laboratory. Sechenov's first impressions of the laboratory, related in a letter to Mechnikov, were full of frustration. The premises were too small, and only two rooms were equipped and adapted to Cyon's research, which mainly involved vivisection. The attendants, 'all former soldiers stand at attention and are coached to stretch the poor dogs on the table at the first nod'. Fortunately, Sechenov acquired an able assistant to help him in preparing lecture demonstrations. Little-by-little, Sechenov borrowed some appliances from other laboratories and bought some instruments on credit.[1] The laboratory still lacked glassware, reagents, measuring tools, and the most common instruments and apparatus used in physico–chemical studies. The meagre funding of the laboratory did not allow acquisition of the necessary equipment, and Sechenov and Ovsiannikov requested an additional thousand roubles from the faculty council.[2] Usual bureaucratic impediments which involved approval, in all instances, from the University and then from the Ministry of Education, left little hope to equip the laboratory in the near future. Sechenov wrote to Mechnikov:

> I am every day at the laboratory, but for my research I do nothing.... Here even fixing a screw takes a week (no exaggeration). And were it not for my teaching, the situation would make me sick.... [B]ecause of such inactivity, I read much....[3]

Although the requested money for basic equipment was finally granted, funding for practical classes remained insufficient to cover expenses for maintenance of equipment, and for materials, utensils, and instruments. The space was limited and the capacity of the laboratory allowed no more than ten students for a practical class.

Year after year, Sechenov continued to report to the University Council on the inadequate conditions for practical classes. His requests to the Council of the University from September 1886 pointed to the same problems as ten years before, when he had started his tenure:

> According to the new University regulations, practical training in all scientific disciplines including physiology is mandatory. The number of students enrolled for practical courses in physiology exceeds the capacity of the laboratory. Therefore, I request additional space for the laboratory by least one more room and an increase in the annual budget of the laboratory by at least 500 roubles to cover expenses for materials, preparations, and apparatus necessary for practical training.[4]

Sechenov's continuous efforts to improve the financial situation of the laboratory proved successful only in 1888, the last year of his stay at the University. The result was a structurally independent physiological laboratory with a special budget.[5] During his tenure at the University, Sechenov introduced some important changes in teaching physiology. According to an established medical tradition, physiology at the University had been taught in the first and second years. Sechenov suggested to the Faculty Council that physiology should be taught in the third and fourth years, after students took theoretical courses in physics and chemistry. His new programme in physiology, introduced in 1887, included a practical course of training in a variety of physical and chemical methods and techniques used in physiological research. The programme was aimed at establishing physiology as a distinct speciality in the University's natural sciences department. Sechenov believed that 'physiology is a discipline that completes the training of a future natural scientist'.[6]

Sechenov trained a number of capable physiologists during his stay at the University, the most distinguished being Nikolai E. Vvedenskii, Sechenov's successor at the laboratory. None of Sechenov's students, however, was involved in the absorptiometric studies of their mentor. One of the students recalled:

> The only person in the laboratory who assisted Sechenov in his studies on salt solutions was attendant Osip. Every morning Ivan Mikhailovich came to the laboratory and in his low voice called 'Osip!' Osip at once left everything and they started working in Sechenov's room.[7]

Sechenov's students usually worked on the problems related to the physiology of the nervous system, in particular on nerve–muscle excitation. On rare occasions, Sechenov supervised biochemical projects, such as research on colorimetry of the blood and colloid state of the protein substances performed by V.P. Mikhailov.[8] Sechenov supplied his students with problems, which could be solved in a limited time by established and predictable methods. He felt that his studies on salt solutions were far beyond the scope of physiology. This area of investigation, although promising for studying particular physiological problems, was too complex for a beginner or even an advanced student. On the other hand, both neurophysiology, which involved classical electrophysiological techniques, and physiological chemistry, which used widely accepted methods, were secure fields in which to master not only research skills but also experimental thinking, so essential for a future independent investigator.

Only once in twelve years at St Petersburg University did Sechenov interrupt his absorptiometric studies. Earlier, in the the 1870s, in Odessa, while working on blood gases, Sechenov occasionally returned to the problem of central nervous inhibition. In 1872, he proved experimentally that the inhibitory process could occur outside of the central nervous system, in the peripheral nerves, but only under a particular type of excitation, 990 impulses per second.[9] In 1875, Sechenov published two papers, in which he refined his arguments based on his experiments of 1863, 1868, and 1872, the arguments which, as he understood, were still lacking decisive evidence. A paper of 1875 was a critical review of Cyon's theory of inhibition, the so-called theory of interference. The same year he published an answer to Ernst Brücke who, as Sechenov believed, was wrong in interpreting the mechanism of central nervous inhibition.[10]

In 1879, Sechenov returned to neurophysiological studies and started a project on electrical phenomena in the spine and medulla oblongata of a frog. The 'galvanic researches', as he called them, lasted two and a half years. Apparently Sechenov became tired from 'long pumping', as he admitted in a letter to Mechnikov, and 'left entirely, for the present, the experiments with CO_2. Now I am working on the electrical properties of the central nervous masses. Something great seems to have turned up.'[11] Sechenov also changed the research projects of his students. Both Vvedenskii and Mikhailov started investigations on the innervation of respiration in a frog. Sechenov believed that there was a similarity between the structure and function of the respiratory and locomotion centres located in the upper part of the medulla oblongata. The equipment of Sechenov's laboratory was, however, modest for such complex experiments as his 'galvanic studies'. He had at his disposal only a Wiedemann galvanometer, which, although widely used in many laboratories, was not sensitive enough to trace rapidly changing fluctuations in the current as compared to the newer capillary electrometer, invented by Gabriel Lippmann (1845–1921). Masterful in electrophysiological technique, Sechenov managed to impart maximum sensitivity to the Wiedemann galvanometer and adapt it to his experiments.[12] In this connection, Sechenov noted to Mechnikov that luckily his 'galvanic studies require little expense, which is why poverty is not a burden to me.'[13]

Sechenov's experimental protocol was quite original. A preparation of the medulla oblongata and an isolated sciatic nerve connected to a piece of spinal cord with small muscles was placed in a humid chamber. Thus, it was an electrophysiological investigation of the central nervous system *in vitro*, since peripheral nervous influences and humoral factors, as well as impulses from the brain could be excluded. At the same time, the isolated 'central nervous system' could be excited by electrical stimulation of the nerves, while the muscles could serve as indicators of the spreading of the excitation from the brain into the spinal cord and then to the periphery. Currents were led through non-polarisable electrodes into a galvanometer. All three forms of electromotor phenomena of the nerves, known as resting currents, electrotonus, and negative fluctuations, could be ascertained in the isolated spinal cord from a frog. Vvedenskii reported that the experiments demanded great care and virtuoso skill.[14] The galvanometer connected to a cross section and a longitudinal section of the medulla oblongata of a frog detected a deflection of the needle, as in the case of the isolated muscle or nerve. As these negative oscillations occurred without apparent reason, Sechenov called them 'spontaneous oscillations or fluctuations', or the 'spontanne Entladungen'. Sechenov attributed these spontaneous fluctuations to the motor impulses spontaneously arising in the medulla oblongata. He believed that the development of spontaneous fluctuations of current in the medulla was in some way connected to the fluctuations of current resulting from the so-called forced movements in the frog, registered in its brain, cut along the medulla. He showed that with intensification of excitation, 'spontaneous fluctuations weaken and reduce, and turn into rest which lasts for some minutes', and vice versa, after rest, spontaneous fluctuations appear again more intensely. It was an important finding, which established a connection between excitation of the nervous centre and the nerve. The changes of activity and rest coincided with the state of excitation and inhibition respectively. Sechenov referred to this phenomenon as to a distinct periodic rhythmic electrical activity in an isolated medulla oblongata and the spinal cord.[15]

Sechenov demonstrated that inhibition of the fluctuations of respiratory impulses occurred during a strong tetanisation of sensory nerves (sciatic nerves), while intensification of the fluctuations of respiratory impulses occurred after the cessation of tetanisation. Thus, the inhibition and intensification of the fluctuations of current could not be attributed to exhaustion or fatigue of the nerve centres, rather, the nerve centres during tetanisations were charged with energy. Finally, the depression of stimulating shocks in the medulla oblongata by tetanisation of the nerves, demonstrated by Sechenov was similar to the inhibition of heart activity by stimulation of the vagi nerves, first demonstrated by Ernst and Eduard Weber in 1845. In 1863, Weber's experiments suggested to Sechenov an idea of the existence of special centres in the mid-brain, which could depress the movements that led to Sechenov's discovery of the central nervous inhibition.

In the 1870s, Rudolf Heidenhain applied Sechenov's methodology in his experiments on the vagus nerve and the heart. Heidenhein demonstrated diastolic stopping of the heart by stimulation of the vagi, and intensification of cardiac activity after cessation of such stimulation. For Sechenov, both depression of stimulating shocks in the medulla oblongata by tetanisation of the sciatic nerves, which he demonstrated in 1882, and inhibition of heart activity by stimulation of the vagi were indisputable cases of the so-called inhibition of the activity of the organ, but not exhaustion or fatigue of the nerve centres.

Sechenov published his *Galvanische Erscheinungen* [*Galvanic Studies*], in *Pflügers Archiv* in 1882. He reviewed the area of research to which he still remained a major contributor. His galvanic studies confirmed, as he believed, the results of his major works on central nervous inhibition, from 1863 and 1868, and convincingly refuted the 'theory of fatigue' by Moritz Schiff, and the 'theory of interference' by Cyon, major opponents to his theory of central nervous inhibition.[16] In his conclusion, Sechenov wrote:

> The major results of my first work on inhibition have sustained a new trial; the effects of the stimulation of the middle section of the brain correspond

to inhibitory actions on its reflexes, and the same applies to the peripheral inhibition.[17]

Apparently, Sechenov ascribed a particular significance to his galvanic studies as a final proof of his theory of central nervous inhibition, which had won him recognition, to which he had devoted ten years of his research career, and which he completely abandoned for his investigation on salt solutions. In a letter to Mechnikov, he wrote: 'Only now, in twenty years, I have finally managed to prove that the so-called inhibition of reflexes is a true... result of the depression of excitation in the nervous centres.'[18] He also had an emotional connection to this work. His discovery of specific centres in the brain embodied for him his youthful passion for science, his bold attempts to explain the psyche in the language of physiology, and those days full of love, hopes, and dreams which had irrevocably passed.

Throughout life, Sechenov maintained that central nervous inhibition is a process *sui generis*, that is, the nature of inhibition is different from the nature of excitation. Therefore, in the central nervous system, the centres that originate impulses responsible for muscular movements, and the centres inhibiting these impulses, are distinctively different. Cyon, Schiff, and later Vvedenskii, held that central nervous inhibition was not a specific process different from excitation. During excitation and inhibition there is only a relative difference in the state of the same centres conditioned by the extent and the way of excitation. Sechenov accepted neither of these concepts and was sceptical about Vvedenskii's early interpretation of the process of inhibition. Later, Vvedenskii formulated his own theory of inhibition, conceptually different from the theory of his mentor. Vvedenskii held that process of inhibition is a process of excitation, which does not spread but remains stationary. That was the so-called catabolic theory of inhibition.[19]

In 1906, in the obituary for his mentor, Vvedenskii noted that, in his later years Sechenov had changed his views on the specificity of central nervous inhibition and preferred to leave that question open for new investigations. There is, however, some discrepancy in Vvedenskii's remarks. Sechenov never returned to neurophysiological studies after he had left St Petersburg in 1889. In his *Autobiographical Notes,* written in 1904, Sechenov expressed his conviction of the theory he had defended throughout his life. What seems interesting in Vvedenskii's picture of his mentor is a parallel to the contemporary history of physiology. Like Sechenov's theory of central inhibition, another landmark theory of the time, that of the electromotive nature of nerve–muscle excitation by du Bois-Reymond, with time underwent changes and began to be viewed from a different prospective. Du Bois-Reymond also defended his theory throughout his life. Vvedenskii worked in the laboratories of both, du Bois-Reymond and Sechenov, and believed that both physiologists, despite all challenges to their theories, were pathfinders in the field.

Notes

1. Sechenov to Mechnikov, letter dated 19 September 1876, St Petersburg, in S.I. Shtraikh (ed.), *Bor'ba za nauku v tsarskoi Rossii: Neizdannye pis'ma I.M. Sechenova, I.I. Mechnikova, L.S. Tsenkovskogo, V.O. Kovalevskogo, i dr.* (Moscow–Leningrad: Gossotsgiz, 1931), 87.
2. F.V. Ovsiannikov and I.M. Sechenov, 'Note to the University Council, Sept. 16, 1876', cited in K.S. Koshtoyants *et al.* (eds), *Nauchnoe nasledstvo: I.M. Sechenov.: neopublikovannye raboty, perepiska i dokumenty* (Moscow: Akademiia Nauk SSSR, 1956), 119.
3. Sechenov to Mechnikov, letter dated 2 November 1876, St Petersburg, in Shtraikh, *op. cit.* (note 1), 88.
4. Sechenov, 'Note to the University Council', in Koshtoyants *et al.*, *op. cit.* (note 2), 122–4; 127–8.
5. Proceedings of the University Council, cited in *ibid.*, 133.
6. A.A. Ukhtomskii, 'I.M. Sechenov v Peterburgskom-Leningradskom universitete', *Fiziologicheskii zhurnal SSSR,* 40, 5 (1954), 527–39: 532, 538.
7. *Ibid.*, 534.
8. Occasionally, as in the case with his student Mikhailov, Sechenov supervised biochemical research, see 'Protokoly zasedanii Soveta St Peterburgskogo universiteta za 1878–80 i 1880–81', in Koshtoyants *et al.*, *op. cit.* (note 2), 122–3.
9. I.M. Sechenov, 'Einige Bemerkungen über das Verhalten der Nerven gegen sehr schnell folgende Reize', *Pflügers Archiv für die gesamte Physiologie,* v (1872), 114–19.
10. I.M. Sechenov, 'Zur Frage über Reflexhemmungen', *Bulletin de la Classe Physico-mathématique de l'Académie Impériale des Sciences de St-Pétersbourg,* ser. 3, xx (1875), 537–42; *idem*, 'Notiz, die reflexhemmenden Mechanismen betreffend', *Pflügers Archiv für die gesamte Physiologie,* x (1875), 163–9.
11. Sechenov to Mechnikov, letter dated 2 February 1880, St Petersburg, in Shtraikh, *op. cit.* (note 1), 103.
12. N.E. Vvedenskii, 'Sechenov: nekrolog', in K.M. Bykov (ed.), *Sechenov, Pavlov, Vvedenskii: fiziologiia nervnoi systemy: izbrannye trudy* (Moscow: Medgiz, 1952), 5 vols, Vol. I, 59–81: 67. Gustav Heinrich Wiedemann (1826–99), a student of Gustav Magnus, was Professor of Physics at Leipzig University.
13. Sechenov to Mechnikov, *op. cit.* (note 11).
14. Vvedenskii, *op. cit.* (note 12), 67; see also A.F. Samoilov, 'Sechenov i ego mysli o roli myshtsy v nashem poznanii prirody', in A.F. Samoilov, *Izbrannye stat'i i rechi* (Moscow-Leningrad: Izd-vo Akademii Nauk, 1946), 43–69: 51.

15. I.M. Sechenov, 'Galvanische Erscheinungen an dem verlängerten Marke des Frosches', *Pflügers Archiv für die gesamte Physiologie,* xxvii (1882), 524–66. Cited from I.M. Sechenov, *Selected Works,* M.N. Shaternikov (ed.) (Moscow: Izd-vo Vsesouznogo instituta eksperimental'noi meditsiny, 1935), 212–43: 235. In the 1890s, the question of spontaneous electrical activity in the brain came sharply into focus. Simultaneously, several researchers succeeded in recording the 'spontaneous oscillations' of brain potentials. Among those engaged in the polemics on the priority in the discovery of the electrical activity of the brain were, Ernst Fleishl von Marxow (1846–92) from Vienna, Victor Horsley (1857–1916) and Francis Gotch (1853–1913) from England, and Adolph Beck (1863–1942) from Warsaw. All these studies, however, were anticipated by the work of Richard Caton in 1875. Interestingly, only Adolf Beck referred, in his work, to Sechenov's articles in *Pflügers Archiv* as the first study on the spontaneous electrical activity of the brain. Beck, Professor of Physiology at Warsaw University was a collaborator and friend of I.R. Tarkhanov, Sechenov's and Cyon's student, and professor of physiology at the Medico-Surgical Academy. On the polemics over the priority, see M.A. Brazier, *A History of Neurophysiology in the Nineteenth Century* (New York: Raven Press), 193–6.
16. Sechenov, *Selected Works, op. cit.* (note 15), 241.
17. *Ibid.,* 242.
18. Sechenov to Mechnikov, letter dated October 1883, St Petersburg, in Shtraikh, *op. cit.* (note 1), 86.
19. On Vvedenskii's theory of inhibition, see I. Arshavskii, 'I.M. Sechenov i N.E. Vvedenskii', in P.G. Kostiuk, S.P. Mikulinskii, and M.G. Iaroshevskii (eds), *Ivan Mikhailovich Sechenov: k 150-letiiu so dnia rozhdeniia* (Moscow: Nauka, 1980), 497–511.

17

The Context to Sechenov's Study of Solution: The Mendeleev–Ostwald Debate on the Theory of Solutions

For twelve years, from 1876 until 1888, Sechenov worked on the absorptiometric properties of salt solutions at his university laboratory in St Petersburg. His 1873–5 studies on the absorption of carbon dioxide by the blood suggested to him a model where a salt solution could represent some notable features of the blood. Sechenov, entering the emerging field of chemistry of solutions, was 'captivated by the thought to find a key to a vast class of phenomena not yet known to anyone.'[1] His research was aimed at establishing a general relationship of the quantity of gases absorbed by salt solutions on the concentration of salts in these solutions. For Sechenov, these studies were a double challenge. As a physical physiologist, Sechenov was particularly interested in physics and chemistry of the blood, the most sophisticated solution in nature. Studying important aspects of the blood as a solution, Sechenov developed a methodology that he believed would work for salt solutions. Conversely, a general dependence or universal law deduced from the absorption of gases by salt solutions could be applied to particular problems in absorptiometric studies of the blood. The major challenge, however, was that relatively few chemists at that time were dealing with the nature of solutions as a problem in its own right. Sechenov was aware that Mendeleev had been developing his theory of solutions when they both studied in Heidelberg in the late 1850s. Later, in 1870, a year-long study in Mendeleev's laboratory proved important in Sechenov's understanding of the basic principles in the chemistry of solutions. Having started independent investigations on solutions, Sechenov used to inform Mendeleev about his results seeking expert advice and approval.

Sechenov's studies here have an interesting context. Mendeleev's influential statement on his hydrate theory of solution, published in the mid-1880s, initiated a debate between Mendeleev on the one side and Arrhenius and Ostwald, proponents of the ionic theory of solution, on the other. Although Sechenov's studies were merely practical and did not concern any theoretical postulates, interpretation of his results and his conclusions still depended on certain theoretical principles, which were hotly debated by the two opposing camps. To put these matters into perspective, it is useful to outline the debate over the two rival theories of

Image 17.1

Dmitrii I. Mendeleev.
Courtesy of the Wellcome Library Collection.

solutions, an important episode in nineteenth-century science relevant to the story.[2]

At the dawn of his career, Mendeleev began to develop his hydrate theory to explain the properties of solutions. For Mendeleev, the investigation of solutions, particularly aqueous solutions, was:

> [E]specially interesting as solutions are present in the earth and water, in plants and animals, in chemical enterprises and industry; solutions also play an important role in chemical transformations.[3]

Mendeleev's first statement on the theory of solutions appeared at a time when the nature of solutions began to be viewed from a different perspective, as new aspects came to the fore as physical and chemical forces became more strongly contrasted. In his doctoral thesis *O soedinenii spirta s vodoiu* [*On the Combination of Alcohol and Water*] (1865), Mendeleev pointed out that solutions are chemical combinations, not mechanical mixtures. Since solutions are ordinary examples of chemical reactions, interaction between solvent and soluble should not be distinguished from other forms of chemical combinations. Dealing with the problem of definite and indefinite combinations in solution, Mendeleev sought to clarify the question whether solutions, particularly water solutions, were to be understood as containing definite chemical species produced by the dissolved substance and water.[4] For Mendeleev, the most plausible explanation for the chemical nature of the combination between solvent and soluble was in terms of the formation of definite combinations, such as hydrates. He pointed out that dissolution in many cases was accompanied by liberation of heat or a sharp change in the properties of the solutions, the phenomena associated particularly with chemical compounds. There also existed hard crystalline solutions and combinations with water of crystallisation.[5] Mendeleev attempted to link the ideas of chemistry of definite proportions, which became prominent in the wake of Dalton's atomic theory, to the ideas of Claude Berthollet, who regarded solutions as compounds in indefinite proportions.[6] Such a link, Mendeleev believed, would enable chemists to draw a distinction between compounds and solutions:

> In my mind solutions are not alien to the atomistic ideas. As with any common definite combinations, solutions are included in now – prevailing studies on the mass action law…[7] on dissociation and on gases. At the same time, solution, I believe, is the most common case of the chemical action (force) which is determined by a relatively weak affinity, and therefore is a fruitful field for further development.[8]

By the 1870s, refinement of techniques and the thermo–chemical studies of Julius Thomsen in Copenhagen and Marcellin Berthelot in Paris, provided an important source of data on the state of substances in solution and on the problem of simultaneous action of several solvents.[9] Marcellin Berthelot (1827–1907) presented an influential and full statement of the hydrate theory in his 1879 paper. He attributed the heat liberated when a substance is dissolved to the chemical combination of the dissolved substance with the water. He further stated that the dissolution of salts in water occurs when definite compounds between the salt and the water are formed, analogous or identical to the hydrates of constant composition known in the crystalline state.[10] Independently of the French chemist, Mendeleev had elaborated basic principles of his hydrate theory by the early 1880s. In particular, he developed a clear theoretical basis of the formation of definite combinations, ie. associations of hydrated molecules in a state of uneven dissociation and dynamic equilibrium, governed by the mass action law.[11]

Outside Russia, Mendeleev's theory of solution attracted attention after publication of his paper 'Über die nach der Veränderungen des spezifischen

Gewichtes beurtheilte chemischen Assoziation Schwefelsäure mit Wasser' ['On the Chemical Association of Sulphuric Acid and Water, Based on the Change of their Specific Weights'] in the journal of the Berlin Chemical Society in 1886.[12] Mendeleev demonstrated the existence of hydrates by using data on the general properties of solutions. Mendeleev's method consisted of plotting the differential of the relative density of the solution (*s*) against the percentage composition (*p*). It resulted in a graph with straight lines, each of which was taken to represent a specific hydrate. When the lines were broken due to increasing dilution, Mendeleev took these points of discontinuity ['Knickes' or 'breaks'] to mean that a complex hydrate had disappeared, leaving a single hydrate containing a lower proportion of water. Mendeleev claimed that the method could be extended to prove the existence of hydrates using other properties of solution, among them electrolytic conductivity. Mendeleev's extensive experimental data supported his hypotheses on the dependence of specific weights on composition. All of these were systematised in a well-built theory and presented in his monograph *Issledovaniia vodnykh rastvorov po udel'nomu vesu* [*Investigation of Water Solutions Based on their Specific Weights*] published in 1887 in St Petersburg.[13] Later, Mendeleev wrote about this work:

> Partly from this, there appeared a kind of vogue for solutions. My ideas on solutions from my youth till now are the same: there is no distinction between the solution and chemical phenomena. I am glad that I have had time to state my ideas.[14]

For Mendeleev, in chemistry, his chief passion after a system of elements, was the nature of solutions, and his monograph on the theory of solutions took pride of place amongst his works. 'I am glad to dedicate this work to my mother to whom I owe everything.' She 'spent her last resources and strength to devote me to science', wrote Mendeleev later. An exceptional woman, Maria Mendeleeva, at the age of fifty-seven, together with Dmitrii, her fourteenth and youngest child, made the long and arduous journey from the Siberian city of Tobolsk to Moscow. Regrettably, Dmitrii Mendeleev, as a Siberian, could not enter Moscow University and so his mother placed him at the St Petersburg Main Pedagogical Institute. Mendeleev studied with Voskresenskii, a student of Liebig and one of the most popular professors of chemistry in Russia. From his youth and throughout his life, Mendeleev adhered to Charles Gerhardt's theory of types, and rejected Berzelius's electrolytic theory of formation of chemical compounds. In his later works, Mendeleev was opposed to linking chemistry with electricity and preferred associating chemistry with physics as a science of mass. Mendeleev's predilection found its most brilliant vindication in the correlation he achieved between the chemical properties and atomic weights of the elements. His periodic system of elements, as well as the voluminous and fascinating textbook *Osnovy khimii* [*Principles of Chemistry*], both first published in 1869, made Mendeleev the most powerful and influential chemist in Russia.[15]

Among the chemists who had intermittently turned their attention to the general problem of solutions were Henry Armstrong (1848–1937) in London and François Raoult (1830–1901) in Grenoble. Although they attacked the problem of solutions from very different angles, they also gave preference to chemical reactions between the solute and the solvent. Since the solvent was usually water, they thought that solutions, including electrolytic solutions involved the formation of hydrates, a crystal formed through the combination of a compound with water. The hydrationists, as they became known, attributed special importance to the role of molecular aggregates, which means that salts in solution exist in the form of complex molecules or molecular complexes.[16] In 1887, with the appearance of the work of Jacobus Henricus van't Hoff (1852–1911) on dilute solutions[17] and Arrhenius's theory of electrolytic dissociation,[18] an alternative view of solution became possible. The solution began to be seen in terms of a physical, rather than a chemical theory. Wilhelm Ostwald (1853–1932), by that time Director of the Physico–Chemical Institute at Leipzig University, originated a new and prolific school of research, which was mainly based on Arrhenius's theory of electrolytic dissociation, van't Hoff's osmotic theory of solutions, and applications to chemistry of the laws of thermodynamics. The Ostwald institute became a stronghold of the Ionists, so named because they believed that chemical reactions in solutions involve only ions and not undissociated molecules. The Ionists were aware that the idea of dissociation required chemists to change the prevailing concepts of the chemical nature of salts and related substances and to step forward from molecular chemistry to the chemistry of ions. For the majority of chemists at the time, however, the fundamental notion of indivisibility of the atom was incompatible with the notion of ions as specific state of matter. In his autobiography, *Lebenslinien*, Ostwald vividly described the reaction of Per Theodor Cleve (1840–1905), Professor of Chemistry at Uppsala University, to Arrhenius's theory of electrolytic dissociation. In a conversation with Cleve, Ostwald acknowledged that the theory had not yet been clearly formulated:

> With iron logic, Cleve refuted one after another all conclusions of Arrhenius's major hypothesis, and at last asked me: 'So, you believe that here in a glass with a solution of sodium chloride [table salt in water] atoms of sodium are floating separately?' And when I answered positively, Cleve glanced at me as if doubting my chemical common sense.[19]

Indeed, it seemed incredible to a chemist that in an aqueous solution of sodium chloride, atoms of sodium, a soft silver-white alkali metal violently reacting with water, and atoms of chlorine, a yellow-greenish suffocating gas, were in a free state.

The dissociation theory was the main target of the attacks on Ostwald's school by Mendeleev and the British hydrationists, who had embraced Mendeleev's method, as stated in his work of 1886, on aqueous solutions and their specific weights. The Ionists were thus confronted, not only with the hydrate theory as an alternative, but also with a method to show that it was correct.[20] Even though Arrhenius and Ostwald, by 1889, had a rich variety of experimental data to support the idea of dissociation, there were still important issues that could not be explained in terms of dissociation, particularly the source of energy necessary for dissociation. Mendeleev started the debate by publishing an article in the *Zhurnal Russkogo fiziko-khimicheskogo obshchestva* [*Journal of the Russian Physico–Chemical Society*], where he rejected 'a specific type of dissociation, into ions, in the electrolytes during formation of weak solutions'. In Mendeleev's view, in the process of dissolution *association* prevailed, that is, the formation of the new complex but with weak and easily dissociated combinations.[21] Arrhenius immediately published his answer to Mendeleev in the *Philosophical Transactions.* He started his article with a critical remark regarding a contradictory manner in which 'the distinguished Russian chemist' presented his theory of electrolytic dissociation in the Russian journal. Arrhenius pointed out that Mendeleev ignored a great part of what had been accomplished by the theory of dissociation. He also argued that theories of dissociation, and of osmotic pressure in application to solutions, made it possible for the first time to calculate the numerical values of thousands of observations with no contradiction between theory and experiment. For Arrhenius, the hydrate theory was useless since not a single numerical datum had been deduced from it. Finally, Arrhenius argued, dissociation theory did not accept Mendeleev's statement that hydrates existed in dilute solutions with a large quantity of water.[22]

Arrhenius's arguments, however, did not convince the Russian chemist. Mendeleev was firmly opposed to the very idea of dissociation and to the concept of the ion as an electrically charged molecular fragment. In his view, the history of the development of chemistry had confirmed the unitary theory, which disclaimed 'pre-existence of opposite constituents'. Elsewhere, Mendeleev stated that Arrhenius's theory 'violates common and conventional principles of chemistry.'[23] Finally, and most importantly, Arrhenius's concept of dissociation undermined a theory of solutions that Mendeleev had been developing for many years. Mendeleev's critical attitude to the dissociation theory was even more enhanced by Ostwald's contradictory views on the atomic theory. Ostwald strongly opposed the atomic theory of matter until 1909. The irony is that the ionic theory would find its ultimate justification in the atomic theory of matter.[24]

In its turn, the hydrate theory was strongly opposed by the German physico–chemical school. Walter Nernst (1864–1941), one of Ostwald's most enthusiastic disciples, argued that hydrate theory could not be even called a 'theory of solution' as it had no theoretical foundation whatsoever and did not result in the determination of any regularity. In his *Theoretische Chemie*, Nernst gave a short description of the hydrate theory. He disputed the correctness of the experimental as well theoretical foundation of Mendeleev's work of 1887, and doubted the existence of hydrates in the solutions.[25] For his part, Ostwald hardly ever mentioned hydrate theory in his publications. Reviewing the works of the hydrationists in his *Zeitschrift für physikalische Chemie,* he was sceptical of both their results and theoretical conclusions. Discussing the report of P.S. Pickering at the meeting of the British Chemical Association in 1890 in Leeds, Ostwald remarked:

> Even if the view that hydrates could exist in solutions is widely accepted, nevertheless, the method by which they have been found is rather peculiar, and not satisfactory.[26]

Later, in his autobiography, explaining '*die Lehre von den Knicken*' [Mendeleev's theory of the breaks], Ostwald wrote that a debate on the problem of the nature of solutions:

> [W]as stimulated by the interference of the well-known Russian chemist Mendeleev.... Like many chemists of the time, Mendeleev believed that between the solvent and dissolved substance there occurred a chemical binding, and to prove that he devised a method, which was as original as it was wrong.[27]

Nevertheless, in his letter to Mendeleev in 1888, Ostwald politely admitted the general scientific importance of Mendeleev's *Investigation of Water Solutions Based on their Specific Weights,* even though its results were doubtful and not significant for him: 'Your book contains an immense effort. I hope that elaboration of this problem will be extremely fruitful for science.'[28]

Mendeleev's influential standing created an intellectual climate for the hydration theory to prevail among Russian scientists. Arrhenius's letter of 1890 to van't Hoff has a fascinating passage on the possible reception of the ionic theory in Russia:

> The Russians are in general very sympathetic towards the new direction, it corresponds so much to the Russian imagination. Only the great Mendeleev

> strongly holds back and the majority dare not reply to the great master and patriot. They are in a state of hypnotism in which they have to obey the hypnotist without resistance. And the analogy between gases and solutions… lies deep in their sensibility.[29]

In his *Solutions,* published in Britain in 1891, Ostwald gives a good picture of the way solution theory looked from inside. In the 'Preface' Ostwald pointed out that the theory of solution, founded by van't Hoff, had made remarkable advances in recent years, both theoretical and practical.[30] Ostwald's detailed account on solution studies omits one important detail, without which the picture is incomplete. Ostwald does not refer to contributions by Mendeleev or by his British supporters P.S. Pickering and H.E. Armstrong. He says nothing whatsoever about the hydrate theory, as if it did not exist, thus eliminating any attempts to assume that there is a rival theory.[31] For Ostwald, as well as for Mendeleev, the nature of solutions was not a mere chemical problem, although important. For the rising star of Ostwald, his physical theory of solution was the very basis of the new discipline of physical chemistry, which he originated. For Mendeleev, his chemical theory of solution was a vindication of his principles of chemistry based on fundamental concepts of indestructibility of the atom, and the universal Newtonian law of mass, to which Mendeleev remained true throughout his life. In view of the authority of the two champions of theory of solution, their recognition of the significance of Sechenov's investigations is quite remarkable. In his *Solutions* and *Lehrbuch der Allgemeinen Chemie,* Ostwald related in detail Sechenov's results and method.[32] Mendeleev also included an account on Sechenov's contribution in his *Principles of Chemistry.* What attracted both sides was solidity and accuracy of the methodology and results and conclusions, which did not encroach on the theoretical postulates of either of the theories. Sechenov appears to agree with both, regarding one type of salts with Mendeleev, and the other, with the 'ionists.' For Sechenov, however, his physico–chemical path turned out to be a challenging undertaking and less rewarding than he had expected. He devoted more than half of his most productive research career to salt solutions. He remained an acclaimed physiologist, but he did not become a successful chemist who suggested a universal law to 'a whole class of phenomena', as he so much hoped.

Notes

1. I.M. Sechenov, *Autobiographical Notes,* D.B. Lindsley (ed.), K. Hanes (trans.) (Washington: American Institute of Biological Sciences, 1965), 156.
2. On the history of theories of solution, see I.I. Solov'ev's, *Istoriia ucheniia o rastvorakh* (Moscow: Izd-vo Akademii Nauk, 1959). Solov'ev has provided a

comprehensive discussion on the development of Mendeleev's hydrate theory and the controversy over hydration between Mendeleev and Arrhenius, as well as an historical survey of the debate on theories of solution in Russia and continental Europe. On the development of the theory of solution in Britain, see R.G. Dolby, 'Debates over the Theory of Solution: A Study of Dissent in Physical Chemistry in the English-Speaking World in the Late Nineteenth and Early Twentieth Centuries', *Historical Studies in the Physical Sciences,* vii (1976), 297–404: 300. On the theories of solution, see also P. Walden, 'Die Lösungstheorien in ihrer geschichtlichen Aufeinanderfolge', *Sammlung chemischer und chemisch-technischer Vorträge,* xv (1910), 277–454. Paul Walden (1863–1957) taught chemistry in Riga, Rostock, and Tübingen, and was a close friend and collaborator of Ostwald. For their correspondence, see R. Zott (ed.), *Wilhelm Oswald und Paul Walden in ihren Briefen,* with 'Introduction: Paul Walden – Wissenschaftler zwischen den Kulturen?' by R. Zott, 12–63 (Berlin: ERS-Verlag, 1994).

3. D.I. Mendeleev, *Osnovy khimii* (Moscow-Leningrad, 1927) in 2 vols, Vol. I, 26–7, in Solov'ev, *op. cit.* (note 2), 53– 4.
4. D.I. Mendeleev, *Rastvory* (Moscow: Izd-vo Akademii Nauk, 1957), 381–2; on Mendeleev's doctoral dissertation and early development of his theory of solutions, see N.A. Figurovskii, *D.I. Mendeleev* (Moscow: Nauka, 1961), 170–87, see also Solov'ev, *op. cit.* (note 2), 57–9.
5. Solov'ev, *op. cit.* (note 2), 59–60.
6. Claude Louis Berthollet (1748–1822), a French chemist, accompanied Napoleon's expedition to Egypt in 1798, where he made some mineralogical analysis that enabled him to conclude that mass action (concentration) could overcome the usual play of elective affinities between substances. Later, Berthollet proposed that compounds combined together in variable and indefinite proportions and he pointed to solutions and alloys and, what would today be defined as mixtures, as empirical evidence for his claim. He did not, however, make a sharp distinction between compounds and solutions and regarded solutions as compounds in indefinite proportions. See W.H. Brock, *The Norton History of Chemistry* (New York: W.W. Norton, 1993), 144–5; and J.W. Servos, *Physical Chemistry from Ostwald to Pauling: The Making of a Science in America* (Princeton: Princeton University Press, 1990), 13–16; see also M.P. Crosland, *The Society of Arcueil: A View of French Science at the Time of the Napoleon I* (Cambridge: Harvard University Press, 1967), 57–60.
7. The mass action law made it possible to express the working of chemical forces by means of mathematical formulae. It was drawn up in 1867, by the Norwegian brothers-in-law Cato Guldberg (1833–1902) and Peter Waage (1833–1900), both professors – the former a mathematician, the latter a chemist – at the University of Christiania in Oslo. The law had been

virtually unknown until Wilhelm Ostwald (1853–1932), in his doctoral thesis of 1877 on the problem of chemical affinity, referred to it and through that reference van't Hoff (1852–1911) was able to reinterpret Waage's work thermodynamically. See Brock, *op. cit.* (note 6), 379.

8. Mendeleev, *op. cit.* (note 4), 383, cited in Solov'ev, *op. cit.* (note 2), 62.
9. The extensive calorimetric researches pursued by Marcellin Berthelot (1827–1907) in France and Julius Thomsen (1826–1908) in Denmark were stimulated by the earlier investigations, which established that the amount of heat liberated during neutralisation of acids by bases was always the same, no matter how many reaction pathways were used. This 'law' was to be found in the later dissociation theory. See Brock, *op. cit.* (note 6), 360–1.
10. M. Berthelot, *Essai de méchanique chimique fondée sur la thermochimie* (Paris: G. Baillière, 1879), 2 vols, Vol. II, 162; see also Dolby, *op. cit.* (note 2), 302.
11. Solov'ev, *op. cit.* (note 2), 65.
12. D.I. Mendeleev, 'Über die nach der Veränderungen des spezifischen Gewichtes beurtheilte chemischen Assoziation Schwefelsäure mit Wasser', *Berichte der deutschen chemischen Gesellschaft,* xix (1886), 379–89.
13. D.I. Mendeleev, *Issledovaniia vodnykh rastvorov po udel'nomu vesu* (St Petersburg: Izd -vo A.S. Suvorina, 1887).
14. D.I. Mendeleev, *Literaturnoe nasledstvo* (Leningrad: Nauka, 1938), 2 vols, Vol. I, 80.
15. Charles Gerhardt (1816–56) rejected Berzelius's dualistic view of molecules for a unitary one that eventually allowed chemical properties to be ascribed to the arrangements of atoms. Through the influence of his *Traité de Chimie Organique* (1848), came the idea of types, the revision of atomic weights, the idea of valency and the structural theory of organic molecules. On Mendeleev's early studies of the chemical properties of substances, his work on the specific weights and their relationship to atomic and molecular weights, see V.M. Kedrov and D.N. Trifonov, *Zakon periodichnosti i khimicheskie elementy: otkrytiia i khronologiia* (Moscow: Nauka, 1969), 127–40. For English language sources on Mendeleev, see Brock, *op. cit.* (note 6), 349–54. On the role of Mendeleev in Russian science and culture, see A. Vucinich, 'Mendeleev's Views on Science and Society', *Isis,* 58 (1967), 342–51.
16. On Raoult's work on freezing points of solutions, see J.R. Partington, *A History of Chemistry* (London: Macmillan & Co., 1964), 4 vols, Vol. IV, 645–8; on Armstrong and the discussion on electrolysis and solutions by British chemists, see Dolby, *op. cit.* (note 2), 309–21; on Armstrong's studies with Kolbe in Leipzig in 1867–70, see A. Rocke, *The Quiet Revolution: Hermann Kolbe and the Science of Organic Chemistry* (Berkeley: University of California Press, 1993), 284–5.

17. Jacobus H. van't Hoff (1852–1911) taught chemistry in Amsterdam, and at Berlin University. Van't Hoff's studies of osmotic pressure suggested that the analogy between gases and solutions was complete. In 1885, he found that at equal osmotic pressure and temperature, equal volumes of solutions contain an equal number of molecules; the same number which, at the same temperature and pressure, is contained in an equal volume of gas. In his paper 'Die Rolle des osmotisches Druckes in der Analogie zwischen Lösungen und Gasen', published in the first volume of the new *Zeitschrift für physikalische Chemie* in 1887, van't Hoff, at the suggestion of Arrhenius, reinterpreted his work in terms of ionisation. The first Nobel Prize in Chemistry was awarded to van't Hoff in 1901. See Brock, *op. cit.* (note 6), 370–1; see also K.J. Laidler, *The World of Physical Chemistry* (Oxford: Oxford University Press, 1993), 207–19.
18. Svante Arrhenius (1859–1927), Professor of Physics at the University of Stockholm, and later Director of the Nobel Institute for Physical Chemistry, presented his theory of ionic dissociation in the paper 'Über die Dissoziation der in Wasser gelösten Stoffe', *Zeitschrift für physikalische Chemie,* i (1887), 631–48, in which he suggested that the dissociation of certain substances dissolved in water was strongly supported by the conclusions drawn from the electrical properties of the same substance. He was awarded the Nobel Prize in Chemistry in 1903. Later, Arrhenius became interested in immunochemistry, postulating a chemical equilibrium between toxin and antitoxin. The most recent biography of Arrhenius is E.T. Crawford, *Arrhenius: From Ionic Theory to the Greenhouse Effect* (Canton: Science History Publications, 1996); see also I.I. Solov'ev, *Svante Arrhenius: nauchnaia biografiia* (Moscow: Nauka, 1990).
19. W. Ostwald, *Lebenslinien: eine Selbstbiographie* (Berlin: Klasing, 1933), 3 vols in 1, Vol. I, 223. Cleve, one of the notable Swedish chemists, fully investigated some of the rare earth elements. He strongly opposed Arrhenius's appointment as Dozent in physical chemistry at the University of Uppsala. On Cleve and Arrhenius, see Crawford, *op. cit.* (note 18), 59–60.
20. On acceptance of the ionic theory by British chemists, see Dolby, *op. cit.* (note 2), 323–7; see also Crawford, *op. cit.* (note 18), 99.
21. D.I. Mendeleev, 'Zametki o dissotsiatsii rastvorennykh veshchestv', *Zhurnal russkogo fiziko-khimicheskogo obshchestva,* xxi (2) (1889), 175–6.
22. S. Arrhenius, 'Electrolytic dissociation versus hydration', *Philosophical Transactions,* xxviii, 5 (1889), 30–1.
23. On the acceptance of Arrhenius's theory in Russia, see Solov'ev, *op. cit.* (note 2), 139–45.
24. Brock, *op. cit.* (note 6), 379.
25. H.W. Nernst, *Theoretische Chemie vom Standpunkte der Avogadroschen Regel und der Thermodynamik* (Stuttgart: Enke, 1893), in V. Nernst (trans.),

Teoreticheskaia khimiia (St Petersburg: Izd-vo A.S. Suvorina, 1904), 100, 432. Hermann Walter Nernst (1864–1941), Professor of Physics and then of Physical Chemistry at the University of Göttingen was a supporter of the ionic theory. In 1889, Nernst, together with Ostwald, experimentally demonstrated the existence of free ions in an electrostatic field. Nernst was the most successful descendants of the original 'triumvirate' of Ostwald, van't Hoff, and Arrhenius. In 1920, Nernst received the Nobel Price in Chemistry. The most recent biography of Nernst is by D.K. Barkan, *Walter Nernst and the Transition to Modern Physical Chemistry* (Cambridge: Cambridge University Press, 1999).

26. For Ostwald's review of P.S. Pickering's report in Leeds, see *Zeitschrift fur physikalische Chemie*, vii (1891), 416–21, in Solov'ev, *op. cit.* (note 2), 333.
27. Ostwald, *op. cit.* (note 19), Vol. II, 126–7.
28. Ostwald to Mendeleev, letter dated 19 January 1888, Leipzig, in Solov'ev, *op. cit.* (note 2), 331.
29. Arrhenius to van't Hoff, letter dated 1890, Stockholm, in Barkan, *op. cit.* (note 25), 62.
30. W. Ostwald, *Solutions: Being the Fourth Book, with some Additions, of the Second Edition of Ostwald's 'Lehrbuch der algemeinen Chemie'*, M.M. Pattison-Muir, (trans.), (London: Longmans, 1891), vii.
31. Arrhenius, however, devoted a whole chapter (III) to the problem of the existence of the hydrates in the solution and specifically to Mendeleev's method, see S. Arrhenius, *Theories of Chemistry being Lectures delivered at the University of California at Berkeley* (London: Longmans, 1907); in German, S. Arrhenius (trans.), *Theorien der Chemie* (Leipzig: Akademische Verlagsgesellschaft, 1906); in Russian, S. Arrenius (trans.), *Teorii khimii* (St Petersburg: Izd-vo A.S. Suvorina, 1907).
32. Ostwald, *op. cit.* (note 30), 18, 28–31.

18

The Universal Law: Expectations and Disappointments

Sechenov published his first communication on the absorption of carbon dioxide by salt solutions in 1873, in the *Protokoly zasedanii Novorossiiskogo obshchestva estestvoispytatelei* [*Proceedings of the Novorossiisk Society of Naturalists*].[1] By that time, he had determined that carbon dioxide in the blood was in an 'unstable binding' (loose or weak chemical binding) state. Using the solutions of salts which had weak chemical bonding with CO_2 analogous to blood, Sechenov examined the absorption of CO_2 by solutions of these salts. The experiments with the solution of sodium acetate yielded results so interesting to Sechenov that he decided to continue his investigation of salt solutions. After a set of experiments with seven different salts and acids arranged in order by their strength, Sechenov assumed that 'there was already sufficient material to establish the general nature of the *weak chemical absorption* of CO_2 by solutions of these salts.' At this point he could have stopped, as 'salts of strong acids did not promise anything for the chemical absorption of CO_2 by the blood', but he decided otherwise and set up new experiments aimed at clarifying further how the composition of salt solutions affected their absorptiometric characteristics.[2] The results confirmed that weak and medium solutions of analogous salts absorbed an equal quantity of CO_2. Sechenov wrote to Mendeleev: 'Now I am experimenting with chlorides, which do not contain water of crystallisation, and already I have very curious facts about which I have to ask your advice.' Seeking approval of the St Petersburg chemists, Sechenov asked Mendeleev to present these findings at the meeting of the Russian Chemical Society.[3]

The next paper, published in 1875, reported that:

> [E]quivalent weights of the related salts bind equal quantities of CO_2. The absorption curve of any salt that binds CO_2 chemically is the resultant of two other curves, one of which represents the run of the absorption of CO_2, the other, the run of the dissolution following the Dalton law.[4]

This time, Sechenov asked Alexander Butlerov, another distinguished St Petersburg chemist, to review the manuscript before submitting it to the Bulletin of the St Petersburg Academy of Sciences: 'Although I am convinced in the correctness of my conclusions, I still have no peace of mind… I have no doubts in my results… but maybe I am too bold in my conclusions.'[5]

Apparently Sechenov realised the limitations in his one-sided background in chemistry, and his letters to Butlerov and Mendeleev convey his concern and anxiety regarding his research on salt solutions. In a small provincial university, he felt isolated from St Petersburg's vibrant scientific life and chemical community, and for him, whom 'fate had thrown to a foreign field', the support and advice of the St Petersburg chemists was of particular value. He wrote to Butlerov:

> I am sincerely grateful to you for your advice. From the German manuscript you will see that I have made use of your remarks. I read Thomsen's as well as Berthelot's papers, but in Berthelot's paper I missed the point that the liberated heat decreases in saturation of acids by bases to the extent of the dilution of solutions, and the decrease is evident especially in weak acids. For me this fact is essential since all my data could be predictable….[6]

Sechenov was well aware that the exactness and reliability of the results depended in great measure on the carefully worked-out method and the instrument devised for a particular kind of measurement and adapted to the demands of precision. He began to develop his absorptiometric method and constructed his first instrument, an absorptiometer as he called it, in Ludwig's laboratory in the late 1850s. During his tenure at the Medico-Surgical Academy, he improved the apparatus having replaced the rubber tubes with thin metal ones. By the mid-1870s, he had elaborated his methods and refined his absorptiometric apparatus, which, as he informed Butlerov, was 'incomparably more sensitive than the previous one'.[7] 'As for the method, I swear that I have used it in hundreds of experiments. The limits of its error are evident, my results are far beyond the error.'[8] Already in St Petersburg, in 1876, Sechenov was able to order a new instrument from a well-known mechanical technician at the Pulkovo Astronomical Observatory.[9] The technician, brought up to work with astronomical instruments that demanded extraordinary skill and mathematical precision, made an excellent instrument. The cost of the instrument at five hundred roubles, equivalent to a professor's salary for two months, was very expensive for Sechenov, who 'lived almost entirely on his salary'. However, he was happy to have the instrument, which 'met all the conditions stipulated excellently', and which 'allows me to trace accurately the finer things than it was possible with the instrument... in Odessa.'[10]

From 1877–9, Sechenov published several papers[11] and two major monographs, *O pogloshchenii ugol'noi kisloty solianymi rastvorami i krov'iu* [*On the Absorption of Carbon Dioxide by Salt Solutions and by the Blood*] and *Weiteres über das Anwachsen der Absorptionskoeffizient von Kohlendioxyd in den Salzlösungen* [*On the Absorption Coefficient of Carbon Dioxide in the Salt Solutions*] in Russia and Germany. Both monographs were also published in the *Bulletin* and *Mémoire* of the St Petersburg Academy of Sciences with an introduction by Alexander Butlerov and Nikolai Zinin. Sechenov demonstrated that in the solutions of salts, on which carbon dioxide acts chemically (sodium salts Na_2CO_3; $Na_2B_4O_3$; Na_2HPO_4), the absorption of CO_2 by these salts is increased and does not follow the Henry–Dalton law. On the other hand, in the solutions of sulphate and nitrate salts, CO_2 is absorbed in lesser amounts, but does follow the Henry–Dalton law. However, in the latter case, there was still explicit evidence of the chemical interaction between the salt, water, and carbon dioxide. Sechenov admitted that there was a physical interaction between substances in the solution, following the lead of Mendeleev, he attributed a significant role to the chemical interaction between the dissolved substances: 'the greater is the attraction of the salt to water, the stronger is the hydration and the weaker is the dissociation, therefore the absorption of CO_2 should be weak in general.'[12]

In 1883, Sechenov started experiments with the solutions of salts indifferent to CO_2. The experiments enabled him to exclude the chemical interaction which distorted the picture of the absorption of carbon dioxide by solutions. Decreasing dilution of the solutions of electrolytes, such as NaCl and $NaNO_3$, Sechenov showed that 'water dissociates a certain amount of salt with the formation of the products, which absorb CO_2 more strongly than pure water, and the dissociation increases with increasing dilution.'[13] Sechenov realised that he had observed at a general pattern, a regularity, which he expressed by the following equation:

$$y = ae^{-k/x}$$

Here y is a coefficient of the absorption of CO_2 in the solution; a is a coefficient of the absorption of CO_2 in the water; e is the base of natural logarithms; k is a constant of a given salt; and x is the volume in which the salt is dissolved.[14] The problem of distribution of substances in different systems was an intriguing problem for the chemists of the 1880s and 1890s. The first attempt to apply a method of distribution for the study of the state of dissolved substances in solutions was made by Marcellin Berthelot in 1872. Based on the distribution of carbon dioxide between water and salt in the solution, Sechenov showed that distribution depends on temperature, pressure, and concentration of the solution. These results were published in 1886 in *Mémoires de l'Académie Impériale des Sciences de St Pétersbourg.*[15]

In a year, Sechenov published the second part of his monograph *On the Absorption Coefficient of Carbon Dioxide in the Salt Solutions*, in which he proved his 'law of increase of the coefficients of absorption' in the experiments with twelve salts (Na_2SO_4, $CaCl_2$, NH_4Cl, etc.). In total, Sechenov performed more than a hundred experiments with these salts. He confirmed the existence of two classes of salts, those which absorb a part of carbon dioxide independently of the Henry–Dalton law (sodium carbonate, disodium phosphate, borax, etc.), and those which exert no definite chemical action on carbon dioxide, such as nitrates, chlorides, and sulphates. Sechenov also showed that solutions containing equivalent quantities of similar salts have nearly equal absorption coefficients.[16]

Sechenov's work with salt solutions, which involved continuous pumping of carbon dioxide out of the liquid, was extremely monotonous and tedious, particularly during the last two years when the work was close to completion. However, he seemed happy with this work. In a letter to Bokova, he remarks with humour: 'Do not worry, my dear, carbon dioxide will not absorb me entirely… this physical labour is even pleasant for me.'[17] Letters to Bokova during this period sound cheerful, they are full of optimism and quiet confidence in his 'law, a general regularity, which governs the absorption of gases by salt solutions'. He describes his experiments, which confirm 'his law' with excitement and delight: 'Isn't it a joy to work on such experiments?'[18]

> My laboratory work goes splendidly. The other day, experimenting with $CaCl_2$, I had a case, something like the astronomer Le Verrier with his discovery of Neptune. I cannot help trembling with agitation and interest.... The work I am doing now is worth the rest of my life.[19]

Indeed, Sechenov's work on salt solutions brought him many happy moments, but the treatment of his 'law as a special case' by the St Petersburg chemists plunged him into great distress. Even years could not erase his disappointment, as one can feel from the passage in *Autobiographical Notes,* written in 1904:

> The attitude of chemists towards my work, justified to a certain extent, pained me even more. They recognised my results and considered my work worthy of attention, but suggested to corroborate these results with gases other than 'eternal' carbon dioxide.[20]

Sechenov knew, as no one else did, that it was technically impossible, because other convenient gases such as oxygen, hydrogen, and nitrogen, are dissolved only weakly, 'so there was nothing to consider about them'. For Sechenov, in view of such a verdict, the work of many years lost its principal significance. He was also aware of the opinion of some biologists regarding his work on salt solutions. His biologist colleagues believed that for Sechenov, it was a sheer waste of time and effort to study non-physiological problems for so many years. Sechenov's frustration was so great that he felt that his stay at the University became 'pointless and even unpleasant'. To make things worse, the Minister of Education once again rejected him for a membership in the Academy of Sciences, and for the title of Honoured Professor. These only added to his unhappiness. At the end of 1888, he sent in his resignation and left St Petersburg for Bokova's estate.[21]

One can imagine Sechenov's dissatisfaction in view of his utterly unreasonable decision to leave his laboratory in St Petersburg and acquire a junior position as Privatdozent at the medical faculty of Moscow University, which did not have any physiological laboratory. In the autumn of 1889, Sechenov began teaching a course in general physiology. Apparently, a year without a laboratory and an appropriate position frustrated him, and next autumn, he took leave and left to go abroad. A visit to Mechnikov in Paris brought soothing relief. They 'met every day and had lengthy discussions'. They 'preserved as before complete similarity in views'. Mechnikov noticed that although Sechenov 'was only sixty-two at that time, and was still vivid, but his old age already began to be felt'. That impression was obviously enhanced by Sechenov's dispirited mood and disappointment regarding the misfortunes of his latest research and absence of appropriate conditions for work. He advised Mechnikov not to return to Russia, 'where it is hard to live'.[22] Mechnikov had already passed through hard experiences connected with a decision to leave Russia. He resigned from Novorossiisk University in 1882, exhausted by difficulties he had encountered pursuing his science, petty academic intrigues, limited independence of the university, and growing political instability and unrest after the assassination of Alexander II in 1881.

Mechnikov moved to Italy. In Messina, he continued his studies on intracellular digestion and made his seminal discovery on the defensive role of leukocytes. In 1888, Mechnikov, by that time an acclaimed scholar, took up a permanent position as Chef de Service in the newly opened Pasteur Institute in Paris. He was invited back to Odessa to head up the Research Institute, similar to the Pasteur Institute, to carry out bacteriological studies, but he decided to stay in Paris. By 1891, when Sechenov visited him in Paris, Mechnikov had become an acclaimed figure in France as a leader of the school of cellular immunity, and his laboratory facilities occupied an entire floor with two rooms for his own use. Mechnikov introduced Sechenov to Louis Pasteur and to Marcellin Berthelot, the most powerful chemist in France and Pasteur's close associate.

Mechnikov sympathised with Sechenov's perseverance in his scientific views and his resistance to criticism and objections, first against his theory of central inhibition, and later his physico–chemical studies. Some turns in Mechnikov's research career were similar to those of Sechenov. Mechnikov, a trained zoologist and embryologist, was deeply engaged in the problems of pathology associated with inflammation and infectious diseases. His theory of phagocytosis, first presented in 1883, held that phagocytic cells constituted a first line of host defence against invading organisms. Mechnikov's theory was well received by such authorities in cellular pathology as Rudolf Virchow in Berlin and Carl Claus in Vienna, but soon became a target of severe and protracted criticism from pathologists, particularly in Germany, who believed in the dogma of inflammation as a deleterious reaction of no benefit to the host organism. Throughout his life, Mechnikov had to defend his scientific views, first in embryology and later in immunology. He published his *Leçons sur la pathologie comparée de l'inflammation* [*Lectures on the Comparative Pathology of Inflammation*],[23] and as a leader of the school continued a ferocious defence of his cellular theory of immunity against the humoralist school. In 1908, Mechnikov received the Nobel Prize in Physiology and Medicine, together with Paul Ehrlich, proponent of the rival theory of humoral immunity.[24]

After Paris, Sechenov went to Ludwig in Leipzig. Ludwig knowing how important recognition and support of the chemical community were to Sechenov, introduced Sechenov to Ostwald. Sechenov was particularly pleased with this meeting:[25]

> My visit to Leipzig was a success. My dreams regarding carbon dioxide were recognised as true.... Ostwald acknowledged the importance of my absorptiometric method for the study of solutions but regretted that I took salts and CO_2, and hence the phenomena were complicated by chemical reactions. In his opinion, I should have started with gases indifferent to substances in the solution. He was convinced that a dissolved salt reacted chemically with CO_2 without my experiments. The only thing he does not agree with is my interpretation of the absorption of CO_2 by solutions of sulphuric and lactic acids with water. In his opinion, *attraction, affinité* cannot explain the phenomena as neither affinity nor its effects could be measured.[26]

Upon returning to Moscow, Sechenov began to think of leaving Russia to work in Leipzig. In his letter, the kind Ludwig, in his usual sincere and sentimental manner, expressed his concern and understanding to Sechenov's disappointments. Ludwig reminded Sechenov that 'there would always be a room for you in the Leipzig laboratory'. Ostwald too had invited him to work at his laboratory.[27] Fortunately for Sechenov, he unexpectedly received a chair of physiology in the medical faculty at Moscow University after the sudden death of Professor F.P. Sheremet'evskii in 1891.

Sechenov's attitude towards his work on salt solutions seems sensitive to the extreme, and in this respect he appeared a distinctly unusual man. He mentioned in his *Autobiographical Notes,* that even his established scholarly reputation 'could not eradicate the splinter from my heart,' regarding 'the fate of my absorptiometric studies'.[28] The mood is best expressed in Sechenov's letter to Mechnikov: 'As you see, I became like poor parents who wished to settle their beloved child. For them he is dear and handsome, but strangers treat him with indifference and suspicion.'[29] Obviously, Sechenov exaggerated. His work on salt solutions was well received: it was included in mainstream texts and published in important journals. Mendeleev included some of the results in a chapter on solutions in the fifth edition of his famous *Osnovy khimii* [*Principles of Chemistry*].[30] Another key figure in the field of theory of solution, Berthelot, favourably commented upon Sechenov's paper and published it in the *Annales de chimie et de physique.*[31] Ostwald too, included Sechenov's method and results in the chapter on solutions in his *Lehrbuch der allgemeinen Chemie*, claiming 'it went deeper into the phenomena of salt solutions' than any other study in the field.[32] Two of Sechenov's articles, published in Ostwald's journal, also contributed to a general recognition of Sechenov's work in the chemical community.[33]

Sechenov perceived, however, that what was missing in a generally favourable reception of his studies was recognition of the universality of the regularity he had discovered. Deep in his heart, he remained convinced that this regularity, which he so strenuously tested in hundreds of experiments represented a unifying law of absorption of carbon dioxide by salt solutions:

> In the first year of my professorship at Moscow University, my torments because of the fate of my work with CO_2 came to an end... as if fate took pity on me.... Moscow chemist A.A. Iakobkin in his studies on chemical statics corroborated my results in a more general form by his investigations. Thus, finally, I obtained the universal key to a wide class of phenomena after all.[34]

Now, at a new place, which happened to be the medical faculty of Moscow University, his *alma mater*, Sechenov found himself 'as in an uncultivated field, in which it is easy and simple to bring great benefit'.[35] He was absorbed by pleasant and familiar problems associated with the arrangement and setting up of a laboratory at the new Physiological Institute, which opened in 1893. Sechenov's family life was also happily settled in Moscow. After twenty-eight years of waiting, in 1888, Maria Bokova received official permission for a divorce from Petr Bokov, and the same year Sechenov and Bokova married. Sechenov spent ten years at Moscow University, and resigned in 1901 – four years before his death. He was happy to have two able assistants, Mikhail N. Shaternikov, his student, and Alexander F. Samoilov, who came to work with Sechenov after his studies with Ivan P. Pavlov in St Petersburg. Shaternikov succeeded Sechenov in the chair, and continued studies on gas exchange and metabolic processes in the organism. Alexander Samoilov, professor at both Moscow and Kazan Universities, became one of most notable electrophysiologists of his time.

Sechenov was renowned as the great old master who introduced the laboratory to Russian physiology and medicine and set up laboratories at the Medico-Surgical Academy and in St Petersburg, Novorossiisk, and Moscow Universities, and who became one of the symbols of his time with its great aspirations for science and its great achievements and disappointments.

Notes

1. I.M. Sechenov, 'O pogloshchenii ugolnoi kisloty shchelochnymy zhidkostiami', *Protok. zased. Novoros. o-va estestvoispytatelei,* i (1873), 24–6.
2. I.M. Sechenov, *Autobiographical Notes.* D.B. Lindsley (ed.), K. Hanes (trans.) (Washington: American Institute of Biological Sciences, 1965), 152.
3. Sechenov to Mendeleev, letter dated 17 October 1873, Odessa, in K.S. Koshtoyants *et al.* (eds), *Nauchnoe nasledstvo: I.M. Sechenov - Neopublikovannye raboty, perepiska i dokumenty* (Moscow: Akademiia Nauk SSSR, 1956), 218–19.
4. I.M. Sechenov, 'O pogloshchenii ugolnogo angidrida rastvorami solei', *Zhurnal russkogo fiziko-khimicheskogo obshchestva,* vii (1875), 214–30, 218.
5. Sechenov to Butlerov, letter dated 14 March 1875, Odessa, in Koshtoyants, *op. cit.* (note 3), 222.
6. Sechenov to Butlerov, letter dated 9 April 1875, Odessa, in *ibid.*, 223.
7. Sechenov to Butlerov, letter dated 16 February 1876, Odessa, in *ibid.,* 224. The letter with the description of a new model of absorptiometer has not been located.
8. Sechenov to Butlerov, letter dated 14 March 1875, Odessa, in *ibid.,* 222.
9. The great telescopes of the nineteenth-century still exist in the Pulkovo Observatory museum, now no longer suited to the needs of modern astronomy but preserved as superb examples of the craft of the nineteenth-century instrument makers.
10. Sechenov, *op. cit.* (note 2), 151.
11. I.M. Sechenov, 'Über die Absorption der Kohlensäure durch Schwefelsäure und deren Gemische mit Wasser', *Bulletin de la classe physico-mathématique de l'Académie Impériale des Sciences de St-Pétersbourg,* xxii, 3 (1877), 102–7.
12. I.M. Sechenov, *O pogloschenii ugol'noi kisloty solianymi rastvorami i krov'iu* (St. Petersburg: Izd-vo L.F. Panteleeva, 1879), 85–6.
13. I.M. Sechenov, 'O narastanii koeffitsientov pogloshcheniia CO_2 razzhizhaemymi vodoi solianymi rastvorami', *Zhurnal russkogo fiziko-khimicheskogo obshchestva,* xviii, 1 (1886), 63–4; *Zhurnal russkogo fiziko-khimicheskogo obshchestva,* xviii, 2 (1886), 124–8: 126.

14. I. Sechenov, 'Über die Absorptionskoeffizienten der Kohlensäure in den zu diesem Gase indifferenten Salzlösungen', *Mémoires de l'Académie Impériale des Sciences de St Pétersbourg,* sér. 7, 34 (1886), 1–24: 19.
15. *Ibid.,* 21.
16. I.M. Sechenov, *Weiteres über das Anwachsen der Absorptionskoeffizient von CO_2 in den Salzlösungen* (St Petersburg: Izd-vo Panteleeva, 1887), 31.
17. Sechenov to Bokova, letter dated 17 October 1886, St Petersburg, in Koshtoyants, *op. cit.* (note 3), 257.
18. Sechenov to Bokova, letter dated 7 April 1886, St Petersburg, *ibid.,* 252.
19. Sechenov to Bokova, letter dated 11 April 1886, St Petersburg, *ibid.,* 252–3. In 1846, the French astronomer Urban Jean Joseph Le Verrier, scrutinising the orbit of Uranus, predicted the existence of the unknown planet Neptune. M. Hoskin (ed.), *The Cambridge Illustrated History of Astronomy* (Cambridge: Cambridge University Press, 2000), 191–5. See also Sechenov's letters to Bokova, dated 29 August 1886; 22 September 1886, and 4 April 1887, St Petersburg, in Koshtoyants, *op. cit.* (note 3), 256–7.
20. Sechenov, *op. cit.* (note 2), 256–7.
21. *Ibid.,* 156, 161.
22. Sechenov to Mechnikov, letter dated 14 April 1891, Leipzig, in S.I. Shtraikh (ed.), *Bor'ba za nauku v tsarskoi Rossii: neizdannye pis'ma I.M. Sechenova, I.I. Mechnikova, L.S. Tsenkovskogo, V.O. Kovalevskogo, i dr.* (Moscow–Leningrad: Nauka, 1931), 110.
23. I. Mechnikov, *Leçons sur la pathologie comparée de l'inflammation faites à l'Institut Pasteur en avril et mai 1891* (Paris: G. Mason, 1892).
24. For Mechnikov's English language biography, see 'Introduction' in A. Tauber and L. Chernyak, *Metchnikoff and the Origins of Immunology* (Oxford: Oxford University Press, 1991), 3–24. On Mechnikov and cellular versus humoral immunity, see A. Silverstein, *A History of Immunilogy* (New York: Academic Press, 1989), 38–54.
25. W. Ostwald, *Lebenslinien: eine Selbstbiographie* (Berlin: Klasing, 1933), 3 vols, Vol. I, 267; on Ludwig and his laboratory, see *idem,* Vol. II, 82–8.
26. Sechenov to Mechnikov, *op. cit.* (note 22), 110.
27. Sechenov to Mechnikov, letter dated 20 May 1891, Teplyi Stan, in *ibid.,* 111. Teplyi Stan was the Sechenov family estate in Simbirsk province on the Volga.
28. Sechenov, *op. cit.* (note 2), 164.
29. Sechenov to Mechnikov, letter dated 20 May 1891, Teplyi Stan, in Shtraikh, *op. cit.* (note 22), 111.
30. D.I. Mendeleev, *Osnovy khimii* (St Petersburg: Izd-vo Suvorina,1889), 60–1; 66.

31. I.M. Sechenov, 'Action de l'acide carbonique sur les solutions des sels à acides forts. Étude absorptiométrique', *Annales de chimie et de physique,* sér. 6, xxv (1892), 226–70.
32. W. Ostwald, *Lehrbuch der allgemeinen Chemie* (Leipzig: Wilhelm Engelmann, 1890), 78.
33. I.M. Sechenov, 'Über die Konstitution des Salzlösungen auf Grund ihres Verhaltens zur Kohlensäure: vorläufige Mitteilung', *Zeitschrift für physikalische Chemie,* Bl. 4, H. 1 (1889), 117–25; *idem,* 'Analogien zwischen der Auflösung von Gas und Salz in einer zu beiden indifferenten Salzlösung', *Zeitschrift für physikalische Chemie,* Bl. 8, H. 6 (1891), 657–60.
34. Sechenov, *op. cit.* (note 2), 164.
35. Sechenov to Mechnikov, letter dated April 1892, Moscow, in Shtraikh, *op. cit.* (note 22), 114.

Bibliography

Archives

Moscow

Russian State Archive of Military History.

State Archive of the Russian Academy of Sciences.

Medical Academy after I.M. Sechenov, Sechenov Museum.

State Central Museum of Musical Culture after M.I. Glinka, Archive and Manuscript Collection.

St Petersburg

Russian State Archive of Military History (St Petersburg Branch).

Fundamental Library of the Military Medical Academy after S.M. Kirov, Division of Manuscripts and Rare Books.

University Archive and Museum.

Museum of the Institute for Experimental Medicine of the Russian Academy of Medical Sciences.

Vienna

Österreichisches Staatsarchive-Kriegsarchiv, Kriegsministerium, Abt., Medizinisch-chirurgische Josephs-Akademie.

Universitätsarchiv.

Josephinum-Museum des Instituts für Geschichte der Medizin.

Berlin

Staatsbibliothek Preussischer Kulturbesitz - Handschriftenabteilung, Autographensammlung.

Universitätsarchive, Humboldt Universität.

Medizinhistorisches Museum der Charité, Medizinische Fakultät der Humboldt-Univerzität, Fotos-Autographensammlung.

Johannes-Müller-Institut für Physiologie der Humboldt-Universität. Historische Instrumentensammlung.

Institute für Geschichte der Medizin der Humboldt-Universität.

Leipzig

Universitätsarchiv.

Universitätsbibliothek – Handschriftenabteilung.

Karl-Sudhoff-Institut für Geschichte der Medizin und Naturwissenschaften, Medizinhistorische Sammlung.

Wilhelm Ostwald-Gesellschaft zu Grossbothen. Ostwald Gedenkstätte.

Stadtmuseum. Portrait Collection.

Heidelberg

Universitätsarchiv.

Institut für organische Chemie, das chemische Universitäts-Laboratorium, historische Instrumentensammlung.

Das fizikalische Universitäts-Laboratorium, historische Instrumentsamlungen.

Munich

Sondersammlungen und Bibliothek des Deutsches Museums.

Feinmechanik und Chemie Instrumentensammlungen.

Handschriftenabteilung, Bayerische Staatsbibliothek.

Primary sources

Anon., *Beitrage zur Anatomie und Physiologie als Festgabe Carl Ludwig* (Leipzig: F.C.W. Vogel Verlag, 1874).

Arrhenius, S., *Theories of Chemistry being Lectures Delivered at the University of California at Berkeley* (London: Longmans, 1907).

Bedson, P.P., 'Lothar Meyer Memorial Lecture', in *Memorial Lectures Delivered before The Chemical Society, 1893–1900* (London: Gurney & Jackson, 1901).

Belogolovyi, N.A., *S.P. Botkin. Ego zhizn' i meditsinskaia deiatel'nost'* (St Petersburg: Tipografiia J.N. Erlikha, 1892).

——— *Vospominaniia i drugie stat'i* (St Petersburg: Tipografiia K. Aleksandrova, 1897), 251–375.

Bernard, C., *Introduction to the Study of Experimental Medicine,* H.C. Green (trans.), (New York: Dover Publications, 1957).

Borodin, A.P. and Butlerov, A.M., 'N.N. Zinin. Vospominaniia o nem i biographicheski ocherk', *Zapiski Akademii Nauk,* 37, 1 (1880), 1–46.

Botkin, S.P., *Die Kontraktilität der Milz und die Beziehung der Infektionsprozesse zur Milz, Leber, den Nieren und dem Herzen* (Berlin: Hirschwald Verlag, 1874).

——— *Kurs kliniki vnutrennikh boleznei,* 3 vols (St Petersburg: Tipografiia Golovacheva, 1867–75), Vol. I, 1867; Vol. II, 1868; Vol. III, 1875.

——— *Medizinische Klinik in demonstrativen Vorträgen: über das Fieber im allgemeinen Flecktyphus* (Berlin: A. Hirschwald Verlag, 1869).

——— *Medizinische Klinik in demonstrativen Vorträgen: Zur Diagnostik, Entwicklungsgeschichte und Therapie der Herzkrankheiten* (Berlin: A. Hirschwald Verlag, 1867).

——— *Pis'ma iz Bolgarii v 1877* (St Petersburg: Izd-vo L.F. Panteleeva, 1893).

——— 'Rech po povody iubileia professora Virkhofa', *Trudy obshch. russ. vrachei v S.-Peterburge, 1881–2 gg* (St Petersburg, 1882), 46–52.

——— 'Über die Wirkung der Salze auf die zirkulierenden roten Blutkörperchen', *Virchows Archiv für pathologische Anatomie und Physiologie und für klinische Medizin,* 15 (1858), 173–6.

——— ' Zur Frage von dem Stoffwechsel der Fette im thierischen Organismus', *Virchows Archiv für pathologische Anatomie und Physiologie und für klinische Medizin,* 15 (1858) 380–2.

Bunsen, R., *Gasometrische Methoden* (Braunschweig: F. Vieweg, 1857).

Chicherin, B.N., *Russkoe obshchestvo 40–50 godov 19 veka,* N.I. Simbaev (ed.), 2 vols (Moscow: Izd-vo Moskovskogo Universiteta, 1991).

———— *Vospominaniia: puteshestvie za granitsu* (Moscow: Sever, 1935).

Cranefield, P.F. (ed.), *Two Great Scientists of the Nineteenth Century: Correspondence of E. du Bois-Reymond and C. Ludwig,* S. Lichtner-Ayèd (trans.), (Baltimore: Johns Hopkins University Press, 1982).

Curtius, T. and Rissom J., *Geschichte des chemischen Universitäts-Labiratoriums zu Heidelberg seit der Gründung durch Bunsen* (Heidelberg: Verlag von F.W. Rochow, Universitäts-Buchhandlung, 1908).

Cyon, I.F., *Die Gefässdrüsen als regulatorische Schuzorgane des Zentral-Nervensystem* (Berlin: Julius Spriger, 1910).

———— *Gesammelte physiologische Arbeiten* (Berlin: Hirschwald, 1888).

———— *Kurs fiziologii* (St Petersburg: C. Ricker, 1873).

———— *Methodik der physiologischen Experimente und Vivisektionen, mit Atlas* (St. Petersburg: C. Ricker, 1876).

———— *Raboty, sdelannye v fiziologicheskoi laboratorii Imperatorskoi mediko-khirurgicheskoi akademii za 1873 god s prilozheniem kriticheskikh statei Professora I. Tsiona* (St Petersburg: C. Ricker, 1874).

Dalton, J., 'On the Absorption of Gases by Water and Other Liquids', *Philosophical Society, Manchester,* 1 (1805) 271–87.

Dianin, A.P., 'A.P. Borodin: Biographicheskii ocherk i vospominaniia', *Zhurnal russkogo fiziko-khimicheskogo obshchestva,* 20, 4 (1888), 367–79.

Dianin, S.A., *A. P. Borodin: Zhizneopisanie, materially i dokumenty* (Moscow: Gos. Muz. Izd –vo, 1960).

———— *Borodin,* R. Lord (trans.), (Oxford: Oxford University Press, 1963).

———— (ed.), *Pis'ma A.P. Borodina, 1857–71,* Vol. I (Moscow: Gos. izd-vo, muzykal'nyi sector, 1927).

———— (ed.), *Pis'ma A.P. Borodina, 1872–77,* Vol. II (Moscow: Gos. muz. izd-vo, 1947).

———— (ed.), *Pis'ma A.P. Borodina, 1878–82,* Vol. III (Moscow: Gos. muz. izd-vo, 1849).

———— (ed.), *Pis'ma A.P. Borodina, 1883–87,* Vol. IV (Moscow: Gos. mus. izd-vo, 1950).

Du Bois-Reymond, Emil, 'Gedächtnissrede auf Johannes Müller, gehalten in der Leibniz-Sitzung der Akademie der Wissenschaften am 8 Juli 1858', *Reden von Emil du Bois-Reymond,* 2 vols (Leipzig: Verlag von Veit, 1912), vol. I, 135–317

———— *Untersuchungen über thierische Elektrizität* (Berlin: Reimer, 1848–84).

———— *Gesammmelte Abhandlung zur allgemeinen Muskel-und Nervenphysik* (Leipzig: Vogel, 1875–7).

———— and Ludwig, Carl, 'Zweiter Beitrag zur Kritik der Blutanalyse', *Zeitschrift für rationelle Medizin,* new series 5 (1854), 101–8.

Du Bois-Reymond, Estelle (ed.), *Reden von Emil du Bois-Reymond,* 2 vols (Leipzig: Verlag von Veit, 1912).

Feierliche Eröffnung der mit allerhöchster Entschliessung von 15. Februar 1854 restaurierten Medizinisch-chirurgischen Josephs-Akademie (Vienna, 1854).

Fernet, Emil, 'Du rôle des principaux éléments du sang dans l'absorption ou le dégagement des gazes de la respiration', *Annales des sciences naturelles: zoologie et biologie animale,* series 4, 13 (1857), 125–52.

Funke, Otto, *Lehrbuch der Physiologie für akademische Vorlesungen und zum Selbstudium,* 2 vols (Leipzig: Leopold Voss, 1858–60).

Glebov, I. T. *Kratkii obsor deistvii Imperatorskoi meditsinsko-khirurgicheskoi akademii za 1857, 1858 i 1859 gody v vidakh ulutsheniia etogo zavedeniia* (St Petersburg: Tipografiia Imperatorskoi akademii nauk, 1860).

Goldie, S. (ed.), *Florence Nightingale: Letters from Crimea, 1854–1856* (Manchester: Mandolin, 1997).

Golubov, N.F., 'O napravleniiakh v russkoi klinicheskoi meditsine' in G.A. Zakhar'in, *Clinicheskie lektsii* (St Petersburg, 1894).

Grigor'ev, V.V., *Imperatorskii St Peterburgskii universitet v techenie pervykh piatideciati let ego sushchestvovaniia* (St Petersburg, 1870).

Heidenheim, R. and L. Meyer. 'Über das Verhalten der Kohlensäure gegen Lösungen phosphorsäuren Natron', *Studien des physiologischen Institut zu Breslau,* 2 vols (Leipzig: Veit und Comp., 1863) Vol. II, 103–24.

Hjelt, E. (ed.), *Aus Jacob Berzelius' und Gustav Magnus's Briefwechsel in den Jahren 1828–1847* (Brauschweig: F. Vieweg, 1900).

———— *Geschichte der organischen Chemie von ältester Zeit bis zur Gedenwart* (Braunschweig: F. Vieweg, 1916).

Hering, E., 'Das physiologische Institut', *Festschrift zur Feier des 500 Jährigen Bestehens der Universität Leipzig* (Leipzig: Verlag von S. Hirzel, 1909), Vol. III, 21–38.

Hermann, Ludimar, *Grundriss der Physiologie des Menschen* (Berlin: A. Hirschwald Verlag, 1874).

——— *Untersuchungen über den Stoffwechsel der Muskeln, ausgehend vom Gaswechsel derselben* (Berlin: Hirschwald Verlag, 1867).

Hoppe-Seyler, Felix, *Handbuch der physiologisch-pathologisch chemischen Analyse für Ärzte und Studirende* (Berlin: Hirschwald Verlag, 1875).

——— *Abbildung des Apparats für physiologischen Chemie* (Berlin: Hirschwald Verlag, 1879).

——— *Anleitung zur pathologisch-chemischen Analyse für Ärzte und Studierende* (Berlin: Hirschwald Verlag, 1858).

——— 'Über Haematokrystallin und Haematin', *Virchows Archiv für pathologische Anatomie und Physiologie und für klinische Medizin,* 17 (1859), 488–91.

——— *Medizinisch-chemische Untersuchungen: Aus dem Laboratorium für angewandte Chemie zur Tübingen,* 4 vols (Berlin: A. Hirschwald, 1866–71).

——— 'Über das Verhalten des Blutfarbstoffes im Spectrum des Sonnenlichtes', *Virchows Archiv für pathologische Anatomie und Physiologie und für klinische Medizin,* 23 (1862), 446–9.

Ivanova, L. (ed.), *A.M. Butlerov: po materialam sovremennikov* (Moscow: Nauka, 1978).

Ivanovskii, N.I. (ed.), *Istoriia Imperatorskoi voenno-meditsinskoi (byvshei medico-khirurgicheskoi) akademii za sto let. 1798–1898* (St Petersburg: Tipografiia Ministerstva vnutrennikh del, 1898).

Karsten, G. (ed.), *Fortschritte der Physik im Jahre 1845: dargestellt von der physikalischen Gesellschaft zu Berlin* (Berlin: G. Reimer Verlag, 1847).

Kliuchevskii, V.O., *A History of Russia,* C. J. Hogarth (trans.), 5 vols (New York: Russel & Russel, 1960)

L. Koenigsberger, *Hermann von Helmholtz,* F. Welby (trans.), (New York: Dover Publications, 1965).

Koshtoyants, *et al,* (eds), *Nauchnoe nasledstvo: I.M. Sechenov. Neopublikovannye raboty, perepiska i dokumenty* (Moscow: Izd-vo Akademii Nauk SSSR, 1956).

Krafft, von F. (ed.), *Bunsen-Briefe in der Universitätsbibliothek Marburg* (Marburg: Universitätsbibliothek, 1996).

Kremer, R. (ed.), *Letters of Hermann von Helmholtz to his Wife, 1847–1859* (Stuttgart: Franz Steiner Verlag, 1990).

Kühne, W., 'Das Vorkommen und die Aussscheidung des Hämoglobins aus dem Blute', *Virchows Archiv für pathologische Anatomie und Physiologie und für klinische Medizin,* 34 (1865), 423–36.

———— *Lehrbuch der physiologischen Chemie* (Leipzig: Engelmann, 1868).

———— *Myologische Untersuchungen* (Leipzig: Verlag von Veit, 1860).

Kutsenko, A.I., *Istoricheskii ocherk kafedry akademicheskoi terapevticheskoi kliniki Imperatorskoi voenno-meditsinskoi akademii, 1810–1898* (St Petersburg: Rikker, 1898).

Leffler, A.C. and Kovalevsky, S., *Sonia Kowalevsky: Biography and Autobiography,* L. von Cossel, (trans.), (New York, 1895).

Ludwig, Carl, *Rede zum Gedächtnis an Ernst Heinrich Weber* (Leipzig: Verlag von Veit, 1878).

———— 'Erwiderung auf Valentin's Kritik der Bemerkungen zu seinem Lehren vom Athmen und Blukreislauf', *Zeitschrift für rationelle Medizin,* 4 (1846), 183–90.

———— 'Zusammenstellung der Untersuchungen über Blutgase, welche aus der physiologischen Anstalt der Josephs-Akademie hervorgegangen sind', *Medizinische Jahrbücher (Wien),* 9 (1865), 145–66.

———— *Lehrbuch der Physiologie des Menschen,* 2 vols (Leipzig: C.F. Winter'sche Verlagshandlung, 1856–58).

———— 'The "Introduction" to the Textbook on Human Physiology', M. Frank and J. Weiss (trans.), *Medical History,* 10, 1 (1966), 76–86.

———— *De viribus physicis secretionem urinae adiuvantibus* (Marburg: Elwert, 1842).

———— *Beiträge zur Lehre vom Mechanismus der Harnsekretion* (Marburg: Akademische Buchhandlung von N.G. Elwert, 1843).

Magnus, G.H., 'Über das Absorptionvermögen des Blutes für Sauerstoff', *Annalen der Physik und Chemie,* 66 (1845), 195–6.

Mechnilov, I.I., 'Vospominaniia o Sechenove', *Stranitsy vospominanii* (Moscow: Izd-vo Akademii Nauk SSSR, 1946), 45–64.

Mendeleev, D.I., 'Dnevniki 1861 i 1862 godov', in S.I. Vavilov *et al.* (eds), *Nauchnoe nasledstvo,* 3 vols (Moscow: Izd-vo Akademii Nauk SSSR, 1951), Vol. II, 95–259.

——— *Osnovy khimii,* 5th edn (St Petersburg: Izd-vo A. S. Suvorina, 1889).

——— *The Principles of Chemistry,* A. Greenaway (ed.), G. Kamensky, (trans.), (New York: Collier, 1901).

Meyer, L., *Die Gase des Blutes* (Göttingen: Dieterich, 1857).

Miliukov, P.N. 'Universitety v Rossii', in F.A. Brokgauz and I.A. Efron (eds), *Rossiia: Entsiklopedicheskii slovar'* (St Petersburg: Izd. Firma brokgauz i Efron, 1902), Vol. LCVIII, 788–800.

Miliutin, D.A., *Vospominaniia general-feldmarshala grapha D.A. Milutina,* 4 vols (Moscow: Rossiiskii fond kul'tury, 1997–2003).

Muehry, A., *Observations on of the Comparative State of Medicine in France, England, and Germany, during a Journey into these Countries in the Year of 1835,* E.D. Davis (trans.), (Philadelphia: Waldie, 1838).

Müller, Johannes, *Handbuch der Physiologie des Menschen für Vorlesungen.* (Coblenz: Verlag von J. Hölscher, 1837).

Nezhdanova, A.V., 'Moi druz'ia', *Materialy i issledovaniia* (Moscow: Iskusstvo, 1967), 59–65.

Ostwald, W., *Lebenslinien: eine Selbstbiographie,* 3 vols in 1 (Berlin: Klasing, 1933).

——— *Solutions, Being the Fourth Book, with some Additions, of the Edition of Ostwald's 'Lehrbuch der algemeinen Chemie',* M.M. Pattison-Muir (trans.), (London: Longmans, 1891).

Pasutin, V.V., *Avtobiografiia* (St Petersburg: K. Ricker, 1899).

——— *Kratkii ocherk Imperatorskoi voenno-meditsinskoi akademii za 100 let eia sushchestvovaniia* (St Petersburg: K. Ricker, 1898).

Pavlov, I.P., *Izbrannye trudy,* I.V. Natochin *et al.* (eds), (Moscow: Meditsina, 1999).

Pirogov, N.I., *Sevastopol'skie pis'ma i vospominaniia* (Moscow: Gos. izd. med. lit-ry, 1950).

——— *Questions of Life: Diary of an Old physician,* G.V. Zarechnak (ed.), (Canton: Science History Publications, 1991).

——— *Otchet o puteshestvii po Kavkazu* (Moscow: Meditsinskaia literatura, 1952).

——— 'Puteshestvie po Kavkazu v 1847', in P.A. Dubovitskii (ed.), *Zapiski po chasti vrachebnykh nauk* (St Petersburg: Izd-vo Jungmeister, 1949).

Pletnev, D.D., *Russkie terapevticheskie shkoly* (Moscow–Petrograd, 1923).

Poggendorff, Johann C., *Biographisch-literarisches Handwörterbuch zur Geschichte der exacten Wissenschaften enthaltend nachweisungen über Lebensverhältnisse und Leitungen von Mathematikern, Astronomen, Physikern, Chemikern, Mineralogen, Geologen usw. aller Völker und Zeiten* (Leipzig: J.A. Barth; Berlin: Akademie Verlag,1863–1904).

Polishchuk, V.R., *Chuvstvo veshchestva* (Moscow: Znanie, 1981).

Popel'skii, L., *Istoricheskii ocherk kafedry fiziologiiv Imperatorskoi voenno-meditsinskoi Akademii za 100 let (1798–1898)* (St Petersburg, 1899).

Protokoly zasedanii Konferentsii Imperatorskoi mediko-khirurgicheskoi akademii za 1872 god (St Petersburg, 1873).

Remak, R., *Galvanotherapie der Nerven- und Muskelkrankheiten* (Berlin: Hirschwald Verlag, 1858).

Rimskii-Korsakov, N.A., *My Musical Life,* C. van Vechten, (ed.), J.A. Joffe, (trans.), (New York: Alfred A. Knopf, 1947).

Roscoe, H., 'Bunsen Memorial Lecture', *Journal of the Chemical Society,* 77 (1900), 513–54.

Sadovskaia, N., *Perepiska S.P. Botkina s N.A. Belogolovym* (Moscow: Izd-vo Akademii Nauk, 1939).

Samoilov, A.F., *Izbrannye stat'i i rechi,* K.S. Koshtoyants (ed.), (Moscow; Leningrad: Izd-vo Akademii Nauk, 1946).

Schilder, N.K., *Imperator Alexander I. Ego zhizn' i tsarstvovanie,* 4 vols (St. Petersburg: Izdatel'stvo Suvorina, 1897).

Schickert, D., *Die militärärztlichen Bildungsanstalte von ihrer Gründung bis zur Gegenwart* (Berlin: Ernst Siegfried Mittler, 1895).

Sechenov, I.M., 'Analogien zwischen der Auflösung von Gas und Salz in einer zu beiden indifferenten Salzlösung', *Zeitschrift für physikalische Chemie,* 8 (1891), 657–60.

——— *Autobiographical Notes,* D.B. Lindsley(ed.), K. Hanes (trans.), (Washington: American Institute of Biological Sciences, 1965).

Sechenov, I.M. (*cont...*),'Beiträge zur Pneumatologie des Blutes', *Sitzungsberichte der mathematisch-naturwissenschaftlichen Klasse der kaiserlichen Akademie der Wissenschaften,* 36, 2 (1859), 293–319.

——— 'Die Kohlensäure des Blutes', *Mémoires de l'Académie Impériale des Sciences de St Pétersbourg,* 7, Vol. 36, No. 13 (St. Petersburg, 1879).

——— *Fiziologiia nervnoi sistemy* (St Petersburg: Tipografiia Golovacheva, 1866).

——— 'Galvanische Erscheinungen an dem verlängerten Marke des Frosches', *Pflügers Archiv für die gesamte Physiologie,* 27 (1882), 524–66.

——— 'Nauchnaia deiatel'nost' russkikh universitetov po estestvoznaniiu za poslednee dvadtsatipiatiletie', *Vestnik Evropy* (1883) 11: 330–42.

——— 'Neuer Apparat zur Gewinnung der Gase aus dem Blut', *Zeitschrift für rationelle Medizin,* 23 (1865), 16–20.

——— *O pogloshchenii ugol'noi kisloty solianymi rastvorami i krov'iu* (St Petersburg: Izd-vo L.F. Panteleeva, 1879).

——— *Physiologische Studien über die Hemmungsmechanismen für die Reflextätigkeit des Rückenmarkes im Gehirne des Frosches* (Berlin: Hirschwald, 1863).

——— *Refleksy golovnogo mozga* (St Petersburg: Tipografiia Golovacheva, 1866).

——— *Selected Works,* M.N. Shaternikov, (ed.), (Moscow: Izd-vo Vsesouznogo instituta eksperimental'noi meditsiny, 1935).

——— *Sobranie sochinenii,* 2 vols (Moscow: Izd-vo Moskovskogo Universiteta, 1907–8).

——— 'Über die Absorption der Kohlensäure durch Salzlösungen', *Mémoires de l'Académie Impériale des Sciences de St Pétersbourg,* Series 7, 22, 6 (1875), 442.

———- 'Über die Konstitution des Sälzlosungen auf Grund ihres Verhaltens zur Kohlensäure: vorläufige Mitteilung', *Zeitschrift für physikalische Chemie,* 4, 1 (1889), 117–25.

——— with Pavlov, I.P. , and Vvedenskii, N.E., *Fiziologiia nervnoi sistemy: Izbrannye trudy,* K.M. Bykov (ed.), 5 vols (Moscow: Gos. izd-vo med. lit-ry, 1948–1952).

Shaternikov, M.N., 'The Life of I.M. Sechenov', in I.M. Sechenov, *Selected Works* (Moscow–Leningrad: State Publishing House, 1935), ix–xxxvi.

Shtraikh, S. Ia., (ed.), Semashko, N.A. (intr.), *Bor'ba za nauky v tsarskoi Rossii: neizdannye pis'ma I.M. Sechnova, I.I. Mechnikova, L.S. Tsenkovskogo, V.O. Kovalevskogo, S.N. Vinogradskogo i dr.* (Moscow-Leningrad: Gossotsgiz, 1931).

Siemens, Werner von, *Recollections,* W. Coupland (trans.), (Munich: Prestel-Verlag, 1983).

Sirotinin, V.N., *S.P. Botkin* (St Petersburg: Izd-vo Panteleeva, 1899).

Solov'ev, S.M., *Istoriia Rossii s drevneishikh vremen,* 29 vols (1851–1879), repr. 18 vols (Moscow: Mysl', 1994).

Stal', A., *Perezhitoe i peredumannoe studentom, vrachem i professorom* (St Petersburg, 1908).

Stasov, V.V., *A.P. Ego zhizn', perepiska i muzykal'nye stat'i, 1834–87* (St Petersburg: Izd-vo Suvorin, 1889).

————— *Selected Essays on Music,* Jonas, F. (trans.), (London: Barrie & Rockliff, 1968).

Svatikov, S.G., *Russische Studenten in Heidelberg,* E. Wieschhöfer (ed.), (Heidelberg: Universitätsbibliothek Heidelberg, 1997).

Tarkhanov, I.R., 'Pamiati professora I. M. Sechenova', *Trudy obshchestva russkikh vrachei,* 73 (1906), 69–75.

Tatishchev S.S., *Imperator Alexander II: ego zhizn' i tsarstvie,* 2 vols (St Petersburg: Izdanie A.S. Suvorina, 1911)

—————- 'Sevastopol' v dekabre mesiatse', *Povesti i rasskasy,* 2 vols (Moscow: Khudozhestvennaia literatura, 1966), Vol. I, 58–72.

Traube, L., *Die Symptome der Krankheiten des Respirations- und Zirkulation Apparats* (Berlin: Hirschwald Verlag, 1867).

————— *Gesammelte Beiträge zur Pathologie und Physiologie,* 2 vols (Berlin: Hirschwald Verlag, 1871).

————— *Über den Zusammenhang von Herz-und Nierenkrankheiten* (Berlin: Hirschwald Verlag, 1876).

Vierordt, K., *Physiologie des Atmens mit besonderer Rücksicht auf die Ausscheidung der Kohlensäure* (Karlsruhe: Groos, 1845).

Virchow, R., 'Errinerung an Ludwig Traube', *Berliner klinische Wochenschrift,* 16 (1876), 200–3.

————— *Gedächtnissrede auf Johann Lucas Schönlein* (Berlin: Hirschwald, 1865).

Voennoe ministerstvo: smeta na 1866 god po Imperatorskoi S.-Peterburgskoi med.-khirurgicheskoi akademii (St Petersburg: Voennoe ministerstvo, 1867).

Vvedenski, N.E., 'I.M. Sechenov: nekrolog', *Trudy C.-Peterburgskogo obshchestva estestvoispytatelei,* 36, 2 (1906), 1–44.

Walden, Paul, 'Die Lösungstheorien in ihrer geschichtlichen Aufeinanderfolge', *Sammlung chemischer und chemisch-technischer Vorträge,* 15 (1910), 277–454.

Wenig, K. (ed.), *Rudolf Virchow und Emil du Bois-Reymond: Briefe 1864–1894* (Marburg-Lahn: Basilisken-Presse, 1995).

Secondary sources

Antonov, V.B. and Georgievskii, A.S., *S.P. Botkin i Voenno-meditsinskaia akademiia* (Moscow: Meditsina, 1982).

Allgemeine deutsche Biographie herausgeben von der historischen Kommission bei der Akademie der Wissenschaften (Leipzig: Duncker, 1875–1912).

Ackerknecht, E., *Medicine at the Paris Hospital, 1794–1848* (Baltimore: Johns Hopkins Press, 1967).

Astrup, P. and Severinghaus, J.W., *The History of Blood Gases, Acids and Bases* (Copenhagen: Munksgaard, 1986).

Barkan, D., *Walter Nernst and the Transition to Modern Physical Science* (Cambridge: Cambridge University Press, 1999.

Baumgart, W., *Peace of Paris 1856: Studies in War, Diplomacy, and Peacemaking* Saab, A.P. (trans.), (Oxford: Clio Press, 1981).

———— *The Crimean War: 1853–1856* (London: Arnold, 1999).

Billroth, T., *The Medical Sciences in the German Universities: A Study in the History of Civilization,* W.H. Welch (trans.), (New York: Macmillan Company, 1924).

Bleker, J., *Die naturhistorische Schule, 1825-1845: ein Beitrag zur Geschichte der klinischen Medizin in Deutschland* (Stuttgart: G. Fischer, 1981).

Bonner, Th., *Becoming a Physician: Medical Education in Britain, France, Germany, and the United States, 1750–1945* (Oxford: Oxford University Press, 1995).

Borell, M., 'Instrumentation and the Rise of Modern Physiology', *Science and Technology Studies,* 5, 2 (1987), 52–62.

Borodulin, F.P., *S.P. Botkin i nevrogennaia teoriia meditsiny* (Moscow: Medgiz, 1949).

Boruttau, H., *Emil du Bois-Reymond* (Vienna: Rikola, 1922).

Brain, R. and Wise, M., 'Muscles and Engines: Indicator Diagrams and Helmholtz's Graphical Methods', M. Biogioli (ed.), *The Science Studies Reader* (London: Routledge, 1999), 51–66.

Brazier, M., *A History of Neurophysiology in the Nineteenth Century* (New York: Raven Press, 1988).

Brock, W.H., *Justus von Liebig: The Chemical Gatekeeper* (Cambridge: Cambridge University Press, 1997).

———— *The Norton History of Chemistry* (New York: W.W. Norton, 1993).

Burdon-Sanderson, J.P., 'Carl Ludwig', in B. Jones (ed.), *The Golden Age of Science: Thirty Portraits of the Giants of Nineteenth Century Science by their Scientific Contemporaries* (New York: Simon and Schuster, 1966).

Cahan, D., 'The Institutional Revolution in German Physics, 1865–1914', *Historical Studies in the Physical Sciences,* 15, 2 (1985), 1–65.

———— (ed.), *Hermann von Helmholtz and the Foundations of Nineteenth-Century Science* (Berkeley: University of California Press, 1993).

———— (ed.), *Letters of Helmholtz to his Parents: The Medical Education of a German Scientist, 1837–1846* (Stuttgart: Franz Steiner Verlag, 1993).

Clelland, C.E., *State, Society and University in Germany, 1700–1914* (Cambridge: Cambridge University Press, 1980).

Coleman, W. and Holmes, F. (eds), *The Investigative Enterprise: Experimental Physiology in Nineteenth-Century Medicine* (Berkeley: University of California Press, 1988).

Cranefield, P.F., 'Carl Ludwig and Emil Du Bois-Reymond: A Study in Contrasts', *Gesnerus,* 45 (1988), 271–82.

———— 'Robert Bunsen, Carl Ludwig and Scientific Physiology', in F. Kao, K. Koizumi and M. Vassale (eds), *Research in Physiology: A Liber Memoralis in Honour of Professor Chandler McCuskey Brooks* (Bologna: A. Gaggi, 1971), 743–48.

———— 'The Organic Physics of 1847 and the Biophysics of Today', *Journal of the History of Medicine and Allied Sciences,* 12 (1957), 407–23.

Crawford, E., *Arrhenius: From Ionic Theory to the Greenhouse Effect* (Canton: Science History Publications, 1996).

Culotta, C.A., *A History of Respiratory Physiology: From Lavoisier to Paul Bert, 1777–1880* (unpublished PhD dissertation: University of Visconsin, 1968).

Cunnigham, A. and Williams, P. (eds), *The Laboratory Revolution* (Cambridge: Cambridge University Press, 1992).

Dolby, R., 'Debates over the Theory of Solution: A Study of Dissent in Physical Chemistry in the English-Speaking World in the Late Nineteenth and Early Twentieth Centuries', *Historical Studies in the Physical Sciences,* 7 (1976), 297–404.

Eimontova, R.G., *Russikie universitety na grani dvukh epokh: ot Rossii krepostnoi k Rossii kapitalisticheskoi* (Moscow: Nauka, 1985).

Faber, K., *Nosology in Modern Internal Medicine, with an Introductory Note by Rufus Cole* (New York: AMS Press, 1923).

Faber V.B., *S.P. Botkin, 1832–1889* (Leningrad: Izd-vo Voenno-meditsinskoi akademii im. S.M. Kirova, 1948).

Feldman, G.E., *Paul Bert 1833–1886* (Moscow: Nauka, 1979).

Figurovskii N.A., *Istoriia estestvoznaniia v Rossii,* 3 vols (Moscow: Akademiia Nauk SSSR, 1962).

————— and Solov'ev, Iu.I., *Aleksandr Porfir'evich Borodin: A Chemist's Biography,* C. Steinberg and G.B. Kauffman (trans.), (Berlin: Springer-Verlag, 1988).

Frieden, N., *Russian Physicians in an Era of Reform and Revolution, 1856–1905* (Princeton: Princeton University Press, 1981).

Fruton, J.S., *Contrasts in Scientific Style: Research Groups in the Chemical and Biological Sciences* (Philadelphia: American Philosophical Society, 1990).

Geison, G. (ed.), *Physiology in the Americal Context, 1850–1940* (Baltimore: American Physiological Society, 1987).

Geison, G. and Holmes, F.L. (eds), *Research Schools: Historical Reappraisals, (Osiris),* (Chicago: University of Chicago Press Journals, 1993).

Gillispie, Ch. C. (ed.), *Dictionary of Scientific Biography* (New York: Charles Scribner's Sons, 1970–present).

Graham, L.R., *Science in Russia and the Soviet Union: A Short History* (Cambridge: Cambridge University Press, 1993).

Grekova, T. I. and Golikov, Iu. P., *Meditsinskii Peterburg* (St Petersburg: Folio-Press, 2001).

Grigoryan, G.G. (ed.), *Pamiatniki nauki i tekhniki v muzeiakh Rossii.* (Moscow: Znanie, 1996).

Henderson, L., 'Blood as a Physicochemical System', *Journal of Biological Chemistry,* 46 (1921), 411–9.

Holmes, F.L., 'The Old Martyr of Science: The Frog in Experimental Physiology', *Journal of the History of Biology,* 26 (1993), 311–28

————— 'Elementary Analysis and the Origins of Physiological Chemistry', *Isis,* 54 (1963, 50–81.

————— and Levere, T.H. (eds), *Instruments and Experimentation in the History of Chemistry* (Cambridge: MIT Press, 2000).

————— and Olesko, K.M., 'The Images of Precision and the Graphical Method in Physiology', N. Wise (ed.), *The Values of Precision* (Princeton: Princeton University Press, 1995), 198–221.

Hoskin, M. (ed.), *The Cambridge Illustrated History of Astromony.* (Cambridge: Cambridge University Press, 2000).

Iaroshevskii, M.G., *I.M. Sechenov* (Leningrad: Nauka, 1968).

Ivanov, B. and Karev, V. (eds), *Istoriia Otechestva s drevneishikh vremen do nashikh dnei. entsiklopedicheskii slovar'* (Moscow: Bol'shaia Rossiiskaia Entsiklopediia, 1999).

Joravsky, D., *Russian Psychology: A Critical History* (Oxford, UK: Blackwell, 1989).

Kassow, S., *Students, Professors, and the State in Tsarist Russia* (Berkeley: University of California Press, 1989).

Katz, V.J., *History of Mathematics* (Reading: Addison-Wesley, 1998).

Kitchen, M., *The Cambridge Illustrated History of Germany* (Cambridge: Cambridge University Press, 1996).

Koblitz, A.H., *A Convergence of Lives: Sofia Kovalevskaia – Scientist, Writer, Revolutionary* (Cambridge: Birkhäuser Boston, 1983).

Kohler, R.E., *From Medical Chemistry to Biochemistry: The Making of a Biomedical Discipline* (Cambridge: Cambridge University Press, 1982).

Koshtoyants, K.S., *Essays on the History of Physiology in Russia,* D.P. Boder, K, Hanes, and N. O'Brien (trans.), (Washington: American Institute of Biological Sciences, 1964).

————— *I.M. Sechenov, 1829–1905* (Moscow: Nauka, 1950).

Kostiuk, P.G., Mikulinskii, S.P. and Iaroshevskii, M.G. (eds), *I.M. Sechenov: K 150-letiiu so dnia rozhdeniia* (Moscow: Nauka, 1980).

Laidler, K.J., *The World of Physical Chemistry* (Oxford: Oxford University Press, 1993).

Lammel, H.-U. and Schneck, P. (eds), *Die Medizin an der Berliner Universität und an der Charité zwischen 1810 und 1850* (Husum: Matthiesen Verlag, 1995).

Lieben, F., *Geschichte der physiologischen Chemie* (New York: Georg Olms Verlag, 1970).

Lenoir, T., 'Models and Instruments in the Development of Electrophysiology, 1845-1912', *Historical Studies in the Physical and Biological Sciences,* 17 (1986) 1–54

——— *Instituting Science: Cultural Production of Scientific Disciplines* (Stanford: Stanford University Press, 1997).

——— 'Science for the Clinic: Science Policy and the Formation of Carl Ludwig's Institute in Leipzig', in W. Coleman and L.F. Holmes (eds), *The Investigative Enterprise, Experimental Physiology in Nineteenth-Century Medicine* (Berkeley: University of California Press, 1988).

Lesky, E., *The Vienna Medical School of the Nineteenth Century* (Baltimore: Johns Hopkins University Press, 1976).

Maulitz, R., 'Rudolf Virchow, Julius Cohnheim and the Program of Pathology', *Bulletin of the History of Medicine,* 52 (1978), 162–82.

Mazumdar, P.M.H., 'Johannes Müller on the Blood, the Lymph, and the Chyle', *Isis,* 33 (1975), 242–53.

——— *Species and Specificity: An Interpretation of the History of Immunology* (Cambridge: Cambridge University Press, 1995).

Meinel, C. (ed.), *Instrument–Experiment historische Studien* (Berlin: Diepholz, 2000).

Mendelsohn, E., 'Physical Models and Physiological Concepts: Explanation in Nineteenth-Century Biology', *British Journal of the History of Science,* 2 (1965), 201–19

Mirskii, B., *Meditsina v Rossii XVI-XIX vekov* (Moscow: ROSSPEN, 1996).

Mladentsev, M.N. and Tishchenko, V. E., *Dmitrii Ivanovich Mendeleev: ego zhizn' i deiatel'nost'* (Moscow: Izd-vo Akademii Nauk SSSR, 1938).

Müller, R.A., *Geschichte der Universität: von der mittelalterlichen Universitas zur deutschen Hochschule* (München: Callwey, 1991).

Neue deutsche Biographie (Berlin: Duncker and Humboldt, 1953–present).

New Grove Dictionary of Music and Musicians, 20 vols, 6th edn (London: Macmillan, 1980).

Otis, L., *Müller's Lab* (Oxford: Oxford University Press, 2007).

Partington, J.R., *A History of Chemistry*, 4 vols (London: Macmillan & Co., 1961–4).

Pollak, K., 'Josephinum und Pépinière. Der Beitrag des Militärsanitätswesens zur Vereinigung von Chirurgie und Medizin', in *Wehrmedizin und Wehrpharmazie*, 3 (1985), 134–8.

Rocke, A.J., *Nationalizing Science: Adolphe Wurtz and the Battle for French Chemistry* (Cambridge: The MIT Press, 2001).

——— *The Quiet Revolution: Hermann Kolbe and the Science of Organic Chemistry* (Berkeley: University of California Press, 1993).

Rothschuh, K.E., *History of Physiology*, G.B. Risse (trans. and ed.), (Huntington: R.E. Krieger Publishing Co., 1971).

——— (ed.), *Von Boerhaave bis Berger: die Entwicklung der kontinentalen Physiologie im 18. und 19. Jahrhundert* (Stuttgart: Fischer, 1964).

Ruff, P.W., *Emil du Bois-Reymond* (Leipzig: Teubner Verlagsgesellschaft, 1981).

Samoilov, V.O., *Istoriia Rossiiskoi meditsiny* (Moscow: Epidavr, 1997).

Schneck, P. and Schultze, W. (eds), *Emil du Bois-Reymond 1818–1896* (Berlin: Gerhardt Weinert GmbH, 1996).

Schröer, H., *Carl Ludwig* (Stuttgart: Wissenschaftliche Verlagsgesellschaft M.B.H., 1967).

Schwabe, K. (ed.), *Deutsche Hochschullehrer als Elite, 1815–1945* (Boppard am Rhein: H. Boldt, 1988).

Servos, John W., *Physical Chemistry from Ostwald to Pauling: The Making of a Science in America* (Princeton: Princeton University Press, 1990).

Shevchenko, Iu. L. (ed.), *Professora Voenno-meditsinskoi (Mediko-khirurgicheskoi) akademii, 1798–1998* (St Petersburg: Nauka, 1998).

Shumeiko, L., *Die Rezeption der Zellularpathologie Rudolf Virchow in der Medizin Russlands und der Sowjetunion* (unpublished PhD dissertation: University of Marburg, 2000).

Slavin, A.J., *The Way of the West*, 2 vols (Lexington: Xerox College Publishing, 1975).

Slonimsky, N. and Kuhn, L. (eds), *Baker's Biographical Dictionary of Musicians,* (New York: Schirmer Books, 2001).

Smith, R., *Inhibition: History and Meaning in the Sciences of Mind and Brain* (Berkeley: University of California Press, 1992).

Solov'ev, Iu.I., *Istoriia ucheniia o rastvorakh* (Moscow: Izd-vo Akademii Nauk SSSR, 1959).

——— *Svante Arrhenius: nauchnaia biografiia* (Moscow: Nauka, 1990).

Spiekermann, P.G., 'Physiology with Cool Obsession: Carl Ludwig – His Time in Vienna and his Contribution to Isolated Organ Methodology', *Pflügers Archiv: European Journal of Physiology*, 432 (1996), 33–41.

Stahnke, J., 'Blutfarbstoff-Kristalle von Reichert bis Hoppe-Seyler', *Sudhoffs Archiv Zeitschrift für Wissenschaftsgeschichte,* 63 (1979), 154–89.

Stangier, H., *Ludwig Traube, sein Leben und Werk* (Freiburg: Buchdruckerei Theodor Kehrer, 1935).

Stochek, A.M., Paltsev, M.A., Zatravkin, S.N., *Meditsinskii fakul'tet Moskovskogo universiteta v reformakh prosveshcheniia perioda pervoi treti XIX veka* (Moscow: Meditsina, 1998).

Thurau, K., Davis, J. and Häberle, D., 'Renal Blood Flow and Dynamics of Glomerular Filtration: Evolution of a Concept from Carl Ludwig to the present Day', in C. Gottschalk *et al.*, *Renal Physiology: People and Ideas* (Bethesda: American Physiological Society, 1987), 31–61.

Todes, D.P., *From Radicalism to Scientific Convention: Biological Psychology in Russia from Sechenov to Pavlov* (unpublished PhD thesis: University of Pennsylvania, 1981).

——— *Pavlov's Physiology Factory: Experiment, Interpretation, Laboratory Expertise* (Baltimore: Johns Hopkins University Press, 2002).

Tuchman, A.M., *Science, Medicine, and the State in Germany: The Case of Baden, 1815–1871* (New York: Oxford University Press, 1993).

Turner, R.S., 'Justus Liebig versus Prussian Chemistry: Reflections on Early Institute Building in Germany.' *Historical Studies in the Physical Sciences,* 13 (1982), 129–62.

Vucinich, A., 'Nikolai Iovanovich Lobachevskii: The Man Behind the First Non-Euclidean Geomentry', *Isis,* 53 (1962), 465–81.

——— *Science in Russian Culture* (Stanford: Stanford University Press, 1963, 1970).

Warner, J.H., *Against the Spirit of System: The French Impulse in the Nineteenth-Century American Medicine* (Princeton: Princeton University Press, 1998).

Weiner, D., *The Citizen–Patient in Revolutionary and Imperial Paris* (Baltimore: Johns Hopkins University Press, 1993)

Weisz, G., *The Medical Mandarins: The French Academy of Medicine in the Nineteenth and Early Twentieth Centuries* (Oxford: Oxford University Press, 1995).

Zaionchkovskii, P.A., *Voennye reformy 1860–1870 godov v Rossii* (Moscow: Izd-vo Moskovskogo universiteta, 1952).

Index

Note: documents are given in italics; p after page numbers indicates a picture; t indicates a table.

C

D

H

I

J

K

L

O

P

R

S